Successful Instrumentation and Control Systems Design

Second Edition

Successful Instrumentation and Control Systems Design

Second Edition

Michael D. Whitt

Notice

The information presented in this publication is for the general education of the reader. Because neither the author nor the publisher have any control over the use of the information by the reader, both the author and the publisher disclaim any and all liability of any kind arising out of such use. The reader is expected to exercise sound professional judgment in using any of the information presented in a particular application.

Additionally, neither the author nor the publisher have investigated or considered the effect of any patents on the ability of the reader to use any of the information in a particular application. The reader is responsible for reviewing any possible patents that may affect any particular use of the information presented.

Any references to commercial products in the work are cited as examples only. Neither the author nor the publisher endorse any referenced commercial product. Any trademarks or tradenames referenced belong to the respective owner of the mark or name. Neither the author nor the publisher make any representation regarding the availability of any referenced commercial product at any time. The manufacturer's instructions on use of any commercial product must be followed at all times, even if in conflict with the information in this publication.

Copyright © 2012 ISA
All rights reserved.

Printed in the United States of America.
10 9 8 7 6 5 4 3 2

ISBN: 978-1-936007-45-5

No part of this work may be reproduced, stored in a retrieval system, or transmitted in any form or by any means, electronic, mechanical, photocopying, recording or otherwise, without the prior written permission of the publisher.

ISA
67 Alexander Drive
P.O. Box 12277
Research Triangle Park, NC 27709

Library of Congress Cataloging-in-Publication Data is in process.

Acknowledgments

I'd like to thank Charlie Thompson for helping me make the transition from I&C to systems integration. Sometimes you need someone who's willing to take a chance on you, and, for me, Charlie was that person.

I also thank my many friends in the professional community here in Knoxville: the folks at Raytheon—now Lauren—with whom I have worked for 15 years, and the wonderful team at Mesa Associates, Inc.

And, of course, to Susan Colwell and my other friends at ISA who helped shepherd this project to its conclusion – thanks.

ACKNOWLEDGMENTS

Dedication

I would like to dedicate this book to my wife Mary and son Elliot, who have continually supported me through the travails of writing this book—I'm truly blessed; my parents, who instilled in me a love for reading, and, by extension, writing; and Jesus Christ, who is my inspiration and through whom all things are possible.

Table of Contents

List of Figures xix

Preface xxix

List of Acronyms – Part I xxxvii

List of Figures – Part I xxxix

Introduction – Part I 1

Part I – Chapter 1: The Project 3
    A. Introduction 3
    B. Project Planning 4
    C. Contracts and their Effects on Project Structure 8
        1. Constraints 9
        2. Contract Types: The Cost-Plus Contract (CP) 10
        3. Contract Types: The Time and Material Contract (T&M) 13
        4. Contract Types: The Time and Material/Not-To-Exceed Contract (T&M/NTE) 14
        5. Contract Types: Lump Sum (Fixed Price) Contract 14
        6. Project Structures: The Turnkey Project 16
        7. Project Structures: The EPC Project 16
        8. Project Structures: The Retrofit and Green-Field Projects 17
        9. Project Structures: The Hybrid Project 17
    D. The Customer/Service Provider Relationship 17
    E. Project Flow for a Controls Project 18
        1. The Owner's Project 21
            a. Owner's Phase 1 – FEL Stage 1: Business Planning 22
            b. Owner's Phase 1 – FEL Stage 2: Project Definition 24
            c. Owner's Phase 1 – FEL Stage 3: Project Planning 24
            d. Owner's Phase 2 – System Design 25
            e. Owner's Phase 3 – Deployment 26
            f. Owner's Phase 4 – Support 27
            g. Owner's Project Deliverables 27
        2. The Control System E/I&C Engineer's Project 30
            a. E/I&C Seller Phase 1 – FEL Stage 1: Business Planning 31
            b. E/I&C Seller Phase 1 – FEL Stage 2: Project Definition 32
            c. E/I&C Seller Phase 1 – FEL Stage 3: Project Planning 33
            d. E/I&C Seller Phase 2 – System Design 33
            e. E/I&C Seller Phase 3 – Deployment 34
            f. E/I&C Seller Phase 4 – Support 34
            g. E/I&C Seller's Project Deliverables 35
        3. The Control Systems Integrator's Project 40
            a. CSI Seller Phase 1 – FEL Stage 1: Business Planning 40
            b. CSI Seller Phase 1 – FEL Stage 2: Project Definition 40
            c. CSI Seller Phase 1 – FEL Stage 3: Project Planning 42
            d. CSI Seller Phase 2 – System Design 42

    e. CSI Seller Phase 3 – Deployment 43
    f. CSI Seller Phase 4 – Support 44
    g. CSI Seller's Project Deliverables 44
  4. The Control Panel Fabricator's Project 50
    a. FAB Seller Phase 1 – FEL Stage 1: Business Planning 51
    b. FAB Seller Phase 1 – FEL Stage 2: Project Definition 51
    c. FAB Seller Phase 1 – FEL Stage 3: Project Planning 51
    d. FAB Seller Phase 2 – System Design 51
    e. FAB Seller Phase 3 – Deployment 52
    f. FAB Seller Phase 4 – Support 52
    g. FAB Seller's Project Deliverables 52
 F. Integrated Control Solutions 52

Part I – Chapter 2: The Project Team 55

 A. The Owner (i.e., Buyer, Customer) 57
  1. Plant Administration 58
  2. Plant Operations 59
  3. Plant Engineering/Maintenance 59
  4. Plant Purchasing 59
 B. The Owner's Engineer (OE) 60
  1. The OE as an Individual 60
  2. The OE as an In-House Team 60
  3. The OE as a Consulting Management Firm 61
  4. The OE as a Prime Contractor 61
 C. The Designer 61
  1. Project Manager (PM) 62
  2. Project Engineer (PE) 63
  3. Discipline Lead Engineer (DLE) 63
  4. Discipline Engineer(s) 63
  5. Discipline Design Supervisor (DDS) 63
  6. Discipline Technical Support (Design) Staff 64
    a. Lead Designer 64
    b. Designer 64
    c. CADD Technician 64
    d. Engineering Aide (EA) 64
 D. The Control Systems Integrator (CSI) 64
  1. The Process Control Team 65
    a. Process Engineers/Specialists 65
    b. PLC/DCS Programmers 66
    c. HMI/SCADA Programmers 66
  2. The Server Setup Team 66
  3. The Network Setup Team 67
  4. The Startup & Commissioning Team 67
 E. The Constructor (i.e., Builder) 67
  1. E&I Construction Superintendent 67
  2. E&I Field Engineer & Coordinator 68
  3. E&I General Foreman 69
  4. Instrument Foreman 69
  5. Instrument Fitter/Mechanic 69
  6. Pipe Fitter 69
  7. Instrument Electrical Foreman 69
  8. Electrician 69
  9. Instrument Electrician/Technician 70

Part I – Chapter 3: The Managed Project 71

A. Key Project Management Tools 73
  1. The Detailed Scope of Work (SOW) 74
  2. The Estimate* 76
     a. Budgetary 76
     b. Bid 76
     c. Definitive 77
  3. The Schedule* 78
  4. The Proposal 79
     a. Executive Summary 81
     b. General Scope of Work 81
     c. Assumptions 82
     d. Inclusions 82
     e. Exclusions (Exceptions) 82
     f. Deliverables 82
     g. Milestone Schedule 82
     h. Safety 83
     i. Price & Payment Schedule 83
     j. Bid Award & Contract Negotiation 83
  5. The Project Execution Plan (PEP) 84
     a. Contact List 85
     b. Existing System Description 85
     c. Disposition of Existing Equipment 85
     d. Addition of New Equipment 85
     e. Company and Applicable Industry Standards 85
     f. Approved Vendors List 86
     g. Vendor-Provided Pre-engineered Subsystems (OEM) 86
     h. Instrumentation Data 86
     i. Quality Control 86
     j. Document Control 86
  6. The Status Report 87

B. Project Management Techniques 90
  1. Assessing Project Status 90
  2. Staff Meetings 91
     a. The Meeting Facilitator 92
     b. The Facilitator's Toolbag 93
     c. The Meeting Agenda 96
  3. Management of Change (MOC) 96

Part I – Summary 101

Part I - References 103

List of Acronyms – Part II 109

List of Figures – Part II 111

Introduction – Part II 115

Part II – Chapter 4: Basic Design Concepts 117

A. Scaling and Unit Conversions 117
  1. Definition of Key Terms 118
  2. Accuracy and Repeatability 120
  3. Resolution Effects on Accuracy 122
  4. Instrument Range versus Scale 123

      5. Instrument Calibration 123
      6. Linearization and Unit Conversions 124
      7. Practical Application 127
  B. Introduction to Information Management 129

Part II – Chapter 5: Design Practice 147
  A. Basic Wiring Practice 148
    1. Inter-Cabinet Wiring 148
      a. Generating a Cable Schedule 149
    2. Intra-Cabinet Wiring 155
      a. Generating a Wiring Diagram 156
  B. Failsafe Wiring Practice 166
  C. Hazardous Area Classification and Effects on Design 170
    1. Hazardous Locations 170
      a. Class I 172
    2. Explosionproofing 174
    3. Intrinsic Safety 175
    4. Purging 175
      a. Class X Purge 175
  D. Connecting to the Control System 177
    1. Discrete (Digital) Wiring 178
      a. Sinking and Sourcing 179
    2. Analog Wiring 184
      a. Circuit Protection (Fusing) 185
  E. Design Practice Summary 192

Part II – Chapter 6: The Control System 193
  A. Introduction 193
  B. The Cognitive Cycle 194
  C. Control System Overview 195
    1. A Historical Perspective 195
    2. PLC versus DCS 197
      a. The Distributed Control System (DCS) 197
    3. Major Control System Elements 200
      a. The Physical Plant 200
    4. Control Modes and Operability 202
      a. Local/Remote (L/R) Mode Selector 204
  D. The Human-Machine Interface 205
    1. The Graphic User Interface (GUI) 205
      a. Action Links 207
    2. The HMI Database 210
      a. Tagnames 210
    3. The HMI Alarm Manager Utility 211
    4. The Historian 212
    5. The Trend Utility 212
    6. Reports 213
  E. Programmable Logic Controller 214
    1. Major PLC Components 214
      a. The Rack Power Supply 215
    2. The PLC Program 216
      a. I/O Map 216
    3. The I/O Interface 222

        a. Physical (Hardware) Address 223
   F. Networking... 230
      1. Optimized/Proprietary Networks 230
      2. Optimized/Non-Proprietary Networks................................ 232
        a. Serial Communications (RS-232) 232
      3. Non-Optimized (Open) Local Area Networks 234
      4. Wireless Local Area Networks....................................... 234
        a. The "Bluetooth" Standard 234
      5. The Ethernet Client/Server Environment 235
        a. "Thick" Client Architecture 236
      6. The Industrial Enterprise-Wide Network 238
        a. The Remote I/O (RIO) LAN...................................... 239
   G. Working with a Control Systems Integrator (CSI)......................... 241
      1. Initial Search 241
      2. Writing a Control System Specification 242
        a. Process Overview... 242
   H. Selecting a Control System 246

References – Part II.. 249

List of Acronyms – Part III 257

List of Figures – Part III... 259

Introduction – Part III 267

Part III – Chapter 7: Piping and Instrumentation Diagrams (P&IDs) 271
   A. General Description 271
   B. Purpose.. 272
   C. Content 272
      1. Symbology (ANSI/ISA-5.1-2009) 273
      2. Symbol Identification 273
        a. Prefix 274
   D. Practical Application 276
      1. Tank Level: LT-10, LSH-10, LSLL-47................................. 276
      2. Tank Fill: HV-13, ZSC-13 277
      3. Tank Discharge: PP-10 277
      4. Pump Discharge Pressure: PIC-48.................................... 277
   E. P&ID Summary 278

Part III – Chapter 8: Links to Mechanical and Civil............................... 279
   A. General Equipment Arrangement Drawing (Civil and Mechanical) 279
      1. Purpose 279
      2. Interfaces... 279
      3. Content 279
      4. Practical Application 280
      5. Equipment Arrangement Summary 280
   B. Piping Drawing (Mechanical) 281
      1. Purpose 282
      2. Interfaces... 282
      3. Content (as related to I&C)... 282
   C. Pump and Equipment Specifications (Mechanical)........................ 282
   D. Links Summary 282

Part III – Chapter 9: Preliminary Engineering 283

A. Development of a Detailed Scope of Work 284
1. Purpose (Project Overview) 284
2. Project Scope—I&C 284
3. Safety Concerns 284
4. Assumptions 284
5. Exclusions 284
6. Deliverables 284
7. Milestone Schedule 285

B. Control System Orientation 285

C. Project Database Initialization 288
1. Initialize Document Control Table 289
   a. Table 290
2. Initialize Instrument and I/O List Table 295
   a. Instrument Table 296
3. Database Summary 302

D. Estimate and Schedule Development* 302
1. Cover Worksheet 303
2. Devices Worksheet 303
3. Count Worksheet 309
4. Labor Worksheet 310
5. Summary Worksheet 314
6. Schedule Worksheet 316
7. Estimate and Schedule Summary 317

E. Preliminary Engineering Summary 319

Part III – Chapter 10: Control Systems Integration (CSI) 321

A. FEL Stage 1 – Business Planning 321
1. Cost/Benefit Analysis 322
2. Control System Specification 323
3. Functional Description 324
   a. Tank Level: LT-10, LSH-10, LSLL-47 324
4. Project Estimate 325
   a. Field Device Control Elements 325
5. Project Proposal 327

B. FEL Stage 2 – Project Definition 327
1. Sequential Function Chart (SFC) 328
2. Continuous Function Chart (CFC) 328
3. Control Narrative 329
4. Sequence Control Detail Sheets (SCDS) 329
5. Device Control Detail Sheets (DCDS) 329
6. Functional Logic Diagrams 329

C. Control Narrative 329
1. Sequential Function Chart (SFC) 329
2. Continuous Function Chart (CFC) 333
3. SFC Control Narrative Fragment 334
   a. Powerup & Initialize 335
4. Sequence Step Detail Sheet (SSDS) 337
   a. Step S02 – Fill Tank 338
5. Device Control Detail Sheet (DCDS) 345
   a. Pump PP-10 Device Logic 346
6. Functional Logic Diagram 350
   a. Tank TK-10 Control Sequence Step 02 353

       7. Logic Diagram Standard ISA-5.1... 362
       8. FEL2 Systems Integration Summary.. 363
  D. Operator Interface Specification Development – The HMI 365
       1. Animation Plan 366
           a. Colors.. 366
       2. Screen Diagrams.. 367
           a. Graphic Screen 369
       3. Tagname Database, Device Driver, and I/O Mapping 374
       4. Finished Graphics Screen 375
       5. Alarm Manager 377
       6. Historian 378
       7. HMI Report Generation 378
  E. Network Single-Line Diagram Generation 378
  F. Other Systems Integration Tasks.. 379
       1. Control System Cabinetry Design and Delivery.......................... 379
       2. I/O Address Assignment (Partitioning)................................... 379
           a. Hardware (HW) Address 380
       3. Factory (or Functional) Acceptance Test (FAT)............................ 381
       4. Site Acceptance Test (SAT).. 384
       5. Commissioning 385
       6. Operations and Maintenance (O&M) Manual............................. 385
           a. Operations.. 386
       7. Onsite Training... 386
  G. Systems Integration Summary 387

Part III – Chapter 11: Information Management................................. 389

  A. Document Control 389
  B. Instrument and I/O List 390
       1. Instrument and I/O List Table.. 391
       2. Preliminary Design Query 391
       3. Plan Drawing Takeoff Query 392
       4. Plan Dwg Takeoff Query Report.. 392
       5. X-Ref Document Cross-Reference Query................................. 393
       6. X-Ref Document Cross-Reference Report 394
  C. Database Summary.. 394

Part III – Chapter 12: Instrument Specifications................................. 397

  A. Purpose.. 398
       1. Mechanical Designers 398
       2. Instrument Designers 399
       3. Other Users 401
  B. Interfaces.. 401
  C. Examples 402
       1. LT/LSH-10 402
       2. PV-48 405
  D. Summary 407

Part III – Chapter 13: Physical Drawings 409

  A. Control Room 409
       1. Environmental Issues 409
           a. Heating, Ventilation and Air Conditioning (HVAC)9 409
       2. Physical Arrangement... 410
       3. Control Room Design Summary 411

- B. Termination Room 411
  - 1. Environmental Issues 411
    - a. Lighting 411
  - 2. Furniture and Equipment Arrangement 413
    - a. Personnel Clearances... 413
  - 3. Termination Room Design Summary 413
- C. Process Area (Instrument Location Plan) 413
  - 1. Why Produce Instrument Location Plan Drawings? 415
  - 2. Anatomy of an Instrument Location Plan 415
  - 3. Design Considerations 416
  - 4. Drawing Production Technique 417
    - a. Step One: Initialize Drawing (Generate drawing background) 417
  - 5. Material Takeoff 429
- D. Instrument Installation Details 433
  - 1. Electrical Installation Details 434
  - 2. Tubing Details 435
  - 3. Mounting Details 436
  - 4. Related Database Activities 436
  - 5. Material Takeoff 438
- E. Summary 438

Part III – Chapter 14: Instrument and Control Wiring 441

- A. Instrument Elementary (Ladder) Diagram 444
  - 1. Motor Elementaries 447
  - 2. AC Power Distribution Schematic 449
  - 3. DC Power Distribution Schematic 451
  - 4. PLC Ladder Diagram (Elementary) 452
    - a. Discrete (Digital) Inputs 455
- B. Loop Sheet (Ref: ISA-5.4-1991)14 458
- C. Connection Diagrams 462
  - 1. Junction Box JB-TK10-1: Initial Layout 463
  - 2. Termination Cabinet TC-2 469
    - a. DC Circuits (TS-2) 470
- D. Wiring Summary 479

Part III – Chapter 15: Panel Arrangements 485

- A. Procedure 486
- B. Junction Box JB-TK10-01 Arrangement Drawing ARR-002 487
  - 1. Set Up a Scale 488
  - 2. Design the Panel 488
  - 3. Generate a Bill of Materials 491
- C. Summary 492

Part III – Chapter 16: Procurement 493

- A. Typical Purchasing Cycle 494
- B. Material Classification 496
- C. Bulk Bill of Materials 496
- D. Detail Bill of Materials 501
- E. Procurement Summary 506

Part III – Chapter 17: Quality Control—The Integrated Design Check 509

- A. Administrative Content – Individual Checks 509

    B. Technical Content – Squad Check 510
    C. Squad-Check Roster 511
    D. Design Check Summary 511

Part III – Chapter 18: Phase 3—Deployment 513

    A. Construction 513
        1. Kickoff Meeting 513
        2. Construction 514
    B. Pre-Commissioning 515
    C. Cold-Commissioning (Site Acceptance) 516
        1. Device Tests 516
        2. Subsystem Tests 517
    D. Hot-Commissioning (Startup) 517
    E. Adjustment of Document Package to Reflect Construction Modifications 517
    F. Issue for Record 518
    G. Phase 3 Summary 518

Part III – Chapter 19: Phase 4—Support 519

    A. Warranty Support 519
    B. Continuing Service Support 519

References – Part III 521

Additional Resources 523

Index 525

List of Figures

Figure 1-1. Typical bid package content 6

Figure 1-2. Success triangle. 9

Figure 1-3. Risk to reward analysis by project type 11

Figure 1-4. Effects of constraints on project structure 13

Figure 1-5. Typical CSP project lifecycle. 19

Figure 1-6. Control system project flow by involvement level 20

Figure 1-7. Sample Owner's capital improvement project plan 22

Figure 1-8. Sample E/I&C seller's project plan. 31

Figure 1-9. Sample CSI seller's project plan. 41

Figure 1-10. Sample FAB seller's project plan 50

Figure 1-11. Typical controls project participants. 56

Figure 1-12. Engineering design team 62

Figure 1-13. Control Systems Integration design team 65

Figure 1-14. Construction team 68

Figure 1-15. Simplified contract award overview. 72

Figure 1-16. Project Execution Plan template for small tasks 87

Figure 1-17. Project status report – data collection and status calculation fields 89

Figure 1-18. Project status report – analysis fields 91

Figure 1-19. Meeting Status Notes form. 94

Figure 1-20. Emerging Issues Notes form 94

Figure 1-21. Needs List form 94

Figure 1-22. Action List form 95

Figure 1-23. Suggestion List form 95

Figure 1-24. Sample project meeting agenda form 97

Figure 1-25. A Management of Change (MOC) process 99

Figure 2-1. Typical error pattern caused by deadband 121

Figure 2-2. Typical error pattern caused by hysteresis. 123

Figure 2-3. Conversion problems 125

Figure 2-4. Data translation process — from field device to HMI 126

Figure 2-5. Signal conversion at PLC input 127

Figure 2-6. Engineering unit calculation at the HMI 127

Figure 2-7. Spreadsheet versus database comparison 130

Figure 2-8. Typical relational database program structure 133

Figure 2-9. ICS-based project flow with database-intensive activities highlighted 135

Figure 2-10. The P&ID takeoff query 136

Figure 2-11. The I/O partitioning query 137

Figure 2-12. The software & logic assignment query 137

Figure 2-13. The cable and conduit schedule query (partially shown) 138

Figure 2-14. The instrument specification query (partially shown) 139

Figure 2-15. The construction checkout query 139

Figure 2-16. The Validation & Verification (V&V) test queries 140

Figure 2-17. The site acceptance test queries 141

Figure 2-18. Typical document handling process 144

Figure 2-19. Typical cabling scheme 148

Figure 2-20. Defining the cable route (wire W1, route C1/T1/T2/C2/C2a) 149

Figure 2-21. Sample cable schedule 150

Figure 2-22. Cable area fill 151

Figure 2-23. Cross-sectional views of cable orientation before, during, and after a conduit bend . 152

Figure 2-24. Conduit facts 152

Figure 2-25. Conduit sizing calculator 153

Figure 2-26. Sample conduit schedule 155

Figure 2-27. Sample instrument arrangement 155

Figure 2-28. Interconnection wiring example 157

Figure 2-29. Form A contact set (SPST – NORMALLY OPEN) 158

Figure 2-30. Form B contact set (SPST – NORMALLY CLOSED) 159

Figure 2-31. Form-C contact set (SPDT) 159

Figure 2-32. 5-pole relay used as a motor starter (shown in shelf state, with interlocks, overloads, and PLC input) 161

Figure 2-33. Common types of switches and their diagrams 162

LIST OF FIGURES

Figure 2-34. Types of contacts 162

Figure 2-35. Interval timer timing diagram 163

Figure 2-36. Time delay on de-energize (TDOD) timer timing diagram 163

Figure 2-37. Time delay on energize (TDOE) timer timing diagram 164

Figure 2-38. Sample ladder elementary format 165

Figure 2-39. Failsafe interlock chain (devices shown in shelf state) 169

Figure 2-40. Hazardous boundaries 171

Figure 2-41. Basic discrete (digital) circuit 178

Figure 2-42. Discrete (digital) circuit wiring technique 179

Figure 2-43. Simple switching 180

Figure 2-44. Sinking and sourcing digital input modules 182

Figure 2-45. Isolated digital output module 184

Figure 2-46. Sinking and sourcing digital output modules 185

Figure 2-47. Analog circuit wiring technique 186

Figure 2-48. Analog wiring methods: 2-wire vs. 4-wire 191

Figure 2-49. The cognitive cycle 195

Figure 2-50. Typical control system 203

Figure 2-51. The Human-Machine Interface (HMI) 206

Figure 2-52. Graphical User Interface with pushbutton configuration template ... 209

Figure 2-53. Trend screen 213

Figure 2-54. Typical PLC rack 215

Figure 2-55. Sequential function chart washing machine sequence control application ... 220

Figure 2-56. Continuous function chart washing machine temperature control application ... 221

Figure 2-57. Control detail sheet 223

Figure 2-58. Suggested program flow of control 224

Figure 2-59. I/O tally worksheet 227

Figure 2-60. Revised I/O tally worksheet reflecting new setup 229

Figure 2-61. I/O tally worksheet with split by I/O type 231

Figure 2-62. Remote I/O network 232

Figure 2-63. Industrial network 240

Figure 3-1. Instrumentation and controls engineering tasks (Phases 1 – 3) ... 269

Figure 3-2. Typical feed tank configuration 270

Figure 3-3. Typical P&ID symbology 274

Figure 3-4. Typical P&ID symbology showing combined automation system functions. 275

Figure 3-5. P&ID presentation of the TK-10 subsystem 276

Figure 3-6. Basic P&ID drawing 278

Figure 3-7. TK-10 feed tank with equipment labels 280

Figure 3-8. TK-10 Feed tank area equipment arrangement 281

Figure 3-9. Detailed Scope of Work. 283

Figure 3-10. Existing control system 287

Figure 3-11. Revised control system 288

Figure 3-12. List of tables 289

Figure 3-13. Document control table structure 290

Figure 3-14. Document control table, datasheet view 291

Figure 3-15. OrderDrawingsQuery (design view) 292

Figure 3-16. Document control table data 292

Figure 3-17. Transmittal query. 293

Figure 3-18. Transmittal query design view (with criteria filter) 294

Figure 3-19. Transmittal query, datasheet view 294

Figure 3-20. Instrument and I/O list table 297

Figure 3-21. Tagname update query, design view 298

Figure 3-22. TagnameUpdateQuery, design view, with criteria filter 298

Figure 3-23. Query tagname display 299

Figure 3-24. Reports. 300

Figure 3-25. Report wizard 301

Figure 3-26. P&ID takeoff query report 301

Figure 3-27. P&ID takeoff query report, design view. 302

Figure 3-28. Finished database products 303

Figure 3-29. Cover sheet for Estimate workbook 305

Figure 3-30. Devices worksheet 306

Figure 3-31. Devices I/O assignment index table 307

Figure 3-32. Devices I/O assignment index, revised 308

Figure 3-33. Devices I/O calculator. 308

Figure 3-34. Count worksheet 309

Figure 3-35. Background data table. 310

Figure 3-36. I/O configuration worksheet 310

Figure 3-37. Labor worksheet. 311

Figure 3-38. Direct engineering labor, Phase 1 312

Figure 3-39. Direct engineering labor, Phase 2 313

Figure 3-40. Indirect engineering labor, Phase 2. 313

Figure 3-41. Engineering and construction labor, Phase 3. 313

Figure 3-42. Engineering summary worksheet 314

Figure 3-43. Project cost summary table. 315

Figure 3-44. Engineering cost summary table 315

Figure 3-45. Phase 1 deliverables summary table. 316

Figure 3-46. Phase 2 deliverables summary table. 316

Figure 3-47. Instrument and I/O summary table 317

Figure 3-48. Schedule worksheet. 318

Figure 3-49. Design schedule and staffing plan 319

Figure 3-50. Project manhour loading chart. 320

Figure 3-51. Systems Integration services checklist 322

Figure 3-52. Existing control system 327

Figure 3-53. New control system. 328

Figure 3-54. TK-10 feed tank control sequence overview 330

Figure 3-55. Sequential function chart fragment. 330

Figure 3-56. Sample sequential function chart logic. 331

Figure 3-57. Sequential function chart (SFC). 332

Figure 3-58. Continuous function chart 333

Figure 3-59. Sequence step 2: "fill tank" sequence 340

Figure 3-60. Sequence step 5: "empty tank" sequence 343

Figure 3-61. Pump PP-10 motor controls elementary wiring diagram 347

Figure 3-62. Pump PP-10 device control detail sheet 348

Figure 3-63. HV-13 fill valve device control detail sheet 351

Figure 3-64. Sample logic diagram format 352

Figure 3-65. Logic diagram showing rat holes 353

Figure 3-66. Naming conventions for this project. 353

Figure 3-67. Timing diagram for a delay timer 354

Figure 3-68. FILL_TK10 control logic 355

Figure 3-69. EMPTY_TK10 control logic 355

Figure 3-70. Device logic for TK-10 fill controls and analog alarms. 357

Figure 3-71. PP-10 device logic 358

Figure 3-72. PP-10 device on/off logic 360

Figure 3-73. One-shot, rising (OSR) edge function block. 361

Figure 3-74. Pump restart inhibit signal processing 361

Figure 3-75. Sample logic diagram 362

Figure 3-76. Selected SAMA symbols now incorporated into ISA-5.1. 363

Figure 3-77. Functional control diagram (ISA-5.1) 364

Figure 3-78. Animation plan. 368

Figure 3-79. Preliminary screen graphics, TK-10 overview screen. 369

Figure 3-80. Sample control overlays 371

Figure 3-81. Pop-up overlays 371

Figure 3-82. Animation detailing. 372

Figure 3-83. Pump PP-10 control overlay. 373

Figure 3-84. Animation chart 373

Figure 3-85. Final screen diagram 374

Figure 3-86. Typical data progression. 375

Figure 3-87. HMI screen, pumping out in manual 375

Figure 3-88. HMI screen, filling in auto 376

Figure 3-89. HMI screen, sequence status 377

Figure 3-90. Sample alarm manager database. 377

Figure 3-91. Historian sampling points 378

Figure 3-92. Simple network single-line diagram 379

Figure 3-93. Hardware addresses 381

Figure 3-94. Software addresses 382

List of Figures

Figure 3-95. Adding the I/O modules 382

Figure 3-96. Instrument and I/O list table, design view 391

Figure 3-97. Instrument and I/O list database, datasheet view 392

Figure 3-98. Preliminary design query 392

Figure 3-99. Plan drawing takeoff query 393

Figure 3-100. PlanDwgTakeoffQuery report 393

Figure 3-101. Plan drawing component schedule (Microsoft® Access to Microsoft® Excel)...... 394

Figure 3-102. Document cross-reference (X-ref) query...................................... 394

Figure 3-103. Document cross-reference report.. 395

Figure 3-104. Ultrasonic level transmitter 403

Figure 3-105. Instrument specification for LT/LSH/LSL-10................................. 404

Figure 3-106. Instrument specification for Control Valve PV-48 406

Figure 3-107. Three termination room configurations 412

Figure 3-108. Sample instrument location plan drawing.................................... 414

Figure 3-109. Initialize drawing... 418

Figure 3-110. Locate major equipment items.. 419

Figure 3-111. Locate instrument items 420

Figure 3-112. Add instrument stations 421

Figure 3-113. PlanDwgTakeoffQuery... 421

Figure 3-114. PlanDwgTakeoffQuery, filtered.. 422

Figure 3-115. 3D to 2D and back 424

Figure 3-116. Add conduit detail... 424

Figure 3-117. Recommended conduit tagging convention 426

Figure 3-118. Instrument arrangement with support data 427

Figure 3-119. Cable code cross-reference chart 427

Figure 3-120. Component schedule.. 428

Figure 3-121. Plan001 component schedule.. 429

Figure 3-122. Cable and conduit takeoff approach.. 430

Figure 3-123. Cable takeoff method 431

Figure 3-124. Conduit sizing calculator results 431

Figure 3-125. Cable takeoff by leg 432

Figure 3-126. Conduit takeoff.. 433

Figure 3-127. Instrument conduit installation detail 434

Figure 3-128. Instrument electrical installation detail 435

Figure 3-129. Instrument mechanical hookup detail 436

Figure 3-130. Instrument mechanical detail with throttling valve........................ 437

Figure 3-131. Instrument mounting detail 437

Figure 3-132. Database log of details.. 438

Figure 3-133. Wiring design basics 441

Figure 3-134. Fabrication.. 442

Figure 3-135. Wiring interconnections 442

Figure 3-136. Elementary wiring diagram 443

Figure 3-137. Typical instrument elementary content 444

Figure 3-138. Four-pole relay coil with contacts 446

Figure 3-139. Four-pole relay coil with cross-references to its contacts................. 446

Figure 3-140. Four-pole relay contacts with cross-reference to its coil 446

Figure 3-141. Motor elementary wiring diagram 447

Figure 3-142. Motor elementary wiring diagram showing fused transformer output 448

Figure 3-143. AC power distribution elementary wiring diagram........................ 450

Figure 3-144. AC power panel loading chart... 451

Figure 3-145. DC power distribution elementary wiring diagram........................ 451

Figure 3-146. Instrument elementary wiring diagram concept............................ 453

Figure 3-147. Traditional ladder elementary—washing machine application.......... 454

Figure 3-148. "Unhide Columns" window 455

Figure 3-149. Instrument and I/O list table, filter by selection............................ 455

Figure 3-150. PLC digital input module elementary wiring diagram 456

Figure 3-151. Filtered on DOI (digital output, isolated).................................... 458

Figure 3-152. Digital output (isolated) PLC output module elementary wiring diagram 459

Figure 3-153. Loop sheet.. 460

Figure 3-154. Advanced filter/sort 460

Figure 3-155. Advanced filter/sort, field selection.. 461

Figure 3-156. Results of advanced filter/sort 461

LIST OF FIGURES

Figure 3-157. Creating a connection diagram 463

Figure 3-158. Terminal strip creation 464

Figure 3-159. Instrument elementary diagram, digital input module 464

Figure 3-160. Instrument elementary diagram, digital output module 466

Figure 3-161. Termination drawing setup 467

Figure 3-162. Termination chart 468

Figure 3-163. Motor elementary fragment 468

Figure 3-164. Finished termination chart 469

Figure 3-165. Junction box wiring diagram 470

Figure 3-166. Inner panel, cabinet TC2 471

Figure 3-167. DC wiring 472

Figure 3-168. Wiring diagram section of TC-1 474

Figure 3-169. AC power distribution 475

Figure 3-170. LT-10 power feed 475

Figure 3-171. Wire runs 476

Figure 3-172. Fuse/terminal numbering sequence 477

Figure 3-173. NFPA wire color scheme 478

Figure 3-174. TC-2 wiring color scheme 478

Figure 3-175. TC-2 PLC cabinet connection diagram 480

Figure 3-176. Partial junction box diagram 480

Figure 3-177. Partial motor elementary wiring diagram 481

Figure 3-178. Power distribution information 481

Figure 3-179. Pressure control loop PIC-48 loop sheet 482

Figure 3-180. Ladder diagram for discrete modules 482

Figure 3-181. Document control table 483

Figure 3-182. Document control table and instrument and I/O list table 483

Figure 3-183. Instrument and I/O list table 484

Figure 3-184. Terminal block 487

Figure 3-185. Setting up a scale 489

Figure 3-186. Initial layout 489

Figure 3-187. Single-door enclosure 490

Figure 3-188. Junction box with bill of materials. 491

Figure 3-189. Finished panel arrangement. 492

Figure 3-190. Typical procurement cycle 495

Figure 3-191. Bulk materials takeoff worksheet 498

Figure 3-192. Wire and cable calculation table 500

Figure 3-193. Terminations and cabinetry 500

Figure 3-194. Conduit and conduit fittings 501

Figure 3-195. Installation detail assignment data 502

Figure 3-196. New detail sheet tally 502

Figure 3-197. Material tabulation by detail 503

Figure 3-198. Consolidated material with detail quantity 503

Figure 3-199. Total item quantities 504

Figure 3-200. Part number and price. 505

Figure 3-201. Final bill of materials worksheet 505

Figure 3-202. Sort by description. 506

Figure 3-203. Engineering bill of materials 507

Preface

This book began long ago when, as a department supervisor at Raytheon Engineers and Constructors, I started a regular "lunch and learn" training program. Over time, my lesson plans evolved into this book, which now encompasses a broad spectrum of design issues.

My purpose in conducting that training was to provide perspective; to help broaden my design group—and myself—by exploring different facets of the I&C design profession. It is my belief that to be efficient, a design team must be able to anticipate troublesome issues before they arise and respond to situations quickly without much conscious thought. In this business, "conscious thought" takes the form of a design meeting or interruption in the flow of the engineering process. How much better would it be if the situation were handled real-time or even ahead of time at the lowest level possible on the design floor?

An effective organization is one in which every member of the team is aware of the issues at hand. Cross-training is expensive and difficult to implement, particularly on projects with tight timelines and budgets. But it is possible to broaden the entire team's perspective, such that their awareness encompasses more than just their particular role in the project. If the design staff has "situational awareness," that staff will consume fewer units of management effort, will be more able to react to emerging issues, and will allow a group of individuals to behave as more of a team. Situational awareness comes only from having perspective beyond one's current level of responsibility.

That is the thought behind this book: to provide perspective and situational awareness. Few books really attempt to describe the art of Instrumentation and Controls design from ground level. This book will do just that. In addition, this second edition reflects the trend toward tighter integration between the traditional engineering process and the systems integration and panel fabrication processes. End users are insisting on single-source service providers that can provide all three of these services under one purchase order. Consulting companies that can provide all three of these services as an organic product (as opposed to partnering) are said to be companies with Integrated Control Systems (ICS) capabilities. The advantages of streamlined information flow and internal process coordination give these new-look companies an advantage in the marketplace.

This book is written from the perspective of the consulting design engineer and/or consulting control technician, but is applicable to the maintenance technologist or the owner's design engineer or technician. For those who do not regularly work on design projects from conception to implementation, this book will be enlightening. For those who do, this will be a second opinion.

The book is divided into three parts:

- Part I provides perspective into the engineering business. What is a project? What are the different elements that make up the project team, and how do they interact? How does a project start? What is involved in planning and estimating? How do you track performance during the execution of the project? What are the deliverables that can reasonably be expected?

- Part II provides many of the key fundamentals of design, from the very basic to the complex. What are some of the industry standards that should be consulted? What is good design practice given certain situations? How does a relay work and when is it appropriate to use one? What is a good wire numbering scheme? Also, background information relating to the control system is given. What is a control system? What is systems integration? How should you go about selecting an integrator? These questions and more will be addressed.

- Part III provides detailed information on the various engineering products and services by expanding upon the tank pumping station example introduced in Part II. To the degree possible, the organization reflects the order of a typical project process flow. Since the end result of a design project is a set of documents that can be used to build and maintain a facility, engineering deliverables are discussed in detail. Their relative utility for construction and/or maintenance applications is discussed, and suggestions are made for how best to produce them. Low-cost alternatives to the typical product are presented where applicable.

Who should read this book? Frankly, this book has something for virtually anyone in the Automation and Instrumentation & Controls business. The book is aimed at the maintenance engineer in a plant who has not been exposed to capital project work; at the process or mechanical engineer who finds it difficult to communicate with the I&C or Automation staff; at the junior designer who needs something extra to put him or her on the path to a successful career; and at the design supervisor, who would like to get some additional tools and ideas about how to manage a project and train people.

The book's topical format—as shown in the Table of Contents—makes it useful as a desktop reference. Some of the sections are very detailed, while others merely hit the high points. It does, after all, reflect the author's personal experience. References to spreadsheet and database tools are made throughout the book. In most cases, the tools are used to teach a topic, though many of them are also practical design tools developed "on-the-fly." For example, in the cabinet arrangement task in Part III, the tool is merely a teaching aid. In that case, free

PREFACE

vendor software is available and should be used when available. Other tools, like the estimating and scheduling package, are useful to the design supervisor, regardless of any upper-level scheduling systems that help report status, but are not effective in helping manage work.

Along those lines, the CD-ROM provided with this book is a great resource for training courses and presentations since most of the figures embedded in the book are presented in their "raw" Microsoft® Excel format, ready to be used as-is or "tweaked" to fit a particular need. Called *Software Tools for Instrumentation and Control Systems Design*, the CD-ROM also includes a Microsoft® Excel-based estimating/scheduling tool.

Design is frequently more of an art than a science. Some may take issue with some of the approaches presented here, having developed other methods of their own that are, perhaps, better. But the design concepts presented here are proven and provide the keys to a successful project. And, if you believe as I do that the best-learned lessons are those learned "in the trenches," then this book is for you!

A couple of clarifications must be made before we begin. Throughout this book, "owner" is synonymous with "customer" and "buyer." The engineering company may be referred to as the "engineer," the "designer," the "service provider,"[*] the "seller" or the "contractor." "Service provider" is an entity that can provide both products and services.

---

[*] Clarification: The term "Service Provider" refers to the provision of products, as well as services.

NOTE TO THE SECOND EDITION:

The Second Edition presents a major revision to Part I. The concept of Integrated Control Systems (ICS) design is presented and described with new sections relating to the Control Systems Integrator (CSI) and the Control Panel Fabricator (CPF), and describing the interplay between those and the Electrical and Instrumentation & Controls (E/I&C) engineer. Several sections have been added to give the reader more tools for either managing a project, or for anticipating the needs of the project manager. Some techniques for leading and participating in project meetings are presented with a view toward turning a potential time-waster into a time saver. The topic of contracts was revised for readability and expanded. Edits were made to align the book better with ISA's new Automation Body of Knowledge, a work released after the first edition of this book, and for which this author was honored to be asked to contribute. Modifications were made in Part II, updating the sections on Industrial Ethernet and practical system design, and updating some of the diagrams and charts. Part III modifications were minor. The CD was updated with the new figures.

Best Regards,
Michael D. Whitt
February 2011

PART I
TABLE OF CONTENTS

List of Acronyms – Part I xxxvii

List of Figures – Part I xxxix

Introduction – Part I 1

Part I – Chapter 1: The Project 3
    A. Introduction 3
    B. Project Planning 4
    C. Contracts and their Effects on Project Structure 8
        1. Constraints 9
        2. Contract Types: The Cost-Plus Contract (CP) 10
        3. Contract Types: The Time and Material Contract (T&M) 13
        4. Contract Types: The Time and Material/Not-To-Exceed Contract (T&M/NTE) 14
        5. Contract Types: Lump Sum (Fixed Price) Contract 14
        6. Project Structures: The Turnkey Project 16
        7. Project Structures: The EPC Project 16
        8. Project Structures: The Retrofit and Green-Field Projects 17
        9. Project Structures: The Hybrid Project 17
    D. The Customer/Service Provider Relationship 17
    E. Project Flow for a Controls Project 18
        1. The Owner's Project 21
            a. Owner's Phase 1 – FEL Stage 1: Business Planning 22
            b. Owner's Phase 1 – FEL Stage 2: Project Definition 24
            c. Owner's Phase 1 – FEL Stage 3: Project Planning 24
            d. Owner's Phase 2 – System Design 25
            e. Owner's Phase 3 – Deployment 26
            f. Owner's Phase 4 – Support 27
            g. Owner's Project Deliverables 27
        2. The Control System E/I&C Engineer's Project 30
            a. E/I&C Seller Phase 1 – FEL Stage 1: Business Planning 31
            b. E/I&C Seller Phase 1 – FEL Stage 2: Project Definition 32
            c. E/I&C Seller Phase 1 – FEL Stage 3: Project Planning 33
            d. E/I&C Seller Phase 2 – System Design 33
            e. E/I&C Seller Phase 3 – Deployment 34
            f. E/I&C Seller Phase 4 – Support 34
            g. E/I&C Seller's Project Deliverables 35
        3. The Control Systems Integrator's Project 40

   a. CSI Seller Phase 1 – FEL Stage 1: Business Planning 40
   b. CSI Seller Phase 1 – FEL Stage 2: Project Definition. 40
   c. CSI Seller Phase 1 – FEL Stage 3: Project Planning 42
   d. CSI Seller Phase 2 – System Design 42
   e. CSI Seller Phase 3 – Deployment 43
   f. CSI Seller Phase 4 – Support 44
   g. CSI Seller's Project Deliverables 44
  4. The Control Panel Fabricator's Project 50
   a. FAB Seller Phase 1 – FEL Stage 1: Business Planning. 51
   b. FAB Seller Phase 1 – FEL Stage 2: Project Definition 51
   c. FAB Seller Phase 1 – FEL Stage 3: Project Planning 51
   d. FAB Seller Phase 2 – System Design 51
   e. FAB Seller Phase 3 – Deployment 52
   f. FAB Seller Phase 4 – Support 52
   g. FAB Seller's Project Deliverables 52
 F. Integrated Control Solutions 52

Part I – Chapter 2: The Project Team. 55
 A. The Owner (i.e., Buyer, Customer). 57
  1. Plant Administration. 58
  2. Plant Operations. 59
  3. Plant Engineering/Maintenance 59
  4. Plant Purchasing 59
 B. The Owner's Engineer (OE). 60
  1. The OE as an Individual 60
  2. The OE as an In-House Team. 60
  3. The OE as a Consulting Management Firm. 61
  4. The OE as a Prime Contractor 61
 C. The Designer 61
  1. Project Manager (PM) 62
  2. Project Engineer (PE). 63
  3. Discipline Lead Engineer (DLE) 63
  4. Discipline Engineer(s) 63
  5. Discipline Design Supervisor (DDS). 63
  6. Discipline Technical Support (Design) Staff 64
   a. Lead Designer. 64
   b. Designer 64
   c. CADD Technician 64
   d. Engineering Aide (EA) 64
 D. The Control Systems Integrator (CSI) 64
  1. The Process Control Team 65
   a. Process Engineers/Specialists 65
   b. PLC/DCS Programmers. 66
   c. HMI/SCADA Programmers 66
  2. The Server Setup Team 66
  3. The Network Setup Team. 67
  4. The Startup & Commissioning Team 67
 E. The Constructor (i.e., Builder). 67
  1. E&I Construction Superintendent. 67
  2. E&I Field Engineer & Coordinator 68
  3. E&I General Foreman 69
  4. Instrument Foreman 69
  5. Instrument Fitter/Mechanic. 69

TABLE OF CONTENTS – PART I

   6. Pipe Fitter 69
   7. Instrument Electrical Foreman. 69
   8. Electrician 69
   9. Instrument Electrician/Technician 70

Part I – Chapter 3: The Managed Project 71
 A. Key Project Management Tools 73
   1. The Detailed Scope of Work (SOW) 74
   2. The Estimate* 76
     a. Budgetary 76
     b. Bid 76
     c. Definitive 77
   3. The Schedule* 78
   4. The Proposal 79
     a. Executive Summary. 81
     b. General Scope of Work 81
     c. Assumptions 82
     d. Inclusions 82
     e. Exclusions (Exceptions). 82
     f. Deliverables. 82
     g. Milestone Schedule 82
     h. Safety 83
     i. Price & Payment Schedule. 83
     j. Bid Award & Contract Negotiation 83
   5. The Project Execution Plan (PEP) 84
     a. Contact List. 85
     b. Existing System Description 85
     c. Disposition of Existing Equipment 85
     d. Addition of New Equipment. 85
     e. Company and Applicable Industry Standards 85
     f. Approved Vendors List. 86
     g. Vendor-Provided Pre-engineered Subsystems (OEM) 86
     h. Instrumentation Data 86
     i. Quality Control 86
     j. Document Control 86
   6. The Status Report. 87

 B. Project Management Techniques 90
   1. Assessing Project Status 90
   2. Staff Meetings 91
     a. The Meeting Facilitator 92
     b. The Facilitator's Toolbag 93
     c. The Meeting Agenda 96
   3. Management of Change (MOC). 96

Part I – Summary. 101

Part I - References 103

List of Acronyms – Part I

BOM	Bill of Material
CADD	Computer-Aided Design & Drafting
CP	Cost-Plus Project
CPF	Control Panel Fabricator
CSI	Control Systems Integrator
CSP	Control Solutions Provider
DCDS	Device Control Detail Sheet
DCS	Distributed Control System
DDS	Discipline Design Supervisor
DTDS	Data Transfer Detail Sheet
E/I&C	Electrical/Instrumentation & Controls
ECDS	Equipment Control Detail Sheet
EPC	Engineering, Procurement, and Construction
FAT	Factory (or sometimes Functional) Acceptance Test
FEL	Front-End Loading
HazOp	Hazard & Operability Study
HMB	Heat & Material Balance Diagram
HMI	Human-Machine Interface
I/O	Input/Output
ICS	Integrated Control Systems
ISA	International Society of Automation (www.isa.org)
LDE	Lead Discipline Engineer
MOC	Management of Change
MOU	Memorandum of Understanding
NPV	Net Present Value
NTE	Not-to-Exceed
NTP	Notice to Proceed
OE	Owner's Engineer
P&ID	Piping and Instrumentation Diagram
PAC	Programmable Automation Controller
PCDS	Process Control Detail Sheet
PE	Project Engineer
PEP	Project Execution Plan
PFD	Process Flow Diagram
PLC	Programmable Logic Controller

PM	Project Manager
PMT	Post-Modification Test
RFP	Request for Proposal
SAT	Site Acceptance Test
SCADA	Supervisory Control and Data Acquisition
SCDS	Sequence Control Detail Sheet
SOW	Scope of Work
T&M	Time and Material
T&M-NTE	Time and Material, Not-to-Exceed
V&V	Validation & Verification
VFD	Variable-Frequency Drive
VSD	Variable-Speed Drive
WBS	Work Breakdown Structure

List of Figures – Part I

Figure 1-1. Typical bid package content 6

Figure 1-2. Success triangle 9

Figure 1-3. Risk to reward analysis by project type 11

Figure 1-4. Effects of constraints on project structure 13

Figure 1-5. Typical CSP project lifecycle 19

Figure 1-6. Control system project flow by involvement level 20

Figure 1-7. Sample Owner's capital improvement project plan 22

Figure 1-8. Sample E/I&C seller's project plan 31

Figure 1-9. Sample CSI seller's project plan 41

Figure 1-10. Sample FAB seller's project plan 50

Figure 1-11. Typical controls project participants 56

Figure 1-12. Engineering design team 62

Figure 1-13. Control Systems Integration design team 65

Figure 1-14. Construction team 68

Figure 1-15. Simplified contract award overview 72

Figure 1-16. Project Execution Plan template for small tasks 87

Figure 1-17. Project status report – data collection and status calculation fields 89

Figure 1-18. Project status report – analysis fields 91

Figure 1-19. Meeting Status Notes form 94

Figure 1-20. Emerging Issues Notes form 94

Figure 1-21. Needs List form 94

Figure 1-22. Action List form 95

Figure 1-23. Suggestion List form 95

Figure 1-24. Sample project meeting agenda form 97

Figure 1-25. A Management of Change (MOC) process 99

Introduction – Part I

The world of process control is a dynamic one. Design engineering professionals know that it can be a chaotic environment in which order doesn't just happen—it must be imposed. At its most elemental level, a process control system provides a means of communicating process information (e.g., temperature, pressure, level, device status, alarms, etc.) to a user (such as an operator or another process control element). The information must be accurate, repeatable and useful. Then the system must provide a means to let the user modify the manufacturing process as necessary to achieve a desired effect. The system should alert the user if control elements fail, and should react to such a failure in a way that will minimize risk to personnel and equipment. The system's documentation package should provide information to the user that will assist in the troubleshooting and repair efforts. The best way to meet these needs is to ensure that the control system is well-designed during the engineering phase.

Designing a process control system that will meet the requirements of the operational user, the constructor, the maintenance team, as well as the funding authority is a difficult task, indeed, as these needs can conflict at times. Given unlimited resources, however, a "gold-plated" control system can be designed/built/documented. The challenge is to take the limited resources available, and generate a design that is not gold-plated, but is still appropriate to the task at hand. Just as a manufacturing process takes raw material, processes it, and yields a finished product, so must the control system design process function. The raw material, in the case of the controls profession, is minutiae—a multitude of minute bits of information. In order for control systems professionals to design a control system that functions properly, the design process must provide a means of efficiently collecting, managing, and presenting this mass of information to yield a clear, concise set of design deliverables that are useful to the construction team, operations and maintenance.

A process control project includes four key players: the Owner (customer), the Electrical/Instrumentation and Controls (E/I&C) engineering design team, the Control Systems Integration team (CSI), and the Control Panel Fabricator (CPF). Historically, and still today in the majority of projects, each entity was actually a different company that would combine with the others to form a team for a particular project, and then disband afterward. Each new organization adds layers of complexity to the project for the customer and for project management. Today, sellers of engineering and construction services (sellers) are beginning to react to pricing pressures and customer (buyer) stipulations by either absorbing additional facets of the controls business into their basic set of services, or by entering into strategic partnerships with systems integrators, panel fabricators and others that allow them to present a unified business model to the customer. This consolidation process is a trend that is likely to continue, as it offers the customer an option to shop for turnkey control system solutions rather than for a laundry list of independent services.

This book describes the typical elements of a controls engineering package and the design process that creates it. The project management and basic design techniques presented here are merely one approach. But the deliverables discussed comprise the basic elements of most design packages. Some of the documents described are legal documents that must be a part of any engineering project. Some are necessary for continued plant maintenance, while others are only needed during construction. These issues are discussed in detail as appropriate. But, lest we forget, the goal here is not to create the perfect drawing, software fragment, or management report, but rather to provide a basic framework that, if followed, will expand the reader's understanding of how a design project team is organized, how design products are developed, how engineering and integration services are performed, and, finally, how control systems are deployed, thereby contributing to the reader's situational awareness.

So, what characterizes the perfect project? The perfect project satisfies the customer's needs in terms of budget, schedule and quality, and allows all the participants to meet their financial goals. These are frequently mutually exclusive, making perfection, from everyone's point of view, a rare and wondrous thing. And, once perfection is approached, it is all too often short-lived. There are just too many variables. A design package that is perfectly applied in one application may be quite unacceptable in another. The phrase "level of detail" is the bane of all design engineers. A highly detailed package that is the ideal tool for construction might blow the engineering budget, or might be worthless for maintenance. Conversely, a design package that is too sparse might spare the engineering budget, but cause huge cost overruns and delays in construction. So, beyond discussing design content, this book also delves into the *business* of design engineering. For example, how can a design team organize itself to produce information once, and use it for multiple purposes? How should the design process be modified between fixed-cost and cost-plus projects? A clear view of the desired end must exist from the beginning.

Part I begins to build this background of understanding by providing perspective into the engineering business. What is a project? What are the different elements that make up the project team, and how do they interact? How does a project start? What is involved in planning and estimating? How do you track performance during the execution of the project? What are the deliverables that can reasonably be expected?

The process begins with *the project*…

Part I – Chapter 1: The Project

A. Introduction

> ***Project:*** *A project is a temporary activity whose purpose is to create a product or service. Projects usually involve a sequence of tasks with definite start and end points. These points are bounded by time, resources and end results.*[1]

> ***Project Engineering:*** *1. Engineering activities associated with designing and constructing a manufacturing or processing facility. 2. Engineering activities related to a specific objective such as solving a problem or developing a product.*[2]

For our purposes, an engineering design project encompasses the delivery of engineering services, documents and components sufficient to meet customer objectives. In short, it is a means to an end, in existence but for a short time and a specific purpose. Other facets of a project include fabrication, installation, checkout and startup, but this chapter focuses primarily on the design facet. For a project to exist, there must be a perceived need and an expectation that the need can be met with a reasonable investment. The customer must weigh the risks against the rewards and conclude that the project is worthwhile. Although there are established approaches to quantifying risk (using analysis methods such as Net Present Value (NPV) and Payoff Period techniques)[5] making this risk/reward assessment is sometimes more of an art than a science.

Every project involves some level of risk....

> *All pilots take chances from time to time, but knowing—not guessing—about what you can risk is often the critical difference between getting away with it and drilling a fifty-foot hole in Mother Earth. — test pilot Chuck Yeager, 1985.*[4]

More than in most endeavors, the effects of mishandling risk in process engineering projects can be catastrophic. Beyond the economic ramifications of a poor estimate, which are bad enough, the potential risk to process operators and even the public at large can be extensive. Therefore, a well-conceived process of preliminary evaluation and project risk management is necessary. This evaluation begins with a thorough understanding of the issues. As General Yeager stated, the key is to know, not guess. Risk management has three components—assessing, planning and managing risk—that will affect the project timeline, scope, and budget.[5]

The true keys to project risk management are knowledge, experience and forethought. Each member of the project team needs to have a thorough knowledge of the issues. What is the project? What are some of the outside influences on how the project is to be executed? How will the project flow from inception to implementation? What are the typical products and services to expect from an E/I&C service provider? Who are the players?

This chapter attempts to answer these questions by discussing the following major topics:

A. *Project planning.* Defining the project. What type of project is it, deterministic or probabilistic?

B. *Contracts and their effects on project structure.* What are some of the most common contract types?

C. *Project flow for a controls project.* Who are the players, and what are their interrelationships? What deliverables can be expected from each? How can their efforts best be orchestrated? What is meant by the term Front-End-Loading (FEL)? Four organizations are discussed:

- The Owner and Owner's Engineer (OE)
- Electrical/Instrumentation & Controls Engineering & Design (E/I&C)
- Control Systems Integrator (CSI)
- Control Panel Fabricator (CPF)

D. *Integrated Control Systems (ICS).* What are some of the benefits of combining service sets into an integrated whole? What should an owner expect from such a company or group of teamed companies? What advantages does a solutions-based approach provide as opposed to a services-based approach?

B. Project Planning

Most projects fail because of poor planning on the front end. How certain are you that your project plan will survive through implementation? If your original plan is flawed, how able are you to adjust? If it is a poor plan with a lot of uncertainties, how likely is a successful outcome?

The responsibility for a properly planned project is shared between the owner and the service provider. It is the owner's responsibility, initially, to identify or pre-certify a number of qualified bidders. Then he must articulate what he wants his project to accomplish through his Request for Proposal (RFP). The RFP (which consists of an invitation to qualified bidders, followed by the Bid Package) contains (Figure 1-1):

- *The Bid Specification*, a section of the RFP that describes how the bid package should be configured. In competitive bid situations, it is important that the owner receives all bids in the same format in order to facilitate comparison.

- *The Project Scope of Work (SOW)*, a section of the RFP that describes the work to be done, which areas of the plant are affected, etc. Quality-based incentives or penalties are discussed in this section.

- *The Project Milestone Schedule*, a section of the RFP that provides dates of particular interest to the bidder as he develops his proposal. This section should also discuss any schedule-based incentives or penalties that are applicable.

- *The Project Specification* is usually a separate document that is referenced by the RFP and provides information relative to design package content: how the work products are to be formatted, what site and industry standards are applicable, etc. The owner may have guidelines on how to number wires, for example, or he may want the drawing borders to appear a certain way. A Project Specification is usually not written to be specific to a particular project, but describes how any project should be done for that site or that owner. Project Specifications are generally written once and published many times.

- *The Functional Description* is usually a separate document that is referenced by the RFP, and augments the SOW by providing operational detail, primarily for the systems integrator. The Functional Description describes the controls issues relative to the SOW.

After the bid package is released, project planning responsibility passes to the prospective service provider as he uses the bid package to develop his estimate, initial execution plan and subsequent proposal. After release of the proposal, responsibility passes back to the owner as the proposal is evaluated. This provides a series of checks and balances that should eliminate most of the unknowns prior to contract award. If any party fails to execute his due diligence, then the project is at risk. For the owner, as his level of sophistication and ability to communicate his needs increase, so does the likelihood that he will get the results he desires. To that end, it helps if the owner knows as much about how each service provider operates as possible. For the service provider, being able to discriminate

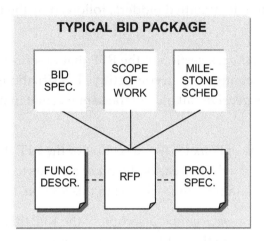

FIGURE 1-1. TYPICAL BID PACKAGE CONTENT

between what you're being asked to do and what the customer really needs is key.

Regardless of the owner's abilities to plan and communicate, the RFP is, at best, an approximation of his wishes. The owner must rely heavily on the service provider(s) to partner with him during the proposal cycle as they interact to further define the project scope, and by extension reduce the financial and physical risks inherent in the project. In this respect, everyone's interests are in complete alignment.

After the proposal is submitted, the bidders depend on the owner to execute due diligence in making sure that every prospective seller has satisfied the stipulations of the RFP. Regardless of what is in the RFP, the seller can only be held accountable for what is stated in his proposal. Note: if the seller plans to omit items the owner has requested in the RFP, Project Specification, or any other document presented for his use in proposal development, then the proposal must take explicit exception to those areas of non-compliance. Silence implies compliance.

For a prospective seller, the process of defining the project has to begin with an analysis of the customer himself, and an evaluation of the customer's ability to articulate what he really wants to achieve. Sometimes this is difficult to do if the project is with a new customer. The seller's tendency is often to assume the customer knows what he wants. Only after working closely with the customer will the seller really know whether that is the case or not, thus making initial projects more risky than subsequent ones. This is one more reason to make every attempt to generate repeat business, as the seller's cost of breaking in new customers is significant.

The seller must gain an understanding of the customer's true objectives, both stated and implied. He must then work with the customer to reduce the number of implied objectives by having the customer revise the bid package or issue written clarifications. Only by getting the bid package revised will he be assured that all parties will be bidding on the same scope of work. Then the seller must reduce those often complex, vaguely stated goals to a series of well-defined, simplified lists of products and services that his staff can execute in a predictable, cost-effective manner. He must then build a labor, materials, and expenses estimate that will allow him to satisfy the bid package and still turn a profit.

This list of products and services is the seller's Scope of Work (SOW), which may cover only a subset of the project's SOW activities. From the seller's point of view, the economics of the project and his SOW are inextricably linked—if one changes, so should the other. The seller's SOW becomes the core of his proposal. (For more information on proposal preparation, see Chapter 3.)

After a seller has been awarded the contract, he should re-organize the information to make it more digestible for his internal organizations. The bid package and Project Proposal become the basis for a new internal document called the Project Execution Plan (PEP). (Note: Figure 1-15 depicts the complete contract award process, which is summarized here.) The PEP will restate the project in terms that are most useful to the seller's execution team(s), providing amplification where necessary. If the project is large enough, a separate, more detailed PEP will often be produced that better defines timing relationships and dependencies between internal organizations.

The trick for the customer is to find service providers who understand these processes and can adapt them to specific circumstances. For the service provider, having customers that understand the process streamlines everything. It is in everyone's interest to foster tight communications and thereby expand the success triangle (Figure 1-2). The success triangle is formed at the point where the goods delivered meet the customer's cost and quality expectations, while still allowing the service provider to make a fair profit and maintain his reputation for good quality work.

Fortunately, while project objectives, strategies, and outcomes may vary widely, most industrial project lifecycles are similar in structure (as will be discussed in Figure 1-6 and its associated discussion). In terms of project flow, a process is begun that is, for the most part, predictable from one project to the next. The customer has a predictable process to follow to get the bid process started. The service provider has a predictable process to follow to prepare a bid. Engineering disciplines have predictable processes to follow in generating their work products. These processes are sometimes adjusted to meet unexpected or unusual conditions. But for the most part, the engineering process that takes an idea from conception to implementation is well defined and consistent, project to project.

Such universal processes lead to a *deterministic* management approach, as opposed to a *probabilistic* approach, which is used for unique processes.

Given the deterministic nature of the engineering process, it would be easy to assume that the interrelationships between customer and service provider would be predictable from one project to the next. This is far from the case. Human nature is involved, further complicating an already complex situation. In addition, external pressures, such as financial conditions in the marketplace, may force changes in strategy from project to project, thereby imposing compromises and more complexities. Thus, it is necessary to establish contractual relationships that expose the risks, and set penalties and rewards that are commensurate. These risk/reward relationships must be given full consideration when setting up the project.

A major factor in project definition is the type of contract in use. The following section explains the various contract types and how they affect the execution strategy of a project. Is the project time driven, or cost driven? Is the project fixed-cost or cost-plus? What is the customer/service provider relationship?

Later sections of this chapter deal with the execution of the project: How does a project develop? Where do you start, and what is the best process to follow? What set of deliverables should be expected from the design team? What is the makeup of a design team? What are the members' roles and responsibilities?

C. Contracts and their Effects on Project Structure

The *structure* of a project has a great bearing on the way the project is executed. This is because each member organization of the project team (customer and service provider) defines *success* from its own unique perspective, as viewed through the prism of the project parameters. From the customer's point of view, success is achieved when the desired end is achieved within the time allotted and/or the funds allocated. This is likely to be a narrower interpretation than that of the service provider, who is also interested in making a profit and maintaining a reputation. As we have seen, a truly successful project is one in which both the customer and the service provider are satisfied with the outcome. For this to occur, a zone of success (the success triangle) must be created that is as large as possible (Figure 1-2).

Staying within this comfort zone is sometimes a bit tricky, and it really helps to understand the physics involved. The "physics" of a project are defined by the forces that both drive and constrain it. The driving force(s) behind a project are the things that compel an owner to take action. Project drivers can be market-based, safety-based, regulation-based, convenience-based, or any combination of these. Project constraints are those things that limit what can be done. The inter-

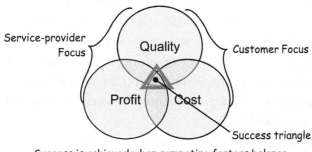

Success is achieved when competing factors balance.

FIGURE 1-2. SUCCESS TRIANGLE

play between these two forces must be managed within the context of the contract.

1. CONSTRAINTS

Time (schedule) and resources (cost) are two parameters that impose limits on the design process. Time can mean either duration or intensity. Duration is measured by units of the calendar (days, weeks, months); intensity is measured by units of labor (e.g., man hours, man-weeks). The term *time driven* in this context implies that the project is constrained by the calendar; the term *cost driven* implies that the project is constrained by the level of labor intensity (in man hours) plus materials and expenses. A third constraint – content (quality) – subjugates both schedule and cost in order to achieve a specific end. A performance specification might define a desired outcome, but might not specify an associated project cost or schedule.

Usually, the mix of these constraints influences major performance goals that might be set for a project. A time driven project, for example, is one that locks in an aggressive schedule. Such a contract sometimes provides a bonus or some other incentive to the service provider if he meets schedule and quality targets and a financial penalty if he does not. In such a project, the ability to control cost is inhibited, as the service provider may be forced to pay overtime premiums, hire additional personnel, or hurry the design, thereby risking rework. The cost control mechanism in this case is the short project duration, which itself tends to limit the amount that can be effectively spent. A cost driven project, on the other hand, is oriented toward limiting cost. The performance incentive would thus be related to meeting project cost and quality targets, while subjugating the schedule, which becomes flexible and can be adjusted for optimum design efficiency. In both cases, the goal is high quality. In practice, quality can be affected negatively in either approach if poorly executed.

Any of these project types can get the job done (Figure 1-4), but the management protocols are very different, as characteristics of each type are somewhat mutually exclusive. If calendar time is compressed, then there will likely be an increased cost per unit of time. If the calendar is not an issue, then expenditures can be better managed and perhaps minimized, provided the project is properly managed. The characteristics of these project types can be applied to any of the project structures listed below.

The contractual relationship between the customer and the service provider (engineer or constructor) requires clear definition. When an equipment vendor provides a quote, then that vendor is bound by it, and must provide the materials or services for the price offered in the quote—even if his suppliers subsequently raise their prices. The same concept applies to a service provider. For the service provider (seller), the act of submitting a proposal constitutes initiation of the contracting process. If the customer (owner/buyer) accepts the proposal, the seller is legally bound to execute per the contract, and the buyer is legally bound to honor it. Following are some of the most common types of contracts:

- Cost-Plus (CP)
- Time and Material (T&M)
- Time and Material, Not-to-Exceed (T&M, NTE)
- Lump Sum (also, Fixed-Price)

Each contract type strikes a different balance of risk/reward for each participant, as noted in the following commentary.

Figure 1-3 depicts the relative risk/reward factors of several of the most common types of contracts. Each contract type is analyzed with respect to the project constraint, and rated on a scale of 0–3 as follows:

0: None
1: Minimal
2: Moderate
3: Maximum

A Risk to Reward ratio of 1:3, therefore, would indicate the condition has minimal risk, with maximum reward ... a very desirable state indeed for the concerned party.

2. CONTRACT TYPES: THE COST-PLUS CONTRACT (CP)

A Cost-Plus contract guarantees a fixed, agreed-upon profit for the seller. In return, the buyer may retain a large measure of control over project content and seller's method of execution. The parameters defined by the contract are unit

Risk / Reward Analysis			
Contract Type	(Seller) Profit	(Shared) Quality	(Buyer) Cost	Project Constraint
Cost-Plus	0 / 1	0 / *	3 / #	Content
Cost-Plus	0 / 1	2 / *	2 / #	Schedule
T&M	0 / 2	0 / *	3 / #	Content
T&M	0 / 2	2 / *	2 / #	Schedule
T&M, NTE	1 / 2	0 / *	1 / #	Content
T&M, NTE	3 / 2	2 / *	1 / #	Schedule
T&M, NTE	3 / 2	2 / *	1 / #	Cost
T&M, NTE	3 / 2	3 / *	1 / #	Cost & Schedule
Lump Sum	1 / 3	0 / *	3 / #	Content
Lump Sum	3 / 3	2 / *	2 / #	Schedule
Lump Sum	3 / 3	2 / *	0 / #	Cost
Lump Sum	3 / 3	3 / *	0 / #	Cost & Schedule

* High quality is the reward for all parties in all cases…
\# Low cost is the reward in all cases for the Buyer…
( ) Indicates concerned party: Seller, or Buyer

FIGURE 1-3. RISK TO REWARD ANALYSIS BY PROJECT TYPE

rates—not project price—plus expenses. The unit rate can be applied either to the various employees' hourly rates or to negotiated rates based on employee classification (e.g., engineer, programmer, designer, clerk, etc.). These unit rates are valid over the life of the contract, which can extend into the future until the scope of work is satisfied. For the seller, this guarantees a minimum negotiated profit, regardless of project constraint (Profit: Minimal risk, Minimal reward). If the seller completes the task below budget, the buyer realizes a windfall. But if the project exceeds budget, the buyer is obligated to pay for the services rendered.

Invoicing typically occurs at regular intervals rather than at pre-designated milestones. Expenses are generally reimbursed at cost, plus a pre-negotiated administrative rate charge of, perhaps, ten percent. Travel expenses include such things as hotels, gasoline, airfare, car rental fees, parking fees, food, etc. For auditing purposes, the buyer usually expects to see labor charges on a weekly or monthly basis, along with an itemized list of expenses. The seller may be asked to provide a document describing work accomplished during the timeframe in question. The rights and responsibilities of the seller with respect to reporting labor charges and expenses need to be clearly defined in the contract.

Refer to Figure 1-3: If the constraint is *content*, the buyer in a cost-plus scenario will eventually get what he wants (Quality: 0; no risk), but with a very real risk of high cost (Cost: 3; high risk). In this scenario, the seller may need to work iteratively until the buyer is satisfied with the result. The seller has no risk, but his reward is usually a small multiplier against his rate, and so profit is predefined and generally small as a percentage of the project cost. But if *schedule* is defined as a constraint, then the risk to quality rises to moderate levels with a corresponding drop in the risk of cost overruns from high to moderate. The seller just has no time to apply an infinite amount of resources. Note that cost is not a constraint factor in this type of contract, even though the buyer always has a budget that was approved internally. The buyer's internal budget has no legal bearing on this type of contract. Though it is in the seller's best interest to know the buyer's target and to try to work within it if he wants repeat business, he is not contractually obligated to do so.

More often than not, the cost-plus contract is not awarded as a result of a bid evaluation for a specific project. Rather, this type of contract is usually negotiated based on hourly billing rates and is most often valid for a period of time, such as one year or three years. Once this type of contract is in place, the customer can request support without further formal negotiation. Usually the customer asks the contracted provider to do a task, they come to an understanding of how many hours the task is likely to take, with the cost based on the contractually negotiated rates, and the project begins. This type of arrangement is best suited for a customer-provider team that has a long-standing relationship. Cost-plus relationships tend to foster a *team* atmosphere between the customer and the service provider, as opposed to the fixed-cost (lump-sum) project that sometimes fosters a contentious relationship, as many sellers low-ball the bids to get the job and then rely on change orders to stay solvent. Cost-plus relieves the pressure on the provider to protect profit since a modest-to-moderate profit margin is assured, depending on how effective he was in protecting profit while negotiating the contract. This frees the provider to concentrate on meeting not only a customer's stated needs, but also his unstated preferences, as long as the number of hours expended does not exceed the amount previously agreed upon and as long as proper communications ebb and flow as needed.

The cost-plus contract's greatest benefit to the customer is that he retains a large measure of control during the life of the project, even to the point of having the provider rip out work and start over if that is what is desired. The benefit to the provider is that his profit anxiety is greatly relieved. Profit anxiety is replaced by a desire to maintain the long-term relationship at a level that will guarantee a contract extension after the current period expires. The trick to a cost-plus arrangement from the provider's standpoint is to satisfy the customer's production and maintenance groups by being flexible in the design (which frequently increases project cost and duration) while simultaneously satisfying the customer's management team's overall budgetary and schedule requirements. This dichotomy can present the provider with conflicting requirements, each tending to place limits on the other. For the service provider, it is unwise to focus on any one of these elements to the exclusion of the others.

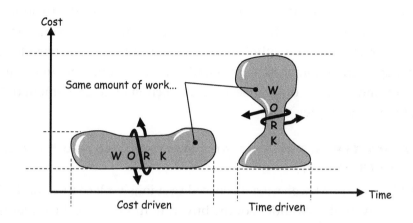

FIGURE 1-4. EFFECTS OF CONSTRAINTS ON PROJECT STRUCTURE

3. CONTRACT TYPES: THE TIME AND MATERIAL CONTRACT (T&M)

The T&M contract and the Cost-Plus contract are very similar. For Cost-Plus, the seller is reimbursed for his labor cost, plus profit and expenses. The T&M contract reimburses for labor cost, plus profit and expenses, plus materials and markup. This kind of arrangement is generally for shorter terms (months, or month-to-month) and is related to a particular project or task. Again, the parameters defined by the contract are unit rates and materials markup—not project price. Therefore, the only major difference between the two is a higher potential profit for the seller due to his markup on materials (see Figure 1-3). Note that the T&M contract's risk/reward scenarios for Quality and Cost are the same as for the Cost-Plus contract.

As in the Cost-Plus scenario, invoicing typically occurs at regular intervals rather than at pre-designated milestones. Expenses are generally reimbursed at cost, plus a pre-negotiated administrative rate charge of, perhaps, ten percent. Materials are generally reimbursed based on an agreed-upon markup. Travel expenses include such things as hotels, gasoline, airfare, car rental fees, parking fees, food, etc. But in this case, materials are included also. Material costs could include fabrication materials, test equipment rental, bulk materials, such as nuts, bolts, screws, certain consumables, etc. For auditing purposes, the buyer usually expects to see labor charges on a weekly or monthly basis, along with an itemized list of expenses and materials procured. The seller may be asked to provide a document describing work accomplished during the timeframe in question, and to provide a bill of material of anticipated purchases. The rights and responsibilities of the seller with respect to reporting charges need to be clearly defined in the contract.

Both Time & Material and Cost-Plus are tried-and-true contract types that encourage stable, long-term business relationships between buyer and seller—

provided the seller guards the buyer's interests and manages the relationship. Good project management techniques are as important for the management of these two contracts as for any other—perhaps more so, since the need for good project management is sometimes not readily apparent, and a tendency toward complacency and laxness could arise. (See the section on Management of Change [MOC] later in this chapter.)

4. Contract Types: The Time and Material/Not-To-Exceed Contract (T&M/NTE)

The Cost-Plus and T&M scenarios discussed previously place most of the budgetary risk squarely on the shoulders of the buyer. In those cases, the buyer is obligated to pay the seller for his expenditures plus profit, regardless of whether the project is under or over budget. The Not-to-Exceed stipulation reverses this by adding additional constraints in the buyer's favor. In this case, the buyer is protected from a cost overrun by virtue of a spending cap. The seller's potential reward remains modest to moderate, as in the straight T&M contract, as long as the project finishes under the buyer's budget cap. But the seller's profit risk quickly climbs to high levels as the NTE constraint is approached. High risk/ moderate reward for the seller causes an increase in his intensity and focus, and decreases his desire to work with the buyer to tweak the product to satisfy buyer preferences. Thus, the buyer loses some control over the way the project is executed, and the likelihood of disagreements between the two organizations increases. It is the seller's responsibility to report impending overruns and to attempt to negotiate contract extensions as the limit is approached. The buyer may not agree to participate in these negotiations, but the seller is still obligated to fulfill the requirements of the contract. The exception to this is scope that is added by the buyer; a proportionate addition to the spending cap should be negotiated immediately when the scope is added. (See MOC for more information on change management.)

5. Contract Types: Lump Sum (Fixed Price) Contract

The terms Lump Sum and Fixed Price are interchangeable. In the lump sum contract, the parameters defined by the contract are a fixed project price for a fixed set of deliverables. Since the price is negotiated before the services are rendered, the fixed-price contract minimizes cost overrun risk for the buyer. Further, the buyer usually signs a contract only after a bidding process in which several potential sellers compete for the contract by providing their lowest bids. The buyer can either accept the lowest bid, or analyze the bids to determine the lowest reasonable bid that will save him the most money while letting him retain a sense of comfort that the seller can actually execute the terms of the contract.

The profit risk to the seller depends upon whether the service or product is deterministic or probabilistic. If the product is widgets, for which little or no new research and development is needed, the product is deterministic, and presumably the seller knows exactly how much resources and funds are required in their production. If the project involves making retrofit modifications to an existing facility in which drawings and/or software listings are out of date, for example, then the project is deemed probabilistic, and his risk is increased. One way for the seller to mitigate his risk, and perhaps maximize his reward, is to add contingency to his bid. Contingency is a factor used to cover normal design development issues that are hard to quantify. If the level of uncertainty is low, then the level of contingency can be low. If not, then contingency should be commensurately higher. Finding the proper amount of contingency that reflects the balance between the level of uncertainty that exists and the level of risk that is deemed acceptable is something of an art.

Contingency can also be reflective of the seller's motivation to win the project. Sometimes the seller may need the work badly and be willing to forgo profit in order to retain staff, leading him to reduce contingency and take on more risk. Conversely, if the seller's staff is busy, it might lead him to raise contingency to reflect the problems inherent in overworking his staff or in having to hire and train new staff members. In either case, if the project is executed with contingency unspent, it becomes profit.

Profit for the seller is at risk depending on the scenarios described above. But the possibility of reward is also high due to the possibility of retaining the contingency (profit: high risk/high reward). Invoicing is generally done based on delivery milestones. Auditing from the buyer's standpoint is minimized, and the seller is rarely obligated to provide backup data to validate his invoices. The products and services being delivered are validation enough.

A fixed price (lump sum) project offers the buyer several benefits, the foremost of which is an enhanced ability to allocate resources. The buyer can set aside project funds (plus a safety buffer—his own contingency) with a high level of confidence that additional funds will not be required. This works to his advantage in planning other projects. The buyer gains these benefits, but in comparison to the cost-plus contract, loses much control over how the project is executed. To submit fixed price bids, bidders generally work at their own expense to clearly define not only the set of deliverables, but the methods they expect to employ to meet the buyer's defined scope of work. These methods cannot be controlled by the buyer except as specified in the Request for Proposal (RFP).

A fixed price project, if properly managed, can be the most efficient and effective format for both organizations. The buyer knows what his cost will be from the beginning, and the seller can make a higher profit if a better execution method is found, or lower-cost materials are found. However, the seller must be ever vigilant in managing scope creep. Once the buyer accepts a bid, the seller has no obligation to adjust the deliverables or methods at all. If the buyer makes a request that is out of bounds with respect to the scope of work, the seller has the right—and obligation—to refuse the request until the buyer approves an engineering change order (see Management of Change), which may also trigger a renegotiation of the price. This defensive posture on the seller's part can lead to a fractious relationship with the customer if a previously agreed-upon method of change control is not employed.

6. Project Structures: The Turnkey Project

> *Turnkey:* Of, pertaining to, or resulting from an arrangement under which a private contractor designs and constructs a project, building, etc., for sale when completely ready for occupancy or operation: turnkey housing, turnkey contract.[1]

In the context of a controls project, a turnkey project is governed by a fixed price contract. The customer wants to have the final product show up with minimal to no participation on his part. The seller's only obligation, once the project is awarded, is to meet the specification and schedule. Once the project is awarded, the customer's input drops to virtually nil with respect to how the project is executed. Projects that provide self-contained equipment that is shipped from the factory fully functioning (a "skid") are examples of turnkey projects. The customer merely mounts the skid somewhere in his facility, hooks up the utilities and makes any process connections that are necessary, and the system is fully operational. The cost and profit for all of the ICS-related services are embedded in the price of the skid.

7. Project Structures: The EPC Project

EPC is an acronym for *engineering, procurement, and construction*. An EPC project is one in which the customer designates a contractor as a single point of contact. This contractor is referred to as the Prime and is the only project participant to be directly paid by the customer. The Prime must obtain bids, award sub-contracts, and provide oversight as necessary to coordinate the activities of all of the sub-contractors. The Prime bears full responsibility for the success of the project, assuming all liability from the standpoint of the customer. To maintain a reasonable risk/reward ratio, the Prime can be expected to have a higher profit target than would typically be the case due to the additional risk. EPC projects may be governed by lump-sum, T&M, T&M NTE, or a combination of each.

---

1. turnkey. (n.d.). *Dictionary.com Unabridged* (v 1.1). Retrieved November 22, 2008, from Dictionary.com website: http://dictionary.reference.com/browse/turnkey

8. Project Structures: The Retrofit and Green-Field Projects

A *Retrofit* project is one in which an existing process facility is being modified. Retrofit projects can be difficult to quantify on the front-end, and so lend themselves more often to the T&M contract or a derivative thereof. A *Green-Field* project is more suited for a lump-sum type of contract because nothing material exists. With everything being new, it is easier to define the project design basis. The exception to that statement is the research & development project, which has no established precursor.

9. Project Structures: The Hybrid Project

Any project can, and most of the large ones do, exhibit several of the characteristics described above. For example, if the scope of work is well defined, the engineering, procurement and construction portions of the project might be done on a lump-sum basis, while unknowns about final operational tweaking might drive startup and commissioning to be handled under a separate T&M agreement. On the same project, pre-engineered turnkey subsystems, such as package boilers and water treatment systems, could arrive and be installed.

D. The Customer/Service Provider Relationship

So, what makes a good service provider from the customer's perspective? For that matter, what makes a good customer from the service provider's perspective? In both cases, it comes down to building a relationship based on some level of verifiable trust. Each party needs to feel that the other, while wanting to win themselves, at least does not want the other party to lose. The contractual configurations described above are really frameworks that define the environment within which a relationship may be built. Both sides may be honorable, and have good intentions, yet if one is not fully aware of that environment, he may misunderstand the actions of the other, or even accidentally do harm.

This relationship begins to be built in the proposal phase. For example, as a service provider, how do you rate a customer? Is your customer sophisticated enough, and is the project well-defined enough, to let you provide a lump sum proposal with minimal contingency, or do you feel compelled to inflate your price to the razor's edge of what you think he can afford in order to cover your potential risk? If the latter is the case, it would be best to let your customer know that your contingencies, in your opinion, are too high—you may either need the funds to survive, which is not good for you, or you may make a profit that is unseemly, which is not good for him. Either the customer needs to refine his RFP, or the project would be best executed as a T&M. This is a good conversation to have, as it lets the customer know you have his interests at heart, as well as your own.

If the customer is unable or unwilling to refine his RFP, and yet is uncomfortable with a straight T&M, then perhaps you can use a hybrid scheme that breaks the project into two segments, the first, smaller segment being the investigation phase done under a T&M/NTE type of contract, the second (detailed design phase) as a lump sum. As a seller, the T&M/NTE arrangement is usually the last thing you want to do, as it places all the advantage on the buyer's side of the ledger. But in this case, it works for you because of the limited scope. At this point, you usually have a short list of items to investigate, trips to take, etc. All you need to do is firm up the scope of work, so it should be easy to arrive at a good NTE number. This strategy gives both parties a breakpoint where they can re-evaluate and tighten up the proposal for the bulk of the project. Then, when enough of the unknowns are removed, a firm price can be agreed upon, and the remainder of the project can be executed under a lump-sum contract.

As the seller, how do you react if the buyer balks at your suggestions to manage your contingency? He may even tell you, at that point, what his maximum price is. If he does that, then it may be assumed that his price is significantly less than yours. Few sellers will walk away at this point, unless the price is one at which no profits are possible. Let's assume the seller takes on the project, knowing that success, from his point of view, is questionable. As the project develops, the buyer can expect the seller to be very rigid in his interpretation of scope—even to the point of being uncooperative. For the seller, building a long-term relationship may become secondary to surviving the project. And if the seller has work already, he may not make himself available the next time this customer has a project up for bid. While the buyer may win on this one project, there is a potential cost later because he will then have to orient a new service provider, opening himself up again to the same first-timer risks. It is always better to work together to arrive at a scope of work that can be executed for the funds available, even if that means the buyer has to compromise.

As the buyer, why not just do the project as a T&M? If the scope is undefined to the point that the seller can't provide a reasonable bid, then it would not be smart to open up your pockets to him. Either the seller is incompetent, or he really needs to have more information. In either case, it would be smart to make a small investment to let the situation sort itself out, and go the hybrid route. Or you could open the project up for competitive bid and get feedback on your RFP from more than one seller. That, too, requires an investment, as it will take both time and money to seek qualified sellers and evaluate multiple bids.

E. Project Flow for a Controls Project

A project may involve many engineering disciplines and organizations. This book focuses on the controls aspects of a project, which limits the discussion to electrical, instrumentation & controls, systems integration, and panel fabrication. For the purposes of this discussion, each organization is a separate company that must be managed by the owner's engineer (i.e., an individual or, for large projects, perhaps a separate engineering company that is contracted to manage the project for the owner), though companies do exist that combine all three major elements. These companies are called Control Solutions Providers (CSP)[7] and will be discussed at the end of this section.

No matter what style of project, whether retrofit, green-field, turnkey, or EPC, most industrial projects are similar in the way they develop. At least four phases are identifiable (Figure 1-5). There is first a period of investigation (Phase 1), followed by a period of design execution (Phase 2), followed by a period of deployment (Phase 3), and then a period of consolidation, closeout, and support (Phase 4).

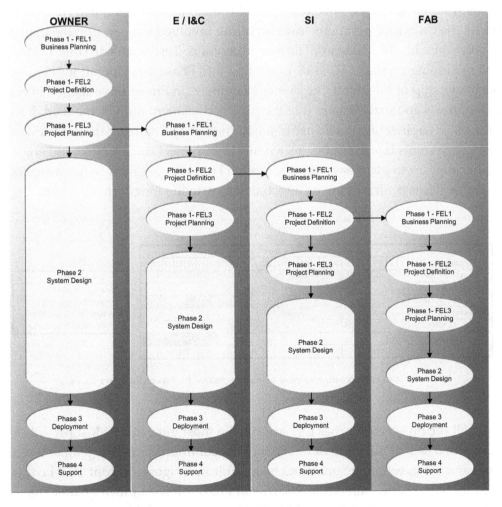

FIGURE 1-5. TYPICAL CSP PROJECT LIFECYCLE

Figure 1-6 shows a thumbnail sketch of a typical project as the involvement intensity of participants ebbs and flows over time. A project generally begins with the owner's operations team identifying a need (Time Period 1). The owner's engineer goes through an evaluation process that results in a Request for Proposal (RFP) for each of the E/I&C, CSI, and CPF sellers (Times 2, 3, and 4). The owner's engineer evaluates the bids and awards contracts (Time 6). The E/I&C seller starts design work and develops product necessary for the CSI (Control Systems Integration) team to begin work (Time 8). The CSI team adds value to the E/I&C team's work to finalize it, and the Panel Fabricator can get started (Time 9). At Time 12, all the design work culminates with a Factory Acceptance Test (FAT), after which all the control system equipment is shipped to the site. The owner's engineer prepares a Construction Bid Package and awards the construction contract (Time 13). For the first stage of construction, the E/I&C seller is closely aligned with the constructor to provide guidance and construction support (Times 13-15). At about Time 15, the CSI seller begins to increase his presence on site, as systems start to be powered up, and functional testing begins. At Times 17 and 18, the owner's Operations team is getting involved in operating the plant in cooperation with the CSI team. The Site Acceptance Test occurs at Time 17, and Commissioning occurs at Time 18, extending into Phase 4 as Operations begins to take ownership of the plant. The owner's engineer, in Phase 4, is engaged in closing loopholes and making sure that the project documentation is finalized. Each of the other engineering participants starts Phase 4 by finalizing their documentation. The Systems Integrator begins to shift from project construction and engineering support to plant operations support (Times 23, 24), after which a separate general services contract may be negotiated to provide long-term support.

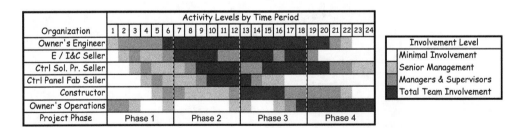

FIGURE 1-6. CONTROL SYSTEM PROJECT FLOW BY INVOLVEMENT LEVEL

Of all the phases of a project, Phase 1 is the most important. Most projects fail because of a lack of planning or not effectively communicating the plan to the customer and the execution team. Projects that follow a rigorous Front-End Loading (FEL)[2] process that eliminates unknowns in Phase 1 have a much greater chance of success than those that do not. FEL, for the purposes of this discussion, is a three-step process that begins at project conception. It involves building a business model and performing initial feasibility studies (FEL1), building a scope of work, developing a preliminary budget, schedule and estimate (FEL2), and building a definitive estimate and project execution plan (FEL3). If the project is still a "go" after the three FEL steps of Phase 1 have been successfully navigated, then the likelihood of the project being both valuable and successful is greatly enhanced.

Figure 1-5 depicts this cycle with all four major participants: the Owner, the Electrical and I&C Engineering supplier, the Systems Integrator, and the Panel Fabricator. Even though each participant starts at a different time in the project's lifecycle, each follows the same predictable four-phase pattern that culminates in project completion. The same model could be extended to the Civil, Construction, or Mechanical organizations, but that is beyond the scope of this book.

For the purposes of the discussion in this section, the project is considered to be a large capital project with multiple contracted organizations, but the steps and concepts discussed are scalable and will also apply to smaller projects. The owner must provide guidance and oversight throughout, as successive organizations come on line at the appropriate time (Figure 1-5). As each subsequent organization gets involved, that organization should perform many of the same FEL investigative steps in their own self-interest that the owner went through. This section steps through each of the processes, giving an overview of what each step means from the perspective of each participant. Deliverables are described in general terms, with more specific examples provided in Part III of this book. It should be noted that the dividing lines between some project phases are fuzzy at best. But this is a good approximation of how a typical project flows. First, the owner…

1. The Owner's Project

As previously discussed, a project is a temporary activity with a purpose. For the owner, it begins with a need, usually one related to a production concern or a profit motive, coupled with the availability of a good market window, and ends either with the need met, or, if due diligence dictates, an intentional abort. Many projects are started and subsequently aborted as the evaluation process evolves to reveal insurmountable problems. On the other hand, projects that are not properly vetted may go on to fail, as those insurmountable problems are not identified until too late. Several hurdles should be crossed internally before funding appropriations can be responsibly made, materials procured and new employees or consultants hired. Figure 1-7 provides a representative sampling of some of the

---

2. "FEL" Front-End-Loading. Construction Industry Institute Best Practices. www.construction-institute.org

activities that are performed by the owner or owner's engineer during the project lifecycle.

FIGURE 1-7. SAMPLE OWNER'S CAPITAL IMPROVEMENT PROJECT PLAN

a. **Owner's Phase 1 – FEL Stage 1: Business Planning**
   Business planning for the owner usually comes down to the development of projects with a suitable return on investment (ROI) considering anticipated lifecycle costs. Projects usually fall into several broad categories:
   - *New Construction Projects:* Large-scale automation projects nowadays usually occur when a new plant is built. The control system in a new "green-field" project is usually a vanishingly small portion of the total project cost, perhaps as small as 3%. This results in the control system aspect of the project receiving scant notice from the capital project planners. The control system is almost treated as a commodity in this

scenario, as various controls manufacturers are consulted and evaluated as they compete to even be included in the discussion. Final selection of a control system probably will not occur until FEL3.

- *Production-Driven Retrofits:* In this type of project, existing control systems are expanded to take on new production lines or are modified as production processes are optimized. Justification for the change is strictly production related, with the cost/benefit analysis heavily tilted toward limiting expense and minimizing production impacts.

- *Control System Maintenance Reliability Upgrades:* Projects that modernize portions of an existing control system with a newer generation of components that do exactly the same thing as the old ones, but more reliably, fall into this category. The upgrade can be software, hardware, or both, and can involve some re-engineering as newer, perhaps smaller equipment form factors (e.g., physical attributes) are deployed.

- *Control System Capability Upgrades:* Projects that modernize portions of an existing control system with a newer generation of components and/or software that do things better fall into this category. Some of the more effective projects, at least in terms of the cost/benefit ratio, replace the older operator interface while leaving the aging PLC (Programmable Logic Controller) or DCS (Distributed Control System) controllers and I/O (input-output) modules in place. Doing this tends to make the plant easier to operate and thus more reliable while giving management more oversight tools, data historian capacity, reports, Web-based tools, real-time business data, etc.

- *Control System Replacement Projects:* Projects to replace an existing control system are difficult to justify, and are usually done as a last resort, when maintenance issues and system failures begin to seriously impact production. Most system manufacturers will rise to the occasion and offer enticements to simply upgrade their system, like low-cost engineering solutions or patch after patch. Replacement projects occur only after a great deal of effort to avoid the action has yielded negative results.

Activities for the owner during this period include such things as developing a preliminary Process Flow Diagram (PFD), a Heat & Material Balance (HMB), and a study of the economics. Whatever is driving the project will drive the business plan and set the constraints. After the range of remedies has been reduced to a manageable number, a best-fit analysis reduces the number of possibilities further. A rough idea of the likely project cost and schedule is developed. Then, a feasibility study and market analysis yield the hard data needed to make a final decision before the appropriation of additional funding to proceed to FEL2 (FEL Stage 2).

b. **Owner's Phase 1 – FEL Stage 2: Project Definition**
At FEL2, enough hurdles have been passed to indicate that pursuing a project might be worthwhile. A small investment will have been approved and funds appropriated to take the next step, which is to better define the project and to prepare to release information for the purpose of obtaining bids. One of the first issues for the owner is to designate an owner's engineer (OE)—whether in-house or hire under contract—to shepherd the project through the preliminary stages and possibly manage the entire project. The owner's engineer investigates the range of technical solutions that were proposed in FEL1 and further validates them. He begins to develop a Scope of Work and a Project Execution Plan for the owner's organization(s) that describe what must be prepared to contribute to the project. Upper tier documents, such as Piping & Instrumentation Diagrams (P&ID), Equipment Arrangements, Equipment Designs, etc., begin to be produced to describe the project, and pertinent standards and procedures are gathered.

A written Functional Description is produced and vetted, and a Control Narrative is written for inclusion in the Project Specification. A Scope of Work and a Project Specification are developed and folded into a Request For Proposal (RFP). After all has been prepared, the RFP is released. Many of these FEL2-generated documents are described further in Chapter 3.

c. **Owner's Phase 1 – FEL Stage 3: Project Planning**
While bidders are responding to the RFP released at FEL2, the owner's engineer continues to validate the design concept. One way to do this is to perform a Hazard and Operability Analysis (HazOp). This early, the HazOp will necessarily be preliminary, but going through the exercise now could prevent problems later. A subsequent HazOp evolution to finalize it could be scheduled for later, after the Phase 1 tasks are completed.

At this point in the process, the OE, working closely with the owner, develops a definitive project estimate, prepares bid packages, and develops information for inclusion in the bid packages, such as Line Lists, I/O Lists, Motor Lists, etc. The OE is usually interacting with the prospective engineering service providers to help them prepare their bids, answering questions, providing site walkdowns (if appropriate), and acting to facilitate information flow. On large projects, this can be a huge undertaking, with multiple engineering and construction companies covering many disciplines. The OE can also be working on milestone schedules and preparing contracts. When the bids are received, the owner and OE spend time and effort evaluating and responding to each bid. Assuming some of the bids meet the financial and schedule requirements, the owner will be able to get funding to proceed with the detailed design phase (Phase 2), make selections, and award contracts. New hiring may be required for project managers and other skilled personnel. A new owner's engineer may even be selected for the execution phase. It is always a good idea to start meeting with the contractors early on to ensure that they understand their roles and responsibilities. The development of an interface matrix is a very good idea at this point. An interface matrix takes each schedule line item on the left side and each participating organization across the top, and lists roles and responsibilities in each cell. This is a living document that can greatly enhance communications between parties. Contractors should be encouraged to provide input to this matrix throughout the project.

d. **Owner's Phase 2 – System Design**
Phase 2 activities begin for the owner's engineer when the first contract is awarded. Other contracts may be in various stages of progress, and so for a time, the owner and OE may have activities across FEL boundaries. But once the first contract is awarded and the contractor begins detailed design, the OE begins Phase 2 for that contractor. In the example depicted in Figure 1-7, the E/I&C contractor is awarded the contract at the end of the Owner's FEL3 process and begins work. Part of the work product of the E/I&C contractor goes toward refinement of a control specification that will be used to finalize the RFP for the systems integrator. Likewise, the work product of the CSI will feed back both into the E/I&C contractor's workflow and into the RFP that will subsequently be released for the control panel fabricator. So, while the owner is engaged in Phase 2 activities with E/I&C, he is, at the same time, still lingering in Phase 1 with the other groups until all Phase 2 contracts have been awarded. The owner also begins to deal with invoices and progress payments at this time.

As factory subassemblies are completed, the owner's engineer is well advised to travel to the various locations and observe as the factory acceptance tests (FAT) are executed. In most cases, FAT procedures must be provided in advance so that they can be vetted by the owner before execution.

In addition to managing Phase 2 contracts, the owner's engineer is also working toward Phase 3. Many of the same steps taken during FEL3 need to be repeated in the final selection of a construction partner for the project. Work product is taken from the various entities working in Phase 2 for the purpose of building a construction specification. Construction work packages are generated and organized by WBS (Work Breakdown Structure) or other criteria that will help the prospective constructor organize his work.

It should be noted that the constructor is probably already known at this point. The owner may already have experience with a local construction company, and the project could already have been tentatively awarded in Phase 1, with the proviso that the constructor's bid can be refined as more information becomes available in Phase 2. This is the difference between a budgetary estimate, made with sketchy information and heavy contingency, and a definitive estimate that is based on adequate detailed information with minimal contingency.

e. **Owner's Phase 3 – Deployment**

Phase 3 activities begin for the owner's engineer when materials begin arriving onsite and/or when the constructor's team begins to mobilize, whichever comes first. Preliminaries such as site preparation, underground work and infrastructure-related work often begin long before the CSP services engineering and construction organizations arrive. The owner's engineer is responsible for maintaining an understanding of materials availability and for managing the construction schedule. The constructor is usually responsible for receiving materials and warehousing them, but the OE closely monitors those activities, clearing obstacles and making sure the constructor is properly managing the flow of materials.

The OE is tasked with providing review oversight of all construction packages (drawings, etc.) as they are released from the engineering disciplines, and with maintaining close supervision of the constructor as the work progresses. The Deployment phase includes the following major steps:

- *Installation:* Buildings are erected, equipment is installed and wired.
- *Continuity Checks:* After wiring is complete, the construction team performs QA audits to ensure that it has been performed per the drawings. These tests are done with power off.

- *Leak Checks/Pressure Checks:* Vessels and pipes are pressurized to check for leaks.

- *Electrical Checkout (Loop Checks, Bump & Stroke):* Power is applied to the equipment. Valves are activated (stroked), and limit switches are adjusted. Motors are briefly energized (bumped) to confirm proper direction of rotation. Pressures and temperatures are simulated to confirm proper input and display.

- *Site Acceptance Tests:* Subsystems are checked for proper operations and sequencing. These checks are usually done "dry" and "cold," meaning without process materials flowing and at ambient temperature.

- *Operator and Maintenance Training:* Plant operations and technical support teams are trained on the systems. Both offsite training to get background information on equipment and onsite training for operational insight are the norm.

- *Startup:* Subsystems begin to be brought online under full load conditions; for example, steam is generated, raw materials are handled and processed. Operations personnel begin to operate sections of the facility.

- *Commissioning:* All systems go into full production with operations personnel in charge. The project team begins to recede into the background as plant operations and maintenance groups take ownership.

f. **Owner's Phase 4 – Support**
After commissioning, the facility is being operated by the operations department. The owner's engineer assumes a support role, remaining on the scene for a time until the plant is in full operation and plant operations and support personnel are capable of autonomous operation. During this time, lingering issues are addressed, such as the collection and archiving of finalized documentation from contractors and subcontractors. Production testing is ongoing as actual production rates and product quality are checked against the original design criteria.

g. **Owner's Project Deliverables**
The owner (or owner's engineer), is responsible for generating several engineering products that are needed by other project participants. In some cases, these products are started by the owner and finished by contractors. Other deliverables listed below may not be owner-generated but

are still required as design input for the controls design process and so should be closely monitored by the owner. Some of the key items from the perspective of the controls groups are:

1. Process Flow Diagram
   The owner typically provides the process flow diagram (PFD). It is a simplified schematic of the system's mechanical configuration. For example, a tank is piped to a pump that is piped to another tank. Very little detail is provided as to tank or pipe sizes, and virtually none of the instrumentation is depicted. The PFD becomes the basis for the heat and material balance (HMB) sheet and the piping and instrumentation diagram (P&ID).

2. Heat and Material Balance Sheet
   The HMB sheet is generally produced by the owner or OE and is not provided to others without signing of a Non-Disclosure Agreement (NDA), and even then the list of valid users is short. The HMB is typically considered "classified," as highly sensitive, confidential, or even secret information is presented. Graphically similar to a PFD, the HMB contains additional information concerning the physics of the project. The drawing shows major equipment items and the piping that connects them. Heat loads, flow rates, and energy expenditures are calculated and charted on the document. Raw materials that are expected to be processed are shown as well.

3. Piping and Instrumentation Diagram
   The P&ID is started very early in the engineering process and originates directly out of the PFD. The P&ID is typically started by the owner or owner's engineer. But once the contracting engineer is selected, ownership of the P&ID passes to him. The primary responsibility for maintaining the P&ID is the seller's process engineering or mechanical engineering department, but responsibility is shared by the electrical and instrumentation departments as well. It is a working document that is continually updated throughout the life of the project as new understanding of the process develops. It provides piping detail in the form of pipe sizes and specifications and shows piping features such as expansion joints, flex hoses, blind flanges and rupture discs. It provides instrument information in the form of instrument "bubbles" to indicate level transmitters, flow switches and the like. It provides some indication of operability (see Point 7 below) by showing hand-switches, indicator lights, gauges, tank-level sight glasses, and annunciator points among other items. It also indicates major con-

trol logic elements by showing interconnected control system functions and interlocks. The P&ID is the primary medium used to communicate cross-discipline information between the electrical, mechanical, I&C and control systems integration groups.

4. Equipment Arrangement Drawings

    Equipment arrangement drawings are produced by the mechanical and civil engineering groups. They are either produced by the owner's engineer or by contract engineers. The basis of these drawings is the civil/structural drawings. These drawings become the basis for the detailed piping arrangement drawings (piping orthographics) and ultimately the instrument area plans.

5. Equipment Specifications

    Process equipment is generally specified by the seller's mechanical engineering group. They, or the process engineering group, will specify vessels, heating, ventilation, and air-conditioning (HVAC) systems, blowers, heat exchangers, and so on.

6. Piping Drawings

    Once the equipment arrangement drawings are done, the mechanical engineering group then begins generating three-dimensional isometric drawings (isos) for each piping section, and two-dimensional orthographic drawings (orthos) showing floor plans and elevation views of the major equipment items, large-bore pipe, and inline instrumentation. The interplay between the isometric drawings and the orthographic drawings continues iteratively until both sets are complete. Access routes, building code requirements, weight considerations, etc., must be taken into account as the drawings are generated. Instruments that must be piped inline are depicted, so the instrument engineer must be prepared to provide the flange-to-flange dimensions of inline instruments.

7. Hazard and Operability Study (HazOp)

    After the production process is defined to a fairly high degree and the HMB's are developed, the envisioned system should be analyzed for hazards and for operability. This is typically an FEL3 activity, and as such is primarily the responsibility of the owner or owner's engineer. Operability concerns relate to the bottom line by perhaps streamlining the process or by optimizing procedures. Safety issues that should be analyzed begin with the various raw materials that are brought onto the site. Handling and storage procedures should be discussed and implemented. Hazards to personnel and equipment should be analyzed for each area of the plant, and for each type of material used and created as part of the process. Finally, production and distribution of the final product should be analyzed.

The HazOp provides a structure for analyzing each element of the process facility. This structure is in the form of a checklist that presents a series of process states. Each state listed is a desired state. The series of states listed in Hazop and Hazan[6] by Trevor Kletz is as follows:

If, for example, you are analyzing pressure changes in a vessel:

- None: What happens if the pressure falls to zero?
- More of: What happens if the pressure rises?
- Less of: What happens if the pressure falls?
- Part of: What if the pressure changes when the agitator is off?
- More than (or as well as): What if the pressure drops but the temperature rises?
- Other than: What other concerns are there relative to pressure changes?

Each of these questions should be asked, and each member of the team should be allowed to provide input. Each question and response should be recorded, along with any remedial action that is defined and operator responses that are needed.

The HazOp is a structured brainstorming session. Free-wheeling discussion of the topics should be encouraged within the boundaries set by the facilitator. The HazOp should be all-encompassing and thorough. The time spent will be more than recouped by the improvements that will result. To gain the most benefit from it, the analysis should be a team effort that includes representatives from every department that is affected by the project. Finally, owner management must commit to a reasonable effort to implement the suggestions; otherwise, the whole thing is a waste of time.

2. The Control System E/I&C Engineer's Project

A four-phase project model applies to the Control System E/I&C contractor just as it does the owner. The E/I&C contractor must evaluate the owner's RFP and make a business decision as to whether to invest in a proposal effort in order to pursue the project. Due diligence can be achieved by following the three-step FEL process described previously. The following is an overview of a typical controls project from the E/I&C perspective:

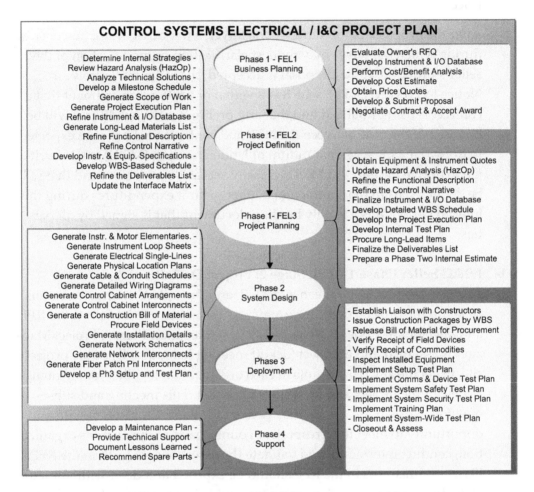

FIGURE 1-8. SAMPLE E/I&C SELLER'S PROJECT PLAN

a. **E/I&C Seller Phase 1 – FEL Stage 1: Business Planning**

Business planning for the E/I&C engineering contractor (seller) asks several questions: Do we have the expertise to meet this specification? Do we have the manpower to execute it? Does this customer have a history of difficult projects, or do they work as a team with their contractors? Do they pay their bills? Have they done their due diligence? Do we have a strategic interest in pursuing this project even if it is marginal in several areas?

If the answer to most of those questions is yes, then the likelihood is that the prospective contractor will invest in the bid process and generate a response to the RFP. The contractor will appropriate funds for travel expenses and manpower as needed to do a proper investigation and build a proposal. Part of this process will be to obtain budgetary estimates from materials vendors that can be rolled up into an overall project materials price.

After the proposal is submitted to the buyer, a period of evaluation ensues during which the buyer may request clarification or modification of the seller's proposal. If the proposal is accepted, the seller may receive a Notice to Proceed (NTP), which is a verbal or written commitment that a contract is being finalized and that any preliminary expenditures will be honored. If the time between the NTP and contract finalization is expected to draw out a bit, a Memorandum of Understanding (MOU) can be written to bridge the gap. An MOU typically sets aside a specific fund that will be deducted from the contractual budget to limit expenditures during this interim period. It is a legally binding document that is signed by all parties.

b. **E/I&C Seller Phase 1 – FEL Stage 2: Project Definition**
At FEL2, the project has been awarded, and the E/I&C contractor begins to set up the project, dedicating staff and resources. He interprets the RFP and develops a Scope of Work and Project Execution Plan. A Project Manager is assigned and a project kickoff meeting held, usually at the project site, to get the key participants in each organization together for an intense review of the project and for a site walkdown. This meeting and subsequent walkdown are key, as they present the contractor's team with an opportunity to meet their respective counterparts in the owner's organization, conduct interviews, and validate the assertions set forth in the RFP. Flaws or omissions in the RFP should be exposed and dealt with immediately. In extreme cases, the contractor could elect to remove himself from the project at this point, with only a minimal impact to his reputation and to either his or the owner's bottom line. If the choice is between backing out early and failing later, the former option is preferable, regardless of the fallout.

Another top priority should be to eliminate any assumptions set forth in the proposal and to further refine the project schedule. Document issuance and change management procedures should be clearly defined. Upper-tier deliverables such as the I/O List, Drawing List, Instrument List, Network Block Diagram, and so on, should begin to be produced, either from scratch or by further refining documents provided by the owner during the bid process. A checklist of some of the key tasks to be accomplished at this point is shown in Figure 1-7.

Upon completion of this stage, both the contractor and the owner should have a thorough understanding of the potential problems and risks inherent in the project, and strategies should be in the process of being developed to mitigate them. Both organizations should be gaining confidence that the task is doable, and the personnel should be starting to align themselves with the goals of the project. The product design should be sufficiently developed to allow the owner to start negotiating with potential systems integration contractors.

c. **E/I&C Seller Phase 1 – FEL Stage 3: Project Planning**
After most of the tasks are identified and the deliverables list is completed, it will be possible to complete the planning stage by developing a detailed Project Schedule and Project Execution Plan (PEP). At this point, it is a good idea to revisit the HazOp in light of new information developed since the original iteration, and, if changes are not required, to reaffirm the current course of action. Upper-tier documents such as the P&ID and I/O List should be finalized during this timeframe in preparation for Phase 2, System Design. Procurement activities can begin for long-lead equipment, or items needed for testing in Phase 2. Internal test and oversight procedures should be revisited and adapted to the project plan. Additional staffing requirements should be defined as the project team expands beyond the senior planners to the rank and file. Training and orientation should be considered for new project team members as the scope of effort widens.

d. **E/I&C Seller Phase 2 – System Design**
For the E/I&C contractor, Phase 2 activities are focused on the production of the agreed-upon design deliverables. If Phase 1 was properly executed, there will be very few unknowns remaining. In a perfect world, management tasks would relate only to monitoring actual progress versus expectations. However, interfacing between the E/I&C contractor, the owner's engineer, and the systems integrator takes a higher priority as the plans are adjusted for design discoveries and for various preference issues that may begin to appear. Plant operations groups tend to begin asking questions and injecting opinions that the owner's engineer may respond to, forcing the E/I&C design team to react by invoking change control procedures. The plan, carefully developed in Phase 1, rarely survives implementation, so it pays to build flexibility into the plan from the beginning. Careful use of contingency funds and the judicious implementation of the pre-approved change control policy help to maintain good customer relations while staying on task, thereby minimizing change orders.

As factory sub-assemblies are completed, the owner's engineer may ask the E/I&C contractor to act as his agent and to travel to the various locations and observe as the factory acceptance tests (FAT) are executed. This is a service that could be included in the proposal or added under a change order.

Deliverables generated in this phase of the work are described in section g below.

e. **E/I&C Seller Phase 3 – Deployment**
Phase 3 activities begin for the E/I&C contractor when the constructor begins organizing his work. Construction Work Packages help facilitate this process, and the E/I&C contractor generally repackages his deliverable sets into drawing and document bundles organized by WBS. Bills of Material (BOM) for commodities such as cable, conduit, and fiber and for wiring components, such as terminals, relays, and fuses, are turned over to the constructor, who reviews them, augments them as necessary, and orders the materials. Engineered items, such as motors, instruments, analyzers, etc., will have already been ordered either by the E/I&C contractor or by the owner's engineer, depending on where that responsibility rests contractually.

At this stage in the project, the E/I&C contractor can begin staffing down as the bulk of his work is complete. Key representatives remain attached to the project with their level of involvement variable, based on the level of involvement by the OE's organization. Construction oversight and technical support are the E/I&S contractor's primary roles during this period.

f. **E/I&C Seller Phase 4 – Support**
After commissioning, the E/I&C contractor must consolidate the design package and issue it for record. Construction teams generally do not install exactly per the original drawing set, either from being poorly managed or due to problems with the design package. In either case, it is important that the final drawing set reflect the as-built condition of the facility. The E/I&C contractor must collect all the marked up drawings used during installation and formally incorporate the modifications. This could entail additional site walkdowns, interviews, and investigations. Some of the documents are legal records, and inaccuracies may be punishable by fines or jail time.

g. **E/I&C Seller's Project Deliverables**

Some of the key deliverables from the perspective of the controls groups are listed below. Responsibility for developing these items may be shared between the owner, owner's engineer, or the contractor, or any combination from one project to the next.

1. Piping and Instrumentation Diagram (P&ID)

   The piping and instrumentation diagram is started very early in the engineering process by the owner or owner's engineer, and it originates directly out of the PFD. It is usually owned by the mechanical or process engineering departments, and is started well ahead of the E/I&C department's involvement. However, the P&ID is a multidiscipline product that is shared, and it is to be completed by layering on controls information. It is a working document that is continually updated as new understanding of the process develops. In addition to mechanical information, it provides information in the form of instrument bubbles to indicate level transmitters, flow switches, and the like. It provides some indication of operability by showing hand switches, indicator lights, gauges, tank-level sight glasses, and annunciator points among other items. It also indicates major control logic elements by showing interconnected automation system functions and interlocks. Most of this detail is layered on during E/I&C Phase 1.

2. Document List

   Document management is very important, and it can be time-consuming. Hundreds of documents might be newly generated. Hundreds more might be ordered from the customer to be modified. Managing the documents and tracking their active revision and disposition can be a major undertaking, one that can be supported by the creation and maintenance of a document list. The document list itself is a design deliverable upon completion of the project.

3. I/O List (I/O Configuration and Partitioning)

   The I/O List may or may not be a deliverable in its own right. However, it is used to support other functions that are deliverables. The I/O configuration is a list of I/O points organized around the characteristics of the control system. The P&IDs are generally organized in a way that is intuitive, and it is usually a good idea to pattern the I/O structure likewise. Proper partitioning of the I/Os is important to give the system a measure of fault tolerance. Partitioning rules should be established early in the design process. Some good rules are:

- Assign I/Os for redundant pumps, motors, equipment items, etc., to different I/O modules to prevent one I/O module's failure from killing both.

- Avoid splitting related I/Os across multiple racks if possible.

- Include some percentage of ready-spare I/Os that are already loaded in the I/O racks, preferably one or two I/O points per I/O card.

4. Instrument Database
   An instrument database is key to a properly managed design process. This database, if developed early and then updated religiously during the design process, can be a very useful design tool and a tremendous aid during the design check. Each item touched by the instrument design team should appear as a record in this database. It is initialized by the I/O list as generated during the I/O configuration process, but eventually it should contain records of even unwired instruments, such as tank-level site glasses, rupture discs, and pressure gages that need to be purchased or refurbished.

5. Instrument Specifications
   Given the availability of support information, such as pump specifications, equipment specifications, P&IDs, and an instrument database, the production of instrument specifications can proceed. HMB-generated data, such as flow rates and temperatures, need to be available as well. Instrument specifications must be completed before the instrument elementaries (i.e., schematic diagrams) or installation details can be finalized since these depend on specific model configurations.

6. Instrument Elementary Wiring Diagrams and Loop Sheets
   Instrument elementary wiring diagrams (sometimes also called Schematics) and instrument loop sheets serve the same basic function in the context of the drawing package. Both types of drawings are non-physical in that distances and physical location information are minimized. Instrument elementary wiring information is presented as a complete electrical circuit, not as a particular physical cabinet or field device for example. A loop sheet might present some physical information, but, like the elementary, its main purpose remains to show the complete electrical circuit. The loop sheet depicts current flowing from hot to neutral or from DC+ to DC-. It is rare that a circuit crosses drawings, though it is sometimes unavoidable. This type of drawing differs from the interconnection wiring drawing in that the interconnection

drawing shows wiring for a particular device or physical area. Showing a complete electrical circuit is less important than showing point-to-point wiring. That format requires the technician to follow the signal across various paths, possibly crossing several different drawings, to assemble a picture of the entire circuit.

7. Marshalling and Control Panel Arrangements; Field Junction Boxes
   Marshalling panels are typically large enclosures that provide a place to mate field cables to control system cables. Generally, field I/Os are not organized in a way that is convenient for mass-termination into the control system. Field cables are oriented toward floor location, while I/O cables are oriented toward I/O module groupings. A marshalling panel provides a place to cross-connect the cabling systems and to perform any necessary circuit protection and power distribution activities. The marshalling panel arrangement drawing is a fabrication drawing that consists of a bill of materials and a component arrangement.

   Control panels house electronic equipment and sometimes have controls (e.g., switches, lights) that are presented on the front panel. The arrangement for this type of panel provides interior panel arrangement and possibly a hole-punch guide for the panel front.

   Field junction boxes are panels that house wiring interconnection components, such as terminals and sometimes relays and fuses. Usually, these boxes are simply "breakout" boxes in which multi-conductor cables are broken out into single-pair cables.

   Each of these panels has associated wiring diagrams described as follows:

8. Wiring Diagram (Interconnection Drawing)
   Wiring diagrams depict interconnections within a specific cabinet or panel. The purpose of a wiring diagram is to provide information that will allow the installer to execute the wiring task with a minimum of investigation. The wiring diagram makes no attempt to show a complete electrical circuit, but provides point-to-point wiring detail only. Cables are shown entering and leaving the enclosure. Any internal wiring is shown in total. A wiring diagram is generally produced for each and every marshalling panel, field junction box, control panel, relay panel, or any other enclosure in which wiring must be terminated.

9. Instrument Installation Details

   Instrument installation details are drawings (or sketches) that provide hookup information for the installer. They provide a pictorial representation of the components in their proper relationship to each other and a bill of materials chart that describes such things as part numbers and item numbers. Installation details usually have multiple uses (i.e., they are generic drawings). For example, there might be several control valves in different parts of the plant that are to be installed the same way. One installation detail will suffice for all provided there is a cross-reference between the instrument and its installation detail. For this, the instrument database is a good tool. It is also a good idea to put that information on the instrument specification as a note. There are usually three categories of such details for each installation: mechanical connections, electrical connections, and mounting.

   (a) Mechanical Installation Detail
   The mechanical installation detail shows a device and its tubing or piping connections and the connection fittings; for example, the instrument air hookup to a control valve. Any fittings that might be needed, including gaskets and bolts, are called out on the detail.

   (b) Electrical Installation Detail
   The electrical installation detail provides information relating to the conduit connections. All conduit fittings are described, such as condulets and flex hoses, but only within about a 2-ft. envelope around the instrument. The instrument location plan takes care of material outside that envelope.

   (c) Mounting Installation Detail
   Mounting information such as instrument pipe-stand fabrication, mounting bracket fabrication, and other mounting particulars are described here.

10. Mechanical and Mounting Bill of Materials

    After the mechanical and mounting installation details are created and assigned to each instrument, a mechanical bill of materials can be produced. This bill of materials lists all items needed for the mechanical installation of the instrument. This includes all tubing, tube fittings, brackets, clamps, and mounting hardware necessary to connect the instrument to the process. All materials called for by the mechanical details are compiled into a single bill of materials. The bill of materials is then submitted to purchasing, where it is broken into packages by vendor and issued for purchase.

11. Electrical Bill of Materials

    After the electrical installation details are created and assigned to each instrument, an electrical bill of materials can be produced. This bill of materials lists all items needed for the electrical installation of the instrument. This includes all conduit, conduit fittings, flex fittings, and drain plugs. All materials called for by the electrical details are compiled into a single bill of materials. The bill of materials is then submitted to purchasing, where it is broken into packages by vendor and issued for purchase.

12. Floor Plans

    There are several types of plan drawings in an engineering design package:

    (a) A site plan is typically included in the Phase 1 study to assist in physical orientation. It is sometimes just a sketch showing the intended future location for equipment, junction boxes, etc. Formal site plans, as engineering deliverables, are considered products of the structural engineering group (if the plan is of a building) or the civil engineering group (if the plan is of a yard area). The E/I&C group can use these drawings as references to generate underground conduit plans, instrument location plans, or other products.

    (b) Process area floor plans, sometimes called instrument location plans, are construction-related E/I&C deliverables rarely maintained past construction. They provide general information about instrument placement on the process floor, the conduit and/or tubing that must be extended to those instruments, and any support equipment, such as field junction boxes and air header connections. The drawings can be scale drawings showing the exact location of the instruments, or they might be done as diagrams depicting each instrument near its expected location. These drawings can be done in two steps if the schedule demands it, though ideally the bulk of the work should happen downstream of the creation of elementaries and somewhat parallel to the creation of wiring diagrams (interconnection drawings). Initial setup of the process area floor plan drawings can occur as soon as the equipment arrangements are essentially complete, which would put this function parallel to the production of the orthographic piping drawings. But, due to the conduit sizing and contents information needed, it may not be possible to finish the floor plans until after the elementaries (or loop sheets) and subsequent wiring diagrams are completed.

13. Control Room and I/O Termination Room Arrangements
    Control room and I/O termination room (I/O room) drawings are closely interrelated. The control room design must be settled before the I/O room design. Sometimes PLC or DCS equipment racks are physically in the control room, with the marshalling panels in an adjacent I/O spreading room. Sometimes all the control system hardware is in the I/O room. Frequently, the deciding factor for placement has to do with the availability of environmental control in the I/O room. If that room is not air-conditioned, then the sophisticated hardware might be found in the control room. Another consideration is the availability of space. Control rooms and I/O rooms are some of the most expensive rooms in the plant in terms of cost per square foot.

3. The Control Systems Integrator's Project

The same four-phase project model applies to the control systems integrator. Figure 1-9 shows an overview of a typical controls project from the control systems integrator's perspective:

a. **CSI Seller Phase 1 – FEL Stage 1: Business Planning**
   Business planning for the control systems integrator is similar to that of the E/I&C contractor, asking many of the same questions. The difference is that the CSI supplier will typically require input from both the owner and the E/I&C contractor in order to develop a tight bid. This brings the CSI team on board a bit later than some of the other team members. Once started, however, the evaluation and business planning tasks are very similar to those described for the E/I&C contractor. Do we have the expertise to meet this specification? Do we have the manpower to execute it? Does this customer have a history of difficult projects, or do they work as a team with their contractors? Do they pay their bills? Have they done their due diligence? Do we have a strategic interest in pursuing this project even if it is marginal in several areas?

   Positive answers to these questions will likely result in bid preparation.

b. **CSI Seller Phase 1 – FEL Stage 2: Project Definition**
   At FEL2, the contract has been awarded, and the CSI contractor begins to set up the project, dedicating staff and resources. He interprets the RFP and develops a Scope of Work and a Project Execution Plan. Using information developed during the E/I&C design process, the CSI team begins to devise control schemes for major control elements. These control schemes include the operability aspects as well as the functional aspects of

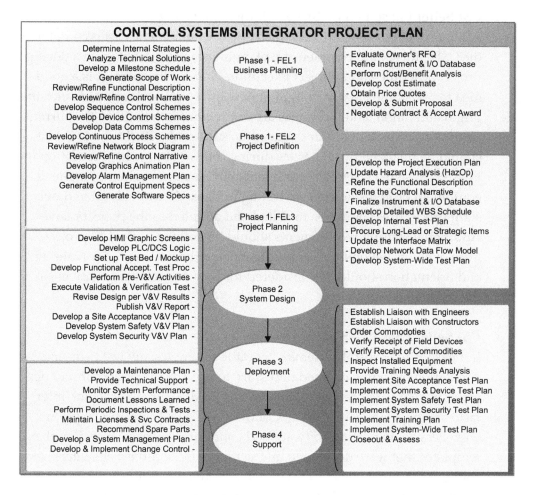

FIGURE 1-9. SAMPLE CSI SELLER'S PROJECT PLAN

each type of control device, such as on/off devices, sequences, and alarms, along with other control and data acquisition topics. Typical examples should be produced and simulated for review and comment by plant operations and by the owner's engineer to ensure that the end result will be acceptable. Templates generated and tested in the FEL2 step will be used throughout the project, which allows the CSI team to plan the remainder of the project in great detail.

Also at this time, the CSI team reviews and confirms the viability of the I/O partitioning work done by the E/I&C as part of the I/O List generation activity, which allows the PLC or DCS hardware configuration to be finalized.

This is the time for the owner's engineer to release the Control Panel Shop RFP and select a control panel fabricator (panel shop).

c. **CSI Seller Phase 1 – FEL Stage 3: Project Planning**

After most of the tasks are identified and the deliverables list is completed, the CSI team should complete the planning stage by developing a detailed Project Schedule and a Project Execution Plan. At this point it is a good idea to revisit the HazOp in light of new information developed since the original iteration, and if changes are not required, to reaffirm the current course of action. Upper-tier documents, such as the P&ID and I/O List, should be finalized by all parties during this timeframe in preparation for Phase 2, System Design. Procurement activities can begin for long-lead equipment or items needed for testing in Phase 2. Internal test and oversight procedures should be revisited and adapted to the project plan. Additional staffing requirements should be defined as the CSI project team expands beyond the senior planners to the rank and file. Training and orientation should be considered for new project team members as the scope of effort widens.

d. **CSI Seller Phase 2 – System Design**

In Phase 2, the control systems integrator assembles any test beds and/or simulation mockups deemed necessary to execute the project and to perform the requisite pre-installation testing. HMI (Human-Machine Interface) and SCADA (Supervisory Control and Data Acquisition) systems engineers begin writing software and configuring screens based on the templates that were developed and vetted in FEL2. Ethernet switches are configured as necessary. VFD (Variable Frequency Drives) are programmed and tested. Networks are temporarily wired and throughputs baselined (i.e., analyzed during ideal conditions). As the project develops, a System Validation & Verification (V&V) Test is being developed and issued by the CSI seller, and typically reviewed by the owner or owner's engineer. Testing activities are typically done at the integrator's workplace, or at the panel shop. It is rarely done at the plant site due to the number of interruptions that are likely.

As the design effort winds down, preparations for V&V Test execution take precedence. The control systems integrator usually executes the V&V Test internally in a pre-V&V to confirm readiness prior to the arrival of the owner's engineer, operations personnel, E/I&C contractor, and any other interested parties that may be involved.

The V&V is a formal test that demonstrates proper system operation as compared to what was published in the control narrative and control system specification. The criteria should be rigorously applied to the exclusion of personal preferences expressed by operations or maintenance personnel. Once the system has been demonstrated to comply with the specifications, then discussions may ensue regarding previously unstated preferences.

Deviations from the specifications, problems, and poor assumptions should be documented and fixed, and the system should be retested. In the end, a V&V report should be generated to describe the tests and their results. This document becomes a valuable historical record for future reference.

At this time, a set of plans for Phase 3 (deployment) can be produced to include checkout, site acceptance, and commissioning.

Deliverables generated in this phase of the work are described in section g below.

e. **CSI Seller Phase 3 – Deployment**
The control systems integrator is the key player in the deployment of the control system. The CSI works closely with the owner's engineer, with construction, and with plant operations to confirm proper installation and to test the software.

- *Loop Checks:* Activities begin for the control systems integrator as the constructor finishes work packages (WBS areas). One of the constructor's responsibilities is to achieve mechanical completion, which entails not only the installation of equipment, but wiring continuity checks, power-up checks, and operational checks called loop checks. Loop checks are sometimes referred to as "bump & stroke," since motors are started just long enough to verify proper rotation, and valves are stroked open and closed to test the pneumatic and electrical actuators and to allow adjustment of the limit switches. Pressures and temperatures are simulated by injecting signals at the sensors and transmitters. The CSI team gets in sync with the constructor at this time to begin checkout of the control system. As each end device is exercised, the team verifies that control can be initiated and that status is properly reported to/from the control system. Work flow is typically driven by the constructor during this time.

- *Site Acceptance Tests (SAT):* After loop checks are completed for a WBS area, the control systems integrator takes charge as subsystems are tested for proper operation and sequencing. These checks are usually done dry and cold. The constructor assigns support personnel to assist with troubleshooting and repair issues as they arise, and the owner's engineer is ever-present, observing the process and signing off on the tests as they are completed.

- *Operator Training:* The control systems integrator begins training plant operations personnel on the systems, both in formal classroom settings and on the job, as they work with the systems integrator during the SAT process.

- *Post-Modification Test (PMT):* PMT's are executed by plant Operations personnel. They can be done "dry," with process conditions being simulated, or "wet," with raw materials flowing, moving, and/or heating/cooling. The CSI team begins to assume a supporting role at this stage.

- *Commissioning:* All systems go into full production with operations personnel in charge. The CSI project team recedes into the background as plant operations and maintenance groups take ownership.

f. **CSI Seller Phase 4 – Support**

After commissioning, the control systems integrator's work continues, migrating toward plant support as opposed to project support. Training is usually an ongoing concern as the plant's personnel cycle in and out of the field of view; maintenance issues arise where additional troubleshooting support may be needed, and software licenses need to be maintained, along with recommended upgrades and patch installation. Many times an ongoing service contract is appropriate, whereby the owner formally shifts the responsibility for the control system to the CSI.

g. **CSI Seller's Project Deliverables**

The CSI team is responsible for generating several engineering products, most of which result in software that runs the plant. Some of the key deliverables from the perspective of the controls groups are:

1. HMI Screen Hierarchy and Navigation Plan

    The HMI screen hierarchy and navigation plan is a document that describes the graphic screens that are envisioned for the project and the navigation rules that will be followed. An intuitive navigation plan employs a multi-tier design, starting with process overviews and ending with control popup overlays. A possible hierarchy could include the following screen classifications and hierarchical level:

    - Master System Overview (Tier 1) – One screen for at-a-glance project status and screen navigation

- Process Area Overviews (Tier 2) – One screen for at-a-glance status of each WBS area

- Subsystem Overviews (Tier 2) – One or more screens for at-a-glance status of each subsystem

- Historical Trends Overview (Tier 2) – One screen for trends navigation

- Reports Overview (Tier 2) – One screen for report selection

- Control Popup Overlays (Tier 3) – One per end device (e.g., a valve, motor, vfd, conveyor, machine, etc.)

- Historical Trend Screens (Tier 3) - One per process control scheme

- Report Screens (Tier 3) – One screen per report

Navigation is usually facilitated horizontally within a particular tier, and vertically between overview tiers.

2. PLC/DCS I/O Partitioning Plan
   I/O point assignments should be made with a view toward fault tolerance. The I/O Partitioning Plan must be tailored for each project, but the rules for partitioning are pretty consistent from project to project. For example, a single I/O card failure should not cause problems in more than one process zone. Further, for critical subsystems where redundant equipment is in place, the I/O on each redundant item should be split to different I/O modules.

3. Data Communications Plan
   The data communications plan describes the organization of data flows across various network structures. The following is one possible organization:

   - PLC-to-Remote I/O: RS485 RIO LAN

   - PLC-to-PLC Data: Modbus-Plus or DH+, etc.

   - PLC-to-SCADA Server Data: Fault-Tolerant Industrial Ethernet Ring #1

   - SCADA Server to Terminal Server Data: Fault-Tolerant Industrial Ethernet Ring #2

   - Terminal Server to Thin Client Data: Fault-Tolerant Industrial Ethernet Ring #3

The data communications plan should detail every communications link, most of which can probably be grouped into one of the five levels shown above. But there will probably be outlying OEM-based communications links, and this would be a good place to describe those as well.

4. Data Security Plan (cyber-security, data integrity)
   Cyber-security and data integrity are topics that are surging in importance as the demand for real-time information grows. The need for instant access to critical performance and market data drives the integration team towards more open architectures, and thus more risk exposure. A data security plan should address the following topics:

   - Physical security
   - User rights and privileges
   - Authentication (passwords, biometrics, etc.)
   - Virus countermeasures
   - Digital signatures and certificates
   - Intrusion prevention and detection
   - Forensics

5. System Safety Response, Archival & Recovery Plan
   Catastrophic events, such as control room fires, acts of sabotage, or natural disasters, can cause major upsets that could take months to recover from. Procedures should be developed and processes established to reduce the upset and provide a reasonable opportunity for restoring operation in a timely, organized manner.

6. Training Plan
   Operators and maintenance personnel will require in-depth training prior to commissioning. In addition, periodic refresher training should be considered. Training materials could include:

   - Operations & Maintenance Manual(s)
   - Practical exercises
   - A training testbed
   - Others

7. Network Single-Line Diagram
   The network single-line diagram is a drawing that shows the major control system components and the various data networks that interconnect them. This type of drawing does not show detailed connections but focuses more on the equipment items themselves, with general interconnection information. Though this convention is somewhat fluid and could change, depending on the division of labor agreed upon for a particular project, for the purposes of this discussion, detailed wiring interconnections for fiber and copper are provided under the E/I&C set of deliverables.

8. Control Logic Diagrams
   Control logic diagrams ("logic diagrams") may be provided to the control systems integrator as a part of the design basis, accompanied by the control narrative. Or the integration team may be given only a narrative and must generate their own logic diagrams. Either way, as software development progresses, the logic diagrams will probably need to be modified, as they are generally only an approximation of the moves that need to be made in the program. Logic diagrams provide a means of communicating control logic at a fairly high level.

9. Device Control Detail Sheets (DCDS)
   Programs organized by control element are usually well-organized programs that are easy to follow. Device control detail sheets are similar to instrument data sheets in that they describe the attributes of a particular device such as a valve or motor. The difference is that instead of physical attributes called out on an instrument data sheet, the control detail sheet describes the software attributes of the device. DCDS deal primarily with on/off type devices, and describe the devices' start/stop or open/close conditions and the actions that occur, and list the interlocks and permissives (see Definition of Terms in Part 2).

10. Equipment Control Detail Sheets (ECDS)
    The operation of conveyors, dust collectors, screw feeders, weigh feeders, pumping stations, and other self-contained equipment subsystems can be described using an ECDS. Similar to the DCDS, these sheets provide start/stop conditions, start/stop actions, interlocks, and permissives.

11. Process Control Detail Sheets (PCDS)
    Continuous processes, such as PID (proportional-integral-derivative) control functions, analog inputs, analog outputs, ratio controllers, and

other such control schemes, are described in PCDS. These sheets provide a place to document control parameters, alarm setpoints, and other information.

12. Sequence Control Detail Sheet (SCDS)
Every project has sequences, whether they are startup and shutdown, batch operation, emergency overwatch, or some other activity. The SDCS describes each sequence in detail.

13. Data Transfer Detail Sheets (DTDS)
Data blocks that are passed from PLC to PLC are described by the DTDS. It allows the data array being transferred to be described, with the base address of the source PLC and the base address of the destination PLC shown. Handshaking and data transfer monitoring methods may also be described.

14. PLC/DCS Programming
Control software can be developed remotely and downloaded to the controller. Multiple programmers can work at the same time in different sections of the program, and then have their work merged at a later time. The programming effort should not begin until the logic diagrams and/or control detail sheets are in an advanced state of readiness.

15. HMI Programming
Visualization software is usually developed on the platform on which it will reside. For that reason, mockups are frequently built up using the actual control elements that will be later installed in the plant. It is possible to write the software elsewhere, and then import it onto the platform of choice—which is the normal method for retrofits. But it is best to develop the software on the same equipment that will later use it, whenever possible.

16. Server Setup Services
There can be several groups of servers on a project. SCADA servers serve up data from the PLC or DCS to visualization equipment. Thin client servers, for example, may take data from the SCADA servers and serve that data up to the thin client workstations, or dumb terminals. A historical data server may just sit on the network and pull data from the SCADA database. Regardless of the actual configuration, all of the servers must be configured and the setup stored for later retrieval. They also require software licenses, sometimes multiple layers of them.

17. Validation & Verification (V&V) Test Procedure
    The V&V procedure is a document written to guide the participants through a regimented program of tests that progressively exercises the software, testing each HMI screen and each major control scheme. These tests are normally executed in the shop environment, away from the eventual installation site. Field device responses are simulated using simulation "loopback" logic or other means of simulation instead of the actual field devices. This allows device logic to be tested for proper response to commands from the HMI, for simulated loss of interlocks, for automatic and manual mode switching, and for other modes of operation in order to gain confidence that the system will function as advertised when connected to the equipment. Some industries, such as pharmaceuticals and nuclear power, require a 100% mockup, including simulated field devices. Loop-back simulation (e.g., simple software simulation that feeds output commands back to inputs with, perhaps, a time lag) is suitable for most industrial customers.

18. Site Acceptance Test (SAT) Procedure
    The SAT procedure is similar to the V&V procedure, except that the test is executed in the field with the control system connected to the field devices. Valves are stroked and monitored for proper limit switch setup. Motors are bumped to confirm proper rotation and to make sure that feedback status is detected at the PLC and properly displayed at the HMI. Process simulators are used to inject simulated temperature and pressure signals at the transmitters so that wiring and scaling can be verified at the HMI. Most of these tests are done dry and cold.

19. Startup (Pre-Commissioning) Plan
    Startup generally happens in pulses of activity that occur as systems are turned over to Operations by the constructor. As each process area is completed and the SAT procedure executed, it becomes an operational issue as to when materials will begin to be processed and PMTs started. For example, a material handling system might begin moving materials and filling dump trucks instead of feeding the materials to the next process area. Fluid-based processes may fill tanks and piping with water to circulate and test pumping systems. Startup is where Operations begins to get more involved, getting operators trained and gaining confidence in making operational decisions.

20. Commissioning Plan

    Commissioning is driven entirely by plant operations, so the plan is primarily theirs. The systems integrator will be heavily involved, in many cases even running the plant while the operations team gradually takes over. The commissioning plan itself is not a direct product of the systems integrator, but it should be closely vetted and approved by the integrator.

4. The Control Panel Fabricator's Project

The same four-phase project model applies to the control panel fabricator (FAB contractor). Figure 1-10 is an overview of a typical controls project from the control panel fabricator's perspective:

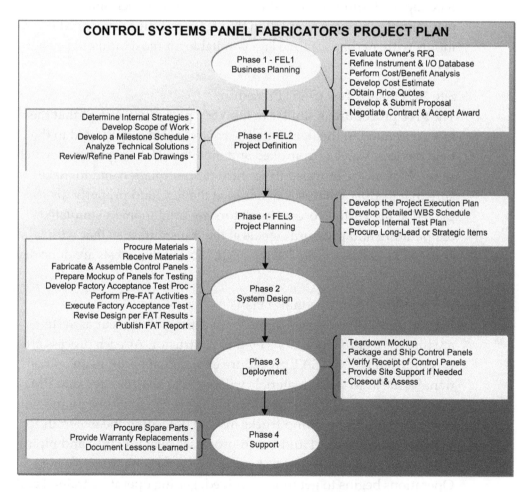

FIGURE 1-10. SAMPLE FAB SELLER'S PROJECT PLAN

a. **FAB Seller Phase 1 – FEL Stage 1: Business Planning**
   The RFP for control panel fabrication is generally issued when the E/I&C design is at a high stage of completion, at least with regard to panel fabrication and wiring drawings. I/O partitioning decisions can affect I/O module distribution, which cascades into panel design, so it is a good idea to get input from the control systems integrator prior to finalizing the I/O partitioning.

   Business planning for the panel fabricator includes many of the same questions as the other contractors: Do we have the expertise to meet this specification? Do we have the manpower to execute it? Does this customer have a history of difficult projects, or do they work as a team with their contractors? Do they pay their bills? Have they done their due diligence? Do we have a strategic interest in pursuing this project, even if it is marginal in several areas?

   Positive answers to these questions will likely result in bid preparation.

b. **FAB Seller Phase 1 – FEL Stage 2: Project Definition**
   At FEL2, the project has been awarded, and the FAB contractor begins to set up the project, dedicating staff and resources. He interprets the RFP and develops a Scope of Work and a Project Execution Plan. Using bills of material developed during the E/I&C design process, the FAB team begins to procure materials and organize the shop floor to accommodate the project.

c. **FAB Seller Phase 1 – FEL Stage 3: Project Planning**
   Project planning for the FAB team revolves around scheduling. The orchestration of materials procurement and flow is important so that materials arrive when needed, not too soon, clogging up the production floor, and not too late, interrupting production processes. Planning for the Factory Acceptance Test should occur here in anticipation of visiting customers.

d. **FAB Seller Phase 2 – System Design**
   In Phase 2, materials are procured and received, and panel fabrication, assembly, and internal panel wiring occur. Internal testing (pre-FAT) activities are done prior to the execution of the formal Factory Acceptance Test. This internal testing includes adding temporary panel-to-panel wiring and power connections. After the FAT is complete, the fabricator makes any corrections necessary and publishes a FAT report. NOTE: The V&V Test discussed for the CSI could be combined with the Factory

Acceptance Test, with the mockup being extended to include the SCADA layer. This is most likely to occur if the same company is providing both CSI and FAB services, but it can be done otherwise as well.

e. **FAB Seller Phase 3 – Deployment**
For the panel fabricator, deployment involves teardown of the mockup, packaging, and shipment of the panels. The fabricator will likely remain in the mix during the site work of installing and checking out the panels but generally on an on-call basis to provide spare parts or to support modifications that may be implemented onsite. In some cases, the fabricator may be asked to send personnel to the site to make panel modifications in situ.

f. **FAB Seller Phase 4 – Support**
Post-project support effort by the fabricator is minimal.

g. **FAB Seller's Project Deliverables**
For the fabricator, the set of deliverables is mostly hardware. Project drawings that are used in the fabrication process should be marked up if minor modifications need to be made. Major modifications to project drawings should be cycled back to the engineering organization that issued them, fixed, and then re-issued to the FAB during the fabrication cycle. Shop drawings may need to be generated to fill any gaps in the project drawing set, and these become deliverables if required by the contract. Custom cabinets that are manufactured will require shop-generated drawings. Cabinets that are procured from reputable cabinet manufacturers will have drawings provided, most of the time downloadable from the manufacturers' websites. The FAB seller's deliverables also include a complete bill of materials, drawing markups, the FAT report, instructions for unpacking and assembly onsite, and any shop assembly drawings that were generated.

F. INTEGRATED CONTROL SOLUTIONS

The previous sections discussed the participants in a control systems project as if they were separate entities—which, historically, has been the case more often than not. A review of the project stages for each reveals that each group must perform many of the same steps in the early stages if they follow prudent FEL-based guidelines. The owner may be able to construct scenarios that allow him to issue a single purchase order for all these services by designating a prime contractor to manage the others. But the overall cost and complexity of the project may not be reduced, even though the project management aspect may be simplified by reducing the number of purchase orders to issue and invoices to process. On the contrary, adding management layers to the contractors' normal set of services may actually increase project cost, and passing the responsibility for managing the interfaces between contracting companies to a third party may increase risk as well.

A new type of company is emerging that addresses these issues, providing all the key engineering, systems integration, and panel fabrication services as an integral aspect of their operation. These companies are sometimes called Control Solutions Providers (CSP), and they provide Integrated Control Solutions (ICS) products and services. The great benefit of these companies is the effect of pre-negotiated interfaces between the control panel fabricator, the E/I&C engineer, and the control systems integrator. Some of the immediate benefits are:

- A single project manager can be appointed (instead of three).

- The E/I&C design product can be tailored to the panel fabrication shop's needs and capabilities.

- Scheduling can be streamlined.

    - The E/I&C designer can specify materials that are already on the fabricator's shelf.

    - The integrator can get involved earlier in the process, assisting the E/I&C designer with an initial system (Instrument and I/O List) database design and partitioning.

- Proposal costs can be minimized, resulting in lower project costs as one bidder instead of three builds proposal recovery costs into his bid.

- Closer, more informal ties among the three internal organizations allow problem resolution to occur below the owner's radar, thus reducing his oversight workload.

The benefits extend beyond the design and fabrication stage. In Phase 3, when the system integrators are in the lead, they can draw from an instrument design knowledge base that is internal to their own company, knowing that all participants have the same success motivation. E/I&C personnel can be called on to augment the CSI in site support, troubleshooting, and construction liaison. FAB personnel can be called on to make field adjustments in the panels. Overall, a better, simpler, and more flexible workflow can be achieved.

In closing, each and every design project is different. Yet the way in which the design engineer should approach a new project should not change. The same initial questions should be asked every time, including: What is the structure of the project? What are the customer's expectations, and how will he measure success? How much risk is it wise to assume?

Once those questions have been answered, and the design environment is known, the specific content of the design package can be resolved. The mix of products and services can change from project to project, and the design team needs to be able to adapt to changing needs. If the team has a good feel for all the deliverables that may be offered, then modifying the package by deleting deliverables will have little effect on the design process. In contrast, adding deliverables that are new to the team could have a great effect, upsetting the work flow. Therefore, it is in the best interest of all concerned to have all team members fully cognizant of the full range of products and services (as shown in Figures 1-7 through 1-10) and their interrelationships.

Part I – Chapter 2: The Project Team

As noted in the previous chapter, a new project is an extraordinarily complex, and sometimes risky, undertaking. A high level of technical expertise on the part of the participants is a must. But other abilities, perhaps less obvious, are also required. The days of sitting in a swivel chair and working the project for weeks in isolation are over. Projects in today's world require a lot of interaction. The customer is looking for partners who are proactive and engaged.

Several issues should be resolved early in the project. Chief among these is the establishment of good lines of communication among the different players who make up the project team. To achieve this, it is important for all team members to understand their roles in the overall project in order to know what to expect from other members and organizations, and to know what others are expecting from you.

Three different classes of organization make up the project team:

1. The owner and the owner's engineer;

2. The designer, which includes E/I&C (engineering), CSI (systems integration), CPF (panel fabrication), and various OEM organizations

3. The constructor, which includes the various companies that install the equipment, wire it up, and get it running.

These organizations are generally composed of multiple distinct groups (as in a CSP) or separate companies, though the smaller the project, the fewer the distinctions. For example, on very small projects, the designer might be the owner's maintenance technician, who will also install (construct) the system. Nevertheless, the three classes are present in some fashion.

In an ideal world, each major team element would function in perfect concert with the others, performing much like a symphony to produce a work that is harmonious and equally beneficial to all concerned. This is, however, difficult to achieve under the best of circumstances. In reality, the normal design process is quite discordant and chaotic. Sometimes the definition of success itself differs widely among the participants, making it impossible to achieve anything other than a compromise in which none are satisfied. Many times, managing discord and chaos seems to be the team's primary task, and a well-executed project may differ from a poorly executed one only by the level of disharmony that exists.

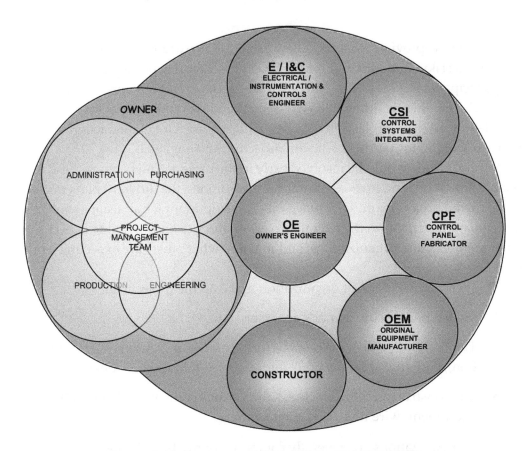

FIGURE 1-11. TYPICAL CONTROLS PROJECT PARTICIPANTS

Though the situation begs for a well-oiled project team, these organizations may never have worked together in the past, making teaming and close cooperation difficult. In addition, some of the participants—particularly those in the owner's organization—may have never been engaged in major projects at all.

All of this complexity and variability impedes good communications, and team building can be a major undertaking. Key stakeholders in each organization need to meet early in an attempt to quickly construct a basis for an effective professional relationship. This pays off in better communications among groups, which results (hopefully) in a more efficient project organization.

The larger the project, the larger the problem, though communications problems on small projects can be equally disruptive. More often than not, a small project's kickoff meeting occurs in a conference room near the project manager's office onsite and is sandwiched between pressing production matters that may take as much of the project manager's mental energy as the meeting. The players probably already know each other, and personal discussions are minimized. Team building exercises are seen as an unnecessary luxury, though the need to achieve some basic understanding between participants is just as great.

Regardless of the size of the project, one way to minimize potential gaps in understanding between organizations is for each participant to understand the other participants and their motivations and to become familiar with the different forces at work behind the scenes. This is particularly important for new engineers or junior designers, who rarely have a "big picture" viewpoint because each project is different, and they see only a few facets at a time. Each new project adds a bit more to their store of knowledge until a sort of critical mass is achieved, whereby reasonably accurate assumptions can be made about activities outside their field of view. This process of assimilation can take several projects before bearing fruit. In an effort to speed up this process, this chapter defines the major organizations and discusses some of their goals and methods.

A. THE OWNER (I.E., BUYER, CUSTOMER)

This book often depicts the owner as some sort of monolithic entity. In reality, however, the owner is an amalgam of individuals and subgroups, all of whom are important and must ultimately be satisfied to some degree. Figure 1-11 shows some of the owner's internal organizations, each with its own set of goals, and each with a wish list of things they'd like to see as outcomes from a project. Part of the owner's Phase 1 FEL process is to build a list of potential projects, nominate a select few to pursue, evaluate the validity and feasibility of those nominated projects, get feedback from each of those internal groups, and either build their feedback into the project plan or reject it. By the time an RFP is produced, this vetting process should be complete, and proposals that result are based on the RFP alone. That doesn't mean that rejected pet ideas and needs will be dropped as possible issues. In fact, those issues frequently arise as the project progresses, and attempts are made by sponsors within the owner's organization to slide them into the agenda, thus causing "scope-creep." It becomes the responsibility of the owner's engineer and every contractor to continually check requests made against the contract. If a contractor performs work outside the contract—knowingly or not—the owner is not obligated to reimburse the contractor. Emerging requests should be dealt with through the Management of Change (MOC) process, which will be discussed later. Management of change is everyone's responsibility, but sloppy management of change is more likely to harm the contractor more than the owner.

Within the owner's organization, each project, whether it is a capital project or a work-order (maintenance) project, is organized in a similar manner. Someone has a need (production representative, perhaps), someone has the funds (business manager or buyer), and someone has the ball. Typically, the one given the ball is the project manager. In some smaller work-order projects, the project manager and the production representative may be the same person.

Work-order projects are generally very small tactical tasks that are funded, and sometimes managed, by the production group (i.e., Operations). The task is generally one that makes a minor change to increase production, add convenience, or improve safety. Generally, someone from plant engineering coordinates this sort of project, acting as project manager and project lead.

Capital projects are those funded from the owner's capital budget that is appropriated during yearly budget meetings. These projects can be thought of as strategic ones that will generate positive long-term effects. Capital project budgets are reported to stockholders, and as such, have high visibility.

Projects differ from one facility to the next and even from one project to the next. But they usually share some similarities. Producers are ultimately driven by the bottom line. If they are spending project money, then projects need to have positive results that increase the profit margin. While owners must frequently rely on outside resources, they need to avoid losing control of the process. Thus, a project management team that understands the owner's goals, understands project management, and can manage contractor and subcontractor relations becomes a necessity.

The owner typically has a group of engineers and production specialists who are tasked as members of the project management team. Often these individuals have other, more production-related duties as well. Depending on the size of the task and the availability of personnel, the project management team could consist of only one individual. In any case, the purpose of this team, under the leadership of the project manager, is to facilitate the work of the contractor(s) by providing oversight, guidance, and information.

The following are some of the primary organizations that will likely have representatives in the owner's project team:

1. Plant Administration

The owner's plant administration staff and officers are concerned with the day-to-day aspects of running the business. This includes funding appropriations, hiring and firing, personnel safety, accounting, proper functioning of the production, maintenance, and marketing departments, procuring raw materials, disposal of waste products, and numerous other concerns that have nothing to do with any particular project that may be in progress. Administration officers are generally willing to fund projects that a) make their or their employees' lives better, b) improve the profit margin, c) improve personnel or equipment safety, or d) improve their product. These concerns leave little room for involvement in the project itself. In fact, ongoing projects tend to interrupt the normal flow of things, and can become irritants, so it should come as no surprise to contracting personnel that meeting times are short, and topics are abruptly dealt with.

2. Plant Operations

The typical project's primary customer is the operations group. Operations staff are held responsible for meeting production goals set forth by higher management and are generally the ones who must justify funding for the project, particularly if the project's goals are for streamlining operations, operational safety, product quality, or process repeatability. The operations group may be ambivalent about projects that change operational procedures for reasons that are outside their purview. It is a good idea for Operations to appoint a representative to interact with the project team, but this frequently does not happen until late in the project. It does seem that more project planners are now making better use of the knowledge base available in the operations group, but this has not always been the case. It is advantageous to get their buy-in, even if their every desire can't be met. The operations representative, ideally, will be the person who is intimate with the production issues and has enough engineering background to be able to both contribute to the conceptual process and communicate decisions and rationale back to the operations group.

3. Plant Engineering/Maintenance

A production facility may or may not have an in-house engineering department, but there will generally be a maintenance department. As in Operations, the primary focus of the maintenance department is not on an ongoing project. The main priority is to cope with problems that affect production. The owner's technical representative to the project team is generally from the maintenance department, but it is usually a sideline function for that individual. That is not to say that the maintenance department isn't interested, it's just a function of resource availability. Ideally, this individual will monitor the project to ensure that the interests of the maintenance department are given proper consideration during the design phase, provide input as appropriate, and report back to the maintenance department. As a result of competing interests, input from this quarter is sometimes not sought or provided early enough to have much impact. Occasionally, rework is the result. It is the project manager's responsibility to make sure that both the Operations' and the maintenance department's concerns are addressed.

4. Plant Purchasing

The plant purchasing group manages the contracts and writes the checks. Purchasing develops a list of qualified bidders, issues the RFP, and manages the bid process. Final bid selection is done in consultation with all other interested parties, but once the selection is made, the purchasing group notifies all the bidders of the decision, works with the losers to provide whatever feedback is deemed appropriate, and works with the winners to finalize contracts. As the project progresses, the purchasing group receives the contractors' invoices, validates the invoices against deliverable milestones or other criteria, and issues payment.

B. The Owner's Engineer (OE)

The function of the owner's engineer is that of liaison between the outside project participants and the interested parties inside the owner's organization. Typically, the OE is focused entirely on the project. The OE could be:

- An individual, either from the owner's organization or a consultant hired for the duration of the project
- A team of individuals from the owner's organization that represents a cross-section of the engineering specialties engaged on the project
- A consulting project management firm
- A prime contractor whose scope of work could include not only owner's engineer responsibilities, but also a major design role.

1. The OE as an Individual

Whether an OE is drawn from the owner's organization, a hired gun brought onboard for the duration of the project, or a firm with a designated team, the OE needs to have a thorough understanding of project management fundamentals plus a broad understanding of the task at hand. Having direct experience on similar projects is a definite advantage. Projects that employ a single individual in this role are usually small projects that involve a minimum of different organizations and disciplines. The person may be working other projects simultaneously, so his involvement will necessarily need to be kept on a high plane, monitoring contractor progress more than content. If specialized skill sets are needed, this person will be able to recognize the need for additional support and will not hesitate to call for temporary assistance. His primary role is that of owner's Project Manager, though other hats to be worn could include Project Engineer, Quality Control Inspector, Construction Manager, and whatever else is needed.

2. The OE as an In-House Team

If a project is more complex than one person can manage, then a team may be assigned to the project. In today's business environment it is rare that a manufacturing company has skilled engineering personnel available to dedicate to a project, though it does happen from time to time. In most cases, there will usually be a designated project manager and some number of project specialists whose job is to interface with their contractor counterparts. Most of the time, these specialists will still have a "day job," spent maintaining equipment or supporting operations, so their support of the project can be a bit fragmented.

3. THE OE AS A CONSULTING MANAGEMENT FIRM

This is a good way to go on large projects. Companies that specialize in project management frequently pay for themselves by delivering projects that are well organized and coordinated. Just a few cases of rework or duplicate work avoided will pay dividends. In addition, project responsibility for project tax and expense audits will be reduced for the owner through the use of a certified project management firm.

4. THE OE AS A PRIME CONTRACTOR

Designating a prime contractor to act as the OE is a very common practice. In this approach, a general engineering firm is selected both to act as the owner's agent in dealing with the array of subcontractors, and to execute a facet of the project themselves.

One example of such an arrangement is common in the municipal water industry. A water utility will hire a civil engineering company as the prime contractor. Generally, civil engineering firms have a solid grasp of municipal building codes, underground conditions, and items specific to the site that a local company would know. The civil engineering firm would then execute the portions of the project related to grading, digging, foundation laying, and tendering the remainder of the work.

Another example is a project in which a number of pre-engineered skids or subsystems need to be installed and started up. A general engineering (A&E) firm may be selected to work with the OEM providers as they build their pre-engineered subsystems, then attend the OEMs' Factory Acceptance Tests, do the Balance of Plant (BOP) engineering that will be required to marry all the subsystems at the plant site, and coordinate checkout and startup.

In both of these cases, the OE is external to the owner's organization, and so internal owner resources will need to be allocated to manage the OE. This internal management entity could be an individual or a team, depending on the need.

C. THE DESIGNER

Like the owner, the designer is an amalgam of several entities that include E/I&C, CSI, CPF and others. For the purposes of this discussion, the designer will be an engineering firm that has been contracted to take the desired end result as envisioned by the owner and engineer—a solution that is both constructible and cost-effective. The execution of a design project is a team effort. Design teams are generally ad hoc groups that are assembled for a specific project and then disbanded. The individuals involved must learn to adapt quickly to new situations and to coalesce into teams on short notice. The success of the project depends on the organization's ability to execute this maneuver in a timely manner.

The designer's project team usually consists of several disciplines or groups with different skills. For example, a design team might require civil engineering to build a structure, electrical engineering to provide lights and power, mechanical engineering to provide plumbing and other services, etc. These different disciplines must be managed and their efforts coordinated to achieve a common purpose.

Figure 1-12 and its accompanying discussion present a top-down description of the project team.

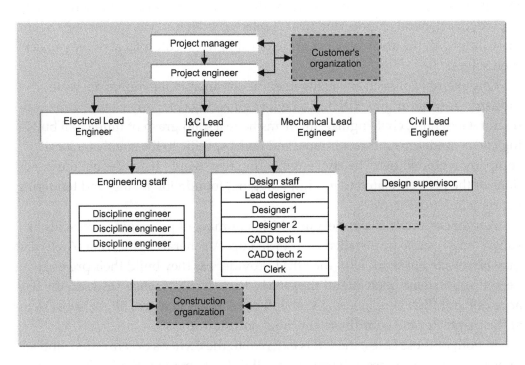

FIGURE 1-12. ENGINEERING DESIGN TEAM

1. Project Manager (PM)

The project manager has overall responsibility for the project. This person typically has developed a business relationship with the customer and is sometimes quite possessive of that relationship since such relationships are difficult to build and easy to disrupt. A project manager guards the customer's interests within the framework of the project scope, budget, and schedule with his own organization, and similarly guards his own organization's interests as they relate to the customer. The PM's primary focus is budget and customer relations.

2. Project Engineer (PE)

The project engineer manages the design team. This person coordinates the efforts of multiple disciplines as they each pursue their particular interests. The project engineer ensures that the various disciplines' engineers are working in concert, using proper procedure and executing their tasks in a manner that is consistent with the goals of the project team. The project engineer also resolves or prevents disputes by enforcing good communication practices among the disciplines. This person is (or should be) an enabler (in the sense of "facilitator"), making sure that all involved are properly trained and have all the support personnel and materials they need. The project engineer's function is sometimes combined with the project manager's function, with the same individual performing both roles, but the PE function is subordinate to the PM function.

3. Discipline Lead Engineer (DLE)

A discipline lead engineer manages a discipline's efforts as they relate to a project. A discipline design team typically consists of an engineering staff and a technical support staff. The engineering staff generates conceptual design documents that the technical support staff uses to build a complete documentation package that is suitable for construction and maintenance. Discipline engineers coordinate their activities with other discipline engineers under the direction of the project engineer. Discipline lead engineers manage their staff to ensure that it is functioning efficiently and effectively.

4. Discipline Engineer(s)

The discipline engineering staff generates products and services under the direction of the discipline lead engineer. These engineers work closely with the design staff to satisfy the scope of work.

5. Discipline Design Supervisor (DDS)

The discipline design supervisor reports to the department head (not shown). The technical support staff reports to the design supervisor administratively. This person is also responsible for the overall quality of the output of the technical support and clerical staff.

6. Discipline Technical Support (Design) Staff

a. Lead Designer
The discipline technical support staff uses the information developed by the engineering staff to develop construction drawings and documents for record. The discipline lead designer is responsible for the technical design content of the engineering package. This person manages the efforts of the technical support staff, which includes the other designers, CADD drafters, and clerical support under the overall direction of the discipline lead engineer. The discipline lead designer also coordinates efforts directly with other technical staffs from the owner and construction organizations.

b. Designer
Designers are technical support personnel with particular skills. Their abilities can range from expertise in a single area of design up to complete competence in all aspects of design. They are responsible for the technical content of the design drawings, bills of materials, databases, etc.

c. CADD Technician
CADD technicians are responsible for the production of the drawing package, reporting directly to the lead designer.

d. Engineering Aide (EA)
The Engineering Aide generates databases and bill-of-material listings and performs other administrative tasks in direct support of design operations. Sometimes a CADD technician will double as the EA.

D. The Control Systems Integrator (CSI)

Trends in the past few years have made it necessary to distinguish between types of Systems Integrators. Control Systems Integrators (CSI) operate in the production environment, working directly with plant-floor equipment and production operations personnel, while Information Technology Systems Integrators (ITSI) work in the business environment.

The original line of demarcation between plant floor control systems and IT has changed from being VERY distinct to being VERY blurred. The IT group used to have responsibility for all the interfaces between a plant and the outside world (e.g., the Web). Nowadays, plant supervisors are calling for real-time plant production data on which to make instantaneous business decisions, and so the trend is more toward piping the data directly to the Web environment rather than going through the IT network. In addition, technologies designed for the IT world, such as Ethernet-based architectures, are quickly finding their way onto the plant floor. It is a rare control systems integration organization today that does not have IT specialists on staff. Still, the overlap is by no means total, and CSI and ITSI remain separate and distinct. Some of the technical aspects of this topic will be discussed in Part II.

The Control Systems Integrator follows much the same project management format as does the engineering design team, with some differences in the skill sets. The project manager and project engineer's relationships and responsibilities are identical to those described previously in the E/I&C section and so will not be discussed here.

In general, there are four divisions of labor within the CSI department, as shown in Figure 1-13.

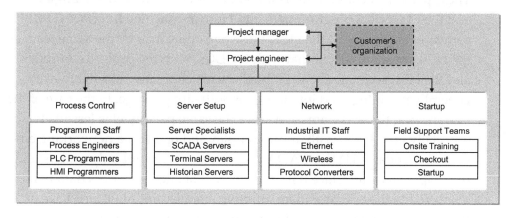

FIGURE 1-13. CONTROL SYSTEMS INTEGRATION DESIGN TEAM

1. THE PROCESS CONTROL TEAM

The process control team works the most closely with the other project organizations. Their work is very interactive with the owner and other project team members as they interpret the project objectives and develop the primary set of deliverables that are tailored to that particular project.

a. **Process Engineers/Specialists**
These are project team leaders who are knowledgeable about the process, whether it be chemical, mechanical, material handling, or whatever the application. These project team leaders understand the principles of operation, equipment, and control schemes necessary to achieve an objective, and can communicate what is needed in terms of control and operation to the other members of the Process Control Team.

b. **PLC/DCS Programmers**

   These personnel convert information provided by the process engineers into Programmable Logic Controller (PLC) or Distributed Control System (DCS) software. This software directly affects plant equipment by energizing or de-energizing it, or by modulating its position or speed. The software used is generally proprietary, produced by the manufacturer of the control processor. The hardware outside of the Central Processing Unit (CPU) can be a mixed bag of proprietary equipment from the same CPU manufacturer, off-the-shelf components, and specialty equipment that may itself need to be programmed or configured. The PLC/DCS programming team must master all these hardware and software combinations.

c. **HMI/SCADA Programmers**

   These personnel convert information provided by the Process Engineer and the PLC/DCS programmers into Human-Machine Interface (HMI) software. In the case of the DCS, these personnel may be the same team as the DCS programmers in that with the DCS, both the process controller and the HMI are manufactured by the same company, and the two are generally pre-integrated into the same software package. In the case of the PLC, having the two teams consist of the same personnel is less likely, especially in today's vastly more complex and open data networking environment. Today's HMI/SCADA programmer must not only understand the graphic user interface (GUI) software that allows process data to be displayed to the operator in an intuitive manner, but must also have a thorough understanding of client/server architecture, networks, and other areas of expertise that traditionally were considered to be more part of the Information Technology (IT) profession. In the past several years, this specialty has seen the most growth in terms of the open technologies available, and in terms of demand, as customers have begun to realize that having real-time, plant-floor information can give them an edge in the marketplace. It is the HMI/SCADA programmer that gets that data where it needs to go, when it needs to be there, in the most effective format possible.

2. THE SERVER SETUP TEAM

Servers are incredibly complex to configure properly and to troubleshoot. While the actual time spent in setting up the servers is very small in terms of the overall project cost, proper setup is crucial to the outcome. These specialists are skilled in server architecture and in understanding the complex licensing requirements. These personnel are typically a subset of the HMI/SCADA programming team.

3. The Network Setup Team

Today's Control Systems Integration teams must have advanced network architecture, implementation, and troubleshooting skill sets. Members of this team are also likely to be a subset of the HMI/SCADA programming team. These personnel understand how to design, configure, and troubleshoot network equipment in the industrial environment that many times includes redundant, fault-tolerant designs. Nowhere has the encroachment of IT been more noticeable than in the industrial network, where Ethernet has essentially taken over as the network protocol of choice. The availability of low-cost, Ethernet-ready equipment has made it convenient and cheap to provide a fully functional network environment. However, the sheer number of choices that are available and the dynamic growth of data management requirements make this team indispensable as a set of specialists.

4. The Startup & Commissioning Team

The startup and commissioning team is made up of individuals with some level of capabilities in each of the specialized areas described above. They may be the same personnel, or they may be an entirely new set of specialists that understand the vagaries of field work. There is a significant shift of mental gears between the office-worker mentality and that of the field engineer, and it is sometimes difficult to find personnel who can make that transition smoothly. The field environment is much more dynamic and stress-filled, as equipment is brought online and raw materials begin to be consumed. Rotating machinery, heating and cooling, and moving materials each contribute to an environment in which the consequences of mistakes made become more dire. The startup and commissioning teams need to be able to react quickly to changing circumstances, identify problems, and deploy solutions quickly in order to minimize waste and equipment/personnel safety.

E. The Constructor (i.e., Builder)

Like the owner and the designer, the constructor is an amalgam of several entities. These entities are called crafts. Electricians, pipe fitters and welders are examples of crafts. The construction team's organizational structure falls more in line with that of electrical and instrumentation (E&I), with controls falling somewhere among them. Figure 1-14 and the following discussion present a top-down description of the construction team.

1. E&I Construction Superintendent

The superintendent's role on the construction team corresponds to the role of the discipline lead engineer on the design team. If the project is large enough, each

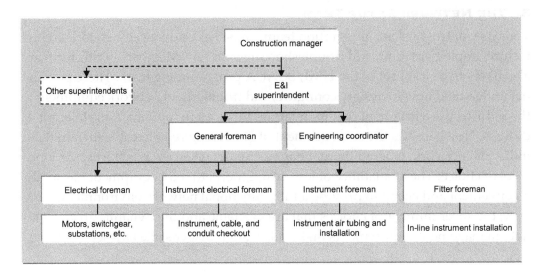

FIGURE 1-14. CONSTRUCTION TEAM

craft could have a dedicated superintendent. The superintendent is responsible for building a good construction team. What follows are two of the most important personnel choices the superintendent makes:

- A team of frontline supervisors
- A good material expediter, coupled with knowledgeable receiving and warehouse personnel, who will inspect and catalog all instruments for tracking purposes.

The construction superintendent is ultimately responsible for the schedule and budget, quality of work, safety of the construction team, and accurate time reporting. In short, the buck stops with the superintendent during the construction process.

2. E&I Field Engineer & Coordinator

The E&I field engineer/coordinator interfaces with the design engineer and the E&I superintendent to confirm the proper field installation of instrumentation. In coordination with the E&I superintendent, the field engineer performs material takeoffs for all instruments and associated items (if not already provided by the design team). This includes material, such as tubing, tube tray/track, tube fittings, and mounting extras such as angle iron, nuts, bolts and miscellaneous parts. This person also generates bills of materials for electrical items, such as conduit, cable, electrical fittings, and associated parts to install the wiring for instrumentation systems. The E&I field engineer is also responsible for initiating field change notices (FCNs) or design change notices (DCNs) as design modification needs are uncovered.

3. E&I General Foreman

The E&I general foreman works with the superintendent and other foremen to develop weekly work plans and daily task plans to ensure that craftsmen know the priority of projects for the day and week.

4. Instrument Foreman

Except for very large projects, this is a position that is usually combined with the E&I General Foreman. The instrument foreman creates daily and weekly work plans and ensures that proper installation procedures are followed by the instrument fitters/mechanics.

5. Instrument Fitter/Mechanic

The instrument fitter/mechanic works within the confines of the instrument air distribution system. This individual installs each air-consuming instrument according to installation details. Copper, stainless steel, or plastic tubing is run from a distribution point to all valves that are automated, either by a solenoid valve or by an I/P (current-to-pressure) converter. Usually this person does not have detailed drawings to work from, as tubing and small-bore pipe are generally not shown on drawings.

6. Pipe Fitter

The pipe fitter installs all inline devices, such as control valves, magnetic flowmeters, and on/off valves, along with any piping associated with pressure gauges, pressure switches, and miscellaneous devices.

7. Instrument Electrical Foreman

The instrument electrical foreman creates daily and weekly work plans and ensures that proper electrical specifications and procedures are followed by the electricians with respect to instrumentation installation. This person supervises the installation of conduit, wire pulls, and termination at the instrument itself and at the DCS/PLC at the instrument's proper address, as well as the point-to-point wiring checkout.

8. Electrician

The electrical craft includes many general service electricians that work across boundaries. Some electricians are specialists that stay within an area, but this usually depends on the size of the project and their own set of abilities. The major divisions in expertise align with the voltage level: High Voltage (500kV), Medium Voltage (4160V), and Low Voltage (480V). Instrumentation experts naturally come out of the Low Voltage area, and specialize in 480VAC and below, including

DC signals and instrument power distribution. During construction and startup, the E/I&C design team generally works in support of the low-voltage electricians to ensure that the instrumentation and PLC/DCS systems are wired per the drawings.

9. Instrument Electrician/Technician

The instrument technician is generally a low-voltage electrician with specialized understanding of the electrical side of instrumentation as opposed to lighting, etc. This person understands the characteristics of each instrument type, when to use shielded cables, where to ground the shield, how to inject simulated signals to the DCS/PLC, and how to use an array of calibrators. During checkout, this person usually works directly with the control systems integrator to test each instrument prior to startup.

Part I – Chapter 3: The Managed Project

A well-conceived, well-managed project is one that has been given a chance to succeed, supposing that the execution team is capable. Conversely, a poorly managed project may fail, regardless of the abilities of the execution team. If a project fails, yet is well managed, then some positive effects can probably be salvaged—even if the outcome is merely a list of lessons learned. In such a project, the problems will likely be well defined, making it possible to avoid similar entanglements in the future. In a poorly managed project, the root causes may never be uncovered.

Well-managed projects do not just happen. They begin with the basics. All members of the project team need to have a clear understanding of the work that needs to be done and a clear view of their role in the process. The days of "mushroom engineering," in which the project manager keeps the design staff in the basement cranking out drawings, are over. Today, design teams need to be fully engaged from top to bottom—it's more of a Velcro® interface. The conversion from mushroom to Velcro means that the support staff needs to be more sophisticated and more attuned to the specific issues being addressed by the project. The more pertinent information they have, the more likely they will be to do the right thing the first time, and deliver the product or service when needed.

But before a project begins, a buyer must select a seller and award a contract. As we have seen, the process that leads to awarding a contract usually begins with the release of the Request for Proposal (RFP). The RFP is usually distributed to a pre-approved set of service providers for their evaluation. The prospective sellers generate a budgetary estimate based on the information provided in the RFP. Most of the time, this budgetary estimate is usually only accurate to ±30%. The buyer may ask for the budgetary estimate in order to pre-validate a prospective seller and to validate his own internal budgetary estimate. If all the bidders respond with estimates that are out of range, the buyer may kill the project right there, or reassess his goals. If one or more respond with a reasonable budgetary estimate, then a pre-bid orientation meeting is generally hosted by the buyer to transmit additional information to the prospective set of bidders (Figure 1-15). The buyer distributes bid packages in the meeting, facilitates site walkdowns, and performs various other tasks to make sure that each bidder's proposal team has enough information to develop a more accurate bid.

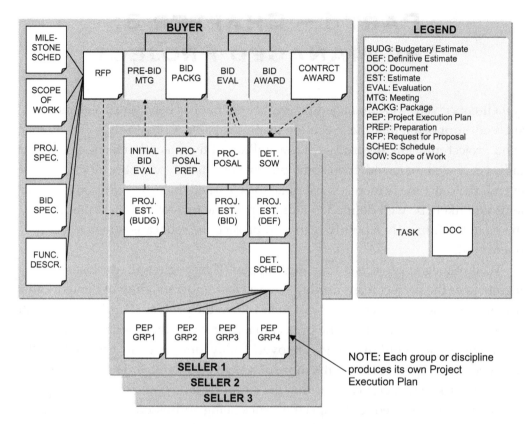

FIGURE 1-15. SIMPLIFIED CONTRACT AWARD OVERVIEW

For each bidder, the proposal preparation process begins by producing a more refined bid estimate, and then preparing a proposal. Once the proposal is completed, it is sent to the buyer, who begins an evaluation process that may cause some recycling of the estimate and proposal as the buyer discovers scope gaps and tweaks the scope of work. Eventually, the buyer selects a seller. That does not necessarily end the estimate/proposal cycle, as the seller may decide to work with the buyer to optimize the project plan and offer alternative approaches. The result is a refined estimate called a definitive estimate, through which the seller provides a final price to the buyer, who plugs the numbers into the contract.

The scene shifts to the seller at that point as all the information collected during the bid process is compiled into a set of documents that can serve as the design basis for the project team(s). Hopefully, the proposal team did a good job of capturing information during the pre-bid site visit and subsequent investigative cycles. The small number of participants exposed directly to that early information makes capturing that information very important (see Figure 1-16 for a reasonably good information-gathering form template, also provided as Form 16 on the enclosed CD). To compound matters, the project specification and RFP provided by the buyer are often inadequate for the purposes of developing a detailed workflow plan. Additional preparatory work is required by the seller in order to ensure project success.

A. Key Project Management Tools

Projects are awarded based on the buyer's evaluation of the proposal (which includes the estimate and a text-based reiteration of the scope of work) produced by a prospective seller. The buyer compares the assertions made by all the bidders and compares those assertions and estimates to expectations established by the scope of work embedded in the RFP and the in-house budgetary estimate. The owner then selects a seller, and a contract is generated that is signed by both parties. Most tend to think of this explicit contract as the only one that matters. But neither the customer's bid package nor the resulting proposal is entirely adequate for detailed design planning. In order to properly set up a project to succeed, a design basis needs to be established that can be used to guide the different disciplines as they execute the project. The design basis should be set early in the project. Fortunately, many of the key elements of the design basis can be developed, at least to a preliminary level, before the project is even awarded. If the information developed during the proposal process is captured, it can form the nucleus of a well-formed design basis. The key inputs for developing a detailed design basis that are started during the proposal process (in order of creation) are:

- The detailed scope of work – the "what" prepared by the buyer for bid purposes.

- The budgetary, bid, and definitive project estimates – the "how much" prepared by the seller for proposal preparation.

- The WBS-based project schedule – the "when" prepared by the buyer for bid purposes, then refined by the seller after winning the bid.

- The proposal – the "who," "what," "where," "when," and "how much" prepared by the seller for delivery to the buyer.

- The project execution plan – the "how" prepared for in-house use by the seller (prepared initially for proposal preparation and in more detail later if the project becomes a reality).

These items do not form a complete design basis, but they do provide a core of information around which adjustments to the design basis can be made while the project is ongoing. But usually these are adjustments only and not wholesale changes; and they are usually documented under a Management of Change

(MOC) process. These inputs to the design basis are discussed in this chapter, along with a suggested format for a Weekly Status Report that will help the project manager ascertain whether the disciplines are performing as expected.

1. THE DETAILED SCOPE OF WORK (SOW)

A scope of work is simply a description of the tasks that need to be accomplished to achieve the desired end. It answers the "what" question in great detail, leaving the other items (who, where, when, how, and how much) for others. The "who," which is individuals or organizations, and the "how," which is dealt with in the Project Execution Plan, are described later. An SOW is usually initiated as part of the bid package. After the project is awarded, the SOW can be more fully developed as new information is gleaned and assumptions resolved. This detailed SOW uses the information provided in the proposal, the RFP, and the project specification.

The SOW should be very detailed, down to the expectations for each subgroup or discipline. A multi-discipline A&E firm will typically have each discipline generate scope sections specific to their areas, and then roll these scope sections up into an overall project scope of work that describes the efforts of the entire company. This can become a very large document. The overall project scope of work will have chapters for each discipline that will be reviewed closely by the project manager. It is the PM's responsibility to make sure the SOW satisfies the requirements of the RFP and meets the assertions of the proposal. The PM will also look at the projected number of man hours and the bill of materials, and satisfy himself that the project can be done as described and still stay within the confines of the contractual budget and schedule. This analysis sometimes causes a cycle of revisions to occur between the disciplines and the PM as the scoping document is refined. Once approved by the PM, however, the SOW forms a sort of internal contract between the PM and each discipline and will be referred to frequently in order to determine whether the workgroups are fulfilling their obligations to the project.

It is recommended that the project scope of work be written in sections that closely reflect the way in which the project will be installed; in other words, it should be written from the point of view of the constructor. Doing this will save a lot of time later on, when the owner's engineer inevitably asks that the information be repackaged to benefit the constructor. It also makes it easier to schedule the workflow to meet realistic timelines, reducing the risk of having to reset priorities after the project gets underway and timelines shift. The Work Breakdown Structure, if one was produced, should help organize the work. For example, if construction will occur in four distinct areas of the plant, then there should be at least four WBS sections and four SOW sections, plus a fifth section dealing with systems that are not specific to a single area. This fifth section will deal with instrument air distribution, process water supply, etc. Each of the sections dealing with a specific area should have subsections devoted to each engineering discipline.

The format of the proposal works well for the SOW; there is just a lot more detail, plus labor estimates and timelines. The following is a suggested format:

- Executive Summary – Reiteration of the Executive Summary in the Proposal with more verbiage about what this particular group is expected to do.
- Scope of Work for Area 1
  - Scope – a narrative describing the details of the project for this specific process area
  - A list of safety concerns
  - A list of assumptions
  - A list of exclusions
  - A list of deliverables
  - A milestone schedule
  - A labor estimate (listed in labor hours)
- Scope of Work for Area 2
  - Scope – a narrative describing the details of the project for this specific process area
  - A list of safety concerns
  - A list of assumptions
  - A list of exclusions
  - A list of deliverables
  - A milestone schedule
  - A labor estimate (listed in labor hours)

The same model should be followed for each WBS area for each discipline. Where appropriate, even the phrase "Not applicable to this discipline" is good information to have in the SOW.

2. THE ESTIMATE*

[* For an in-depth look at how to generate an accurate bid estimate and create a project schedule using Microsoft® Excel, a CD-ROM titled "Software Tools for Instrumentation and Control Systems Design" is included with the book.]

A proper understanding of the project requirements begins with a well-defined scope of work, as previously discussed. But the scope of work is only part of the story. After the "what" is addressed by the scope of work, "how much" needs to be addressed. Therefore, the next step on the path to a successful project is to generate an accurate estimate of material cost (in terms of dollars allocated) and labor costs (in terms of time, measured in man hours and then converted to dollars). These measures define the "physics" of the situation, setting limits that affect the way in which the project will be executed. Dollars spent is the primary tool of the project manager in tracking the project at the strategic level and man hours spent of the design supervisor in managing the work at the tactical level. Only after the number of man hours is known can a schedule and project plan be developed. There are three main types of estimates: budgetary, bid, and definitive.

The following is a brief description of each:

a. **Budgetary**
   The budgetary material cost and labor estimate is usually produced by a group who is familiar with the site and with the plant operation. This group may or may not execute the project. The purpose of this type of estimate is primarily to obtain funding. It is typically "quick and dirty" and is expected to be approximate. In fact, an error margin of ±30% or more is often acceptable for this type of estimate. Prior to this estimate, a formal scope of work has probably not been done, and, quite possibly, the project specification has not been finalized. This type of estimate is generally unfunded by the customer, or at least under-funded. The bidder pulls the numbers together as quickly as possible, with a minimum of research. The expectations for the bid and the definitive estimates, on the other hand, are much higher.

b. **Bid**
   The bid estimate is produced by the various engineering bidders vying for the work. Again, this estimate is typically unfunded by the customer. The raw information for this estimate is the finalized project specification bid package. All prospective bidders are provided identical bid packages on which to base their bids. This bid package is sometimes flawed or incomplete, but the resulting bids give the customer a good picture of the capa-

bilities of each bidder. Usually, all bidders produce a written proposal that accompanies their estimate. The proposal describes the work as understood by the bidder. This gives the customer an idea of the deliverables each bidder plans to provide. It also gives the customer a means for comparing the capabilities of the various bidders.

c. **Definitive**

The definitive estimate is prepared by the engineering firm that was awarded the contract based on its bid. The customer typically includes this as a part of the project, so it is probably fully funded. Once the contract has been awarded and any secrecy agreement issues have been settled, the engineering contractor is given full access to the information developed by the customer during the internal evaluation process. The engineering firm then develops a formal scope of work. This information sometimes alters the picture significantly, and the engineering contractor is given an opportunity to adapt the estimate and schedule to the new information. Of course, this re-estimation process is bypassed if full disclosure was made during the bid process. In that case, the bid estimate becomes the definitive estimate.

The definitive estimate is the baseline document for project management, providing a yardstick by which performance will eventually be measured. It is based on the scope of work and should reflect the work scope on a task-by-task basis wherever possible. Many estimating tools are available. Whichever is chosen, the engineering firm should be able to produce a definitive estimate that is

- *Accurate*, by identifying each task relating to the scope of work and then quantifying the labor and expense of its execution.

- *Timely*, by being repeatable. Once an estimate structure has been optimized for a particular customer, subsequent estimates should simply build upon the first, thus decreasing the cost and time required to generate them. Each project should be approached in a consistent manner, with consistent estimating practices and tools.

- *Verifiable*, by adding notes and amplifications and by making the calculations available for dissection. Remember, the time span between the estimate and project execution can be large, and the estimator may not be the executor.

- *Meaningful*, by relating all the deliverables to hard data. The use of rules of thumb and suppositions should be minimized.

- *Adaptable*, by being easy to modify.

An estimate that is properly executed and formatted will feed directly into the project planning process. A poorly-executed estimate does not capture enough detail to relate to the project plan.

In practice, control systems budgets are often set by simply taking ratios of the project's overall budget, particularly on large capital projects. In those cases the estimate is done "offline," after the bid has been awarded, merely as a way to determine whether the budget available is going to be sufficient for the job at hand. In the case where the budget allocated is not sufficient to the task at hand, the supplier must find creative ways of modifying his normal execution patterns in order to still satisfy scope, and yet do it in a leaner's fashion. This process can get very interesting.

3. THE SCHEDULE*

After the scope of work is complete, a work breakdown structure (WBS) may need to be created. This is usually done by the owner during the bid package preparation stage. A WBS structure could be derived from physical locations if the project entails different sites, or it could derive from natural break points in the production process. On projects that are large enough, entire project teams may be assigned to specific WBS areas. The key piece of information is that a WBS should be used to partition the project into segments that are meaningful and manageable. The WBS provides a detailed, task-oriented strategy for project execution that can be broken down by discipline and by sub-contractor. This, in turn, becomes the basis for the project execution plan and detailed project schedule, which is usually finalized by the winning bidder working in concert with the owner.

Within each WBS section, each task should be evaluated as to its relationship with other tasks. For example, generating loop sheets depends on the P&ID drawings. The P&IDs, therefore, are predecessors to the loop sheets. Such linked relationships should be identified and analyzed. For the most part, these relationships may be defined as follows:

- *Start-to-start:* Task B can't start until Task A starts. This relationship is used when an outside trigger initiates a scheduled chain of events. For example, receipt of a specific material shipment might trigger several tasks at the same time.
- *Finish-to-start:* Task B can't start until Task A finishes.
- *Start-to-finish:* Task A can't finish until Task B starts.

- *Finish-to-finish:* Task B can't finish until Task A finishes.

After the relationships are defined, and the resources have been assigned to each task, the overall project duration may be estimated. Many times, the customer has a timeline in mind already, which constrains the schedule to a particular set of milestones.** When the schedule is compared to this time line, resources may need to be modified to lengthen or shorten the schedule.

[** For an in-depth look on how to generate an accurate bid estimate and create a project schedule using Microsoft® Excel, a CD-ROM titled "Software Tools for Instrumentation and Control Systems Design" is available with this book.]

4. The Proposal

The proposal is a multipurpose document whose prime purpose is sales oriented: to convince the buyer that this particular seller is the right one for the job. After the project is awarded, the proposal's second purpose is technical: to feed into the design team's effort to form a design basis. This means the proposal's value as a project management tool after project award can vary greatly from project to project.

In general, proposals are produced by sellers who are taking calculated financial risks in their production. Proposals are rarely funded by the customer, and the seller understands that not all of these investments will result in the award of a project. So the risks are calculated and the amounts invested carefully controlled. This cost boundary affects the number of participants that can be involved, the amount of time they can spend in the field gathering information or in meetings, and the amount of time it takes to develop the proposal document itself. The personnel involved in these early stages are usually upper level managers and key technical personnel that are able to size up a situation in a minimum amount of time. These company representatives sometimes make business decisions in those early stages that not only affect *what* will be done, but also *how* a project will need to be executed. That "how" may force their design team to make adjustments to their normal design process in order for the project to succeed.

There are three basic ways to approach the proposal process, the choice usually being based on the potential size of the project. A good rule of thumb is to make a budgetary estimate of the probable project size based on the RFP, and select a proposal team budget of something less than ten percent of that estimated project value. The dollar amounts provided below reflect the author's experience as a project manager in the engineering services business, and may not translate to other business types, particularly those that supply a lot of equipment.

- *Small Projects:* Rule-of-thumb bids are frequent in this scenario, with only a bare minimum of key information provided by the seller, and probably one or at most two personnel from the seller's organization present at a one-day pre-bid meeting. The pre-bid meeting can even be done over the phone if practical. These projects are usually single discipline efforts with small budgets. Proposals that result from this sort of off-the-cuff analysis are necessarily very short and to the point, sometimes one or two pages in length, thus making the proposal of small use to the execution team later, except to provide the major boundaries of time and cost. This kind of bid can only be made by senior personnel who have intimate knowledge of what the buyer's expressed and implied wants and needs are, and who know exactly what the capabilities are of the design team who will execute the project.

- *Medium Projects:* The seller forms a small team of key personnel who are experienced in similar projects or are familiar with the buyer. These projects can be multi-discipline projects that have a minimal number of process areas to deal with or single-discipline projects that are more involved. The investment in the proposal depends on the project's size and complexity. This kind of investment will typically support a small team for about a week's work to gather the information and develop the proposal.

- *Large Projects:* The seller must be willing to accept a large financial risk to take on projects of this magnitude. Proposal costs could vary widely, depending on the size of the project and its complexity. Usually, as the size of the project goes ever higher, the percentage that will need to be spent on the proposal will decrease, often in a nonlinear fashion. For such projects, the proposal process can take weeks and is a project in itself. Such proposals tend to be very detailed, and are of great use as fodder for the development of subsequent management tools after the project is awarded.

Regardless of the size of the project, the proposal should, at minimum, answer the "who, what, when, where, how, and how much" questions that are germane to each team subgroup, and should list any assumptions made and any exceptions taken to requests made in the RFP.

The format of the project proposal is often dictated by the RFP, because if the buyer gets multiple bids, they should be in the same format to facilitate comparison. Sometimes, the format may be left to the prospective seller. In that case, it is usually a good idea for the seller to organize his proposal such that the buyer can relate it, point-by-point, back to his RFP.

Proposal content and level of detail are very subjective topics. The level of detail that should be provided in the proposal depends on the size of the task and the amount of detail that is known versus the amount of detail the prospective seller wants to reveal. In most cases, the less said about *how* the project will be executed the better, as having this information will give the buyer more power over the execution process, which can reduce the seller's ability to react to emerging circumstances as the project develops. Deliverables, on the other hand, should be detailed to a high level of precision, with assumptions clearly stated. From the Deliverables List, a Definitive Estimate can be produced that will not be provided to the buyer, but will form the basis of the quoted project price.

At minimum, the proposal should contain the following sections:

- Executive Summary
- General Scope of Work
- Assumptions
- Inclusions
- Exclusions (Exceptions)
- Deliverables
- Milestone Schedule
- Safety
- Price & Milestone-based Payment Schedule (Based on Definitive Estimate)

a. **Executive Summary**
The executive summary can probably be copied from the RFP. It should contain a simple statement of project purpose—a paragraph or so—followed by a cost table or chart and a schedule table or chart. This information can be followed by a short description of some of the key elements of the project.

b. **General Scope of Work**
This section describes in broad terms the purpose and plan for the project. An example of a general scope of work is as follows:

> "The existing plant Auto-call alarm system in Building 504A is defective. This project will replace the existing Auto-call plant alarm system with a new Auto-call system and replace alarm bells as necessary, as identified by field examination. The project

> will be supervised and funded by the facilities department. The new system will be designed and installed by the maintenance department and will be in place by early March."

This statement should include who, what, where, when, and, to a more limited extent, how.

c. **Assumptions**

An assumptions list is very important. Rarely is enough information available to be absolutely sure of the amount of work that will need to be done. Assumptions must often be made to allow the work required to be estimated. Sometimes worst-case scenarios are used, and sometimes best-case scenarios. All assumptions should be noted in this section along with the basis for each assumption. It is also a good idea to state the associated level of certainty (or uncertainty) when describing each assumption.

d. **Inclusions**

If a controversial or non-apparent task or function is intended in this proposal, then it should be clearly noted in the inclusions section. This is in the best interest of all concerned, particularly if this document is being submitted as part of a competitive bid.

e. **Exclusions (Exceptions)**

The exclusions section should note any related work that will be done by others. Perhaps a buyer organization, for example, has elected to do a portion of the project using its own maintenance forces. This section should detail this and any other aspects of the project that are specifically excluded from the scope of the design team.

f. **Deliverables**

Drawings, specifications, studies, databases, and any other physical or intellectual property that will be delivered to the buyer should be listed in detail.

g. **Milestone Schedule**

A milestone-based schedule should be included that describes when major events are to take place. Such events could include the following targets:

- Notice to Proceed (NTP)
- Start design

- Issue Documents for Customer Review (IFR)
- Issue Documents for Construction (IFC)
- Issue construction package(s)
- Procure long-lead items

The dates provided for the milestones should support only key delivery dates specified in the RFP. For example, the RFP might have a line item for Issue Construction Packages, but not for Issue Documents for Review, etc. For more ideas about which line items to include in a milestone schedule, look at Phase 2 and Phase 3 of Figures 1-7, 8, 9, and 10.

h. **Safety**
Any points of concern relating to personnel or equipment safety should be discussed here. If safety training was discussed in the RFP, it should be echoed back here with detail describing compliance or plans to attain compliance.

i. **Price & Payment Schedule**
Data presented in the executive summary should be fully supported in this section. Dollar amounts should be detailed to the level specified in the RFP. The data should provide information as to the total project cost and the total project duration. Also, depending on the type of project, a schedule of payments and a rate schedule may need to be included.

j. **Bid Award & Contract Negotiation**
When the buyer makes his selection, a period of contract negotiation ensues. That period could last minutes or days, depending on the size and complexity of the project. During this time the design team may be working away under a verbal authorization to proceed, which is basically a handshake agreement that can be formalized with a simple email, or under a more binding Memorandum of Understanding (MOU) that can set spending limits and duration until the contract is finalized. The contract negotiation could include refinements to information presented in the proposal or detailed in the RFP. The amount and terms of the assessment of Liquidated Damages (LDs) or incentives might be adjusted, as could the schedule milestones and other items. Changes made during this process need to be captured and made available to the design team if it affects their production planning.

The winning proposal will eventually result in a contract written that refers to or includes the proposal and any interim MOU that may have been enacted as attachments. Sometimes, though, the proposal, coupled with the RFP, *is* the contract. So care should be taken as to the assertions made, guarding against the tendency to write checks during the proposal process that can't be cashed during execution.

For more information on contracts, see Chapter 1, Section C, above.

5. The Project Execution Plan (PEP)

The PEP is an internal document that is produced after the project is awarded. It is usually generated in the FEL 3 stage of Phase 1. It answers the "who (which specific individuals or groups)" and the "how" questions. The PEP actually incorporates the SOW as either a section or an attachment. The PEP could be as simple as a one page document for small projects (Figure 1-16), or a very large and involved one for a big multi-discipline project.

The core function of the PEP is to bridge the gap between the more formal, higher level documents produced earlier on, and the particular personnel that make up the supplier's project team. The project engineer writes the PEP to the level of detail he thinks his design team needs. One of the primary purposes of the PEP is to distribute the work across the staff and orchestrate their efforts, giving them a tool for laying out their individual work plans. Using the detailed scope of work, milestone schedule, and project estimate as a basis, the project engineer builds a manpower plan that will reserve key resources within the company for use at specific times. If rolled up into a departmental report, this manpower plan gives each department head some ability to forecast manpower needs and avoid possible conflicts between the various resource managers who are drawing from a common manpower pool.

A peripheral function of the PEP is to provide a list of the parties involved in the project, internal and external, along with their role and contact information. This facilitates communication and becomes a quick reference for the project team members.

Of all the documents listed in this section, the PEP is the one that the rank and file personnel on the design team will refer to the most. The project engineer should cull information from the RFP, SOW, milestone schedule, and any of the company's internal procedures and guidelines that he wants to stress to his team and reproduce it in the PEP, making this information readily available to them. Much of the information can be clipped from the other documents and dropped into this one. The PEP should be considered a living document that will be revised as conditions change. A read-only copy should be kept in a location that is easily accessible by all team members and updated regularly. Any editing that they think should happen needs to be fed to the project engineer who will approve and incorporate the changes as appropriate. After its original publication, change control measures should be invoked, and when the PEP is revised the entire team should be notified.

Additional information should be in the final PEP version as appropriate. Some of the items that could be included are described below:

a. **Contact List**
   A list of personnel involved. This list should include all personnel who have been identified as part of the owner's team, and of any subcontractors, as well as the internal project execution team. The list should provide the name, role on the project, phone numbers and email addresses of each key player in each organization.

b. **Existing System Description**
   This document should pose questions such as: Which process areas are involved? Where is the equipment located? What is its general state of repair? What is the availability of power and services?

c. **Disposition of Existing Equipment**
   Which existing items will be demolished, refurbished, relocated, or replaced? How will these activities be done? Who will be responsible? How will this activity be documented?

d. **Addition of New Equipment**
   What new equipment will be added? Where? How will the area be prepared? Who will purchase the equipment? How will this activity be documented? How will this addition affect existing power and service capacity?

e. **Company and Applicable Industry Standards**
   If the engineering company has internal standards that apply to this project that are not already listed in either the SOW or the RFP, they should be listed here. Any industry standards that are applicable or that are superseded by the owner's standards should also be noted. Any applicable guidelines should also be listed. The following questions should be anticipated and answered: Are there any unusual requirements or safety concerns? What pertinent standards or guidelines were listed in the project specification?

Impediments that would prevent members of the design team from consulting the standards should be removed. The best way to ensure that standards are easily consulted is to provide copies of the standards, or at least instructions as to how they may be obtained.

f. **Approved Vendors List**
    Does the customer have specific vendors in mind to supply the materials and equipment? This information is probably listed in the project specification and can be copied and placed in this section.

g. **Vendor-Provided, Pre-engineered Subsystems (OEM)**
    If any pre-engineered subsystems are to be installed intact from a vendor or OEM, they should be listed here, along with key sales and technical resources in that organization. Any anticipated interaction with the design team should also be noted. A factory acceptance test may be necessary, and personnel may need to plan to attend. In addition, vendor drawings often need to be reformatted to suit the customer's drawing system.

h. **Instrumentation Data**
    If the customer has an existing instrument and/or I/O list, it should be included here. Otherwise, the customer contact who will facilitate gathering this information should be indicated.

i. **Quality Control**
    What QC/QA guidelines apply to this project? References to internal procedures should be included here, along with any listed in the RFP. The latter can be clipped from the RFP and placed here.

j. **Document Control**
    The customer's document control contact should be listed and any guidelines that may be available provided as attachments. Guidance should be included as to who should interact with this individual, and which internal document control practices should be used. It sometimes proves beneficial to appoint an internal resource as the single point of contact with the customer's document control aide to reduce confusion. Any internal guidelines for document control should also be included in this section.

Figure 1-16 provides a means for collecting a minimum of information and actually works as a PEP on some small projects. This is also a good form to use when on a plant orientation or pre-bid walkdown. The form is provided on the CD that accompanies this book.

Project Execution Plan Worksheet								
Customer:					Date:			
Project:								
Location:								
Plant Contacts					Schedule			
Role	Name	Ph#	Email		Task	From	To	
ProjEngr					Kickoff			
Ld Tech					Walkdown			
					PEP			
					Estimate			
Central Engineering Contacts					Award			
Role	Name	Ph#	Email		Kickoff			
ProjMgr					Design			
ProjEngr								
Ld Dgn								
Project Participants								
Role	Name	Ph#	Email					
ProjMgr					Checkout			
ProjEngr					SAT			
Ld Dgn					Commissioning			
					Complete			

Ref Docs	Scope of Work				

	Execution Plan				

	Estimate (Manhours)				
	Deliverable	Engrg	Design	Cadd	Total
					-
					-
					-
					-
					-
	Total	-	-	-	-

FIGURE 1-16. PROJECT EXECUTION PLAN TEMPLATE FOR SMALL TASKS

6. THE STATUS REPORT

After the project has begun, the schedule should be used as a management tool to help track progress. Periodically, the design team should be asked to provide some feedback as to their execution status. This feedback will answer questions, such as the following: Did you start the project on time? What is your completion percentage? Will you finish the project on time and within budget?

Reporting data is collected for each task (WBS item). The design team provides data for each task as follows:

- *Start Date:* If the task has not started, is the indicated start date still accurate? If the task has been started, on what date did it start?

- *Finish Date:* If the task is not finished, is the indicated finish date still accurate? If the task is finished, what was the date of completion?

- *Percentage Completed:* If the task is active, what is the estimated completion percentage?

The scheduling team then plows that information back into the schedule. For example, if a start date is moved by the design team, the schedule is updated to reflect the change. Successive tasks need to change accordingly depending on their relationship to the task in question.

After data have been collected, the scheduling team begins drawing some conclusions based on the information as follows:

- *Earned Hours:* The completion percentages are used to develop earned hours. This is done by multiplying the reported percentage completed by the total number of hours allocated to the task. For example, if a task has 100 hours allocated and the design team reports 50% completion on the task, then the earned-hours value is 50.

- *Actual Hours:* Actual hours are the number of hours actually expended to date.

- *Efficiency Ratio:* By comparing actual hours to earned hours, it is possible to develop a ratio that describes the performance of the design team as compared to expectations. In our previous 100-hour task example, we might find that 35 hours were expended on that particular task. Therefore, the efficiency ratio of the design team for that task would be 50 divided by 35, or 1.43. If 60 hours were expended, the ratio would be 50 divided by 60, or 0.83. Any ratio under 1.0 should be flagged for further investigation.

[* Milestone: A schedule item that is pegged to a particular date. These usually represent targeted points in the engineering process, such as "issue drawings."]

As a practical matter, relating actual hours down to the task level is difficult because time charges are generally not kept to that degree of resolution. But since the estimate is likely to be broken down to that level, it should be possible to derive *task impact factors* to help relate the percentage-complete values being reported to the actual hours.

Figure 1-17 shows a status data collection form that has ten WBS groupings. Each grouping is subdivided into six subtasks, which in this case equates to the six phases of a project (CAP Model). Of the ten, only WBS-T01 and WBS-T02 have been started. WBS-T01 covers the Railcar Unloading area of the plan, and has 250 man hours allocated. Ideally, the estimate would have been done in this format, with each subtask being estimated individually, and with the hours rolling up into the main task equaling 250. In this case, we started with 250, and distributed them based on Subtask Weight. The Subtask Weight must equal 100%.

WBS Item#	Description	Project Timeline			Total Mhrs	Subtask Weight	Subtask Manhrs	Est. %Comp	Manhrs Earned	Manhrs Actual	Manhrs ETC
		Start	Finish	Float							
T01	Railcar Unloading	4/3	10/1		250.0			42.4%	69.4	106.0	188.0
T01 -1	Feasibility Study	4/3	4/20	5d		10.0%	25.0	100%	25.0	32.0	-
T01 -2	Definition	5/1	6/1	15d		10.0%	25.0	95%	23.8	40.0	8.0
T01 -3	System Design	6/15	7/15	15d		25.0%	62.5	25%	15.6	32.0	40.0
T01 -4	Software	7/1	8/1	15d		20.0%	50.0	10%	5.0	2.0	60.0
T01 -5	Deployment	8/15	8/15	15d		25.0%	62.5	0%	-	-	56.0
T01 -6	Support	9/1	10/1	na		10.0%	25.0	0%	-	-	24.0
T02	Bulk Storage	4/3	10/1		400.0			18.0%	111.0	72.0	260.0
T02 -1	Feasibility Study	4/3	4/20	5d		10.0%	40.0	100%	40.0	16.0	-
T02 -2	Definition	5/1	6/1	15d		10.0%	40.0	95%	38.0	32.0	32.0
T02 -3	System Design	6/15	7/15	15d		25.0%	100.0	25%	25.0	16.0	16.0
T02 -4	Software	7/1	8/1	15d		20.0%	80.0	10%	8.0	8.0	72.0
T02 -5	Deployment	8/15	8/15	15d		25.0%	100.0	0%	-	-	100.0
T02 -6	Support	9/1	10/1	na		10.0%	40.0	0%	-	-	40.0
T03	River Water	4/3	10/1		80.0			0.0%	-	-	80.0
T04	Fire Protection	4/3	10/1		400.0			0.0%	4.0	-	376.0
T05	Material Handling	4/3	10/1		1,000.0			0.0%	-	-	980.0
T06	Crush & Slurry	4/3	10/1		1,500.0			0.0%	-	-	1,490.0
T07	Convey	4/3	10/1		400.0			0.0%	-	-	400.0
T08	Truck Loading	4/3	10/1		80.0			0.0%	-	-	80.0
T09	System Services	4/3	10/1		80.0			0.0%	-	-	80.0
T10	Miscellaneous	4/3	10/1		80.0			0.0%	-	-	80.0
	Project Summary:	4/3	10/1		4,270.0			4.2%	184.4	178.0	4,014.0

Project Status Reporting Structure - by WBS Area

FONT LEGEND: XXX-Calculated Value; **XXX**-Data by Project Manager; *XXX* Data by Design Team Leads
Note: Efficiency rating is calculated by dividing "Apparent % Comp" by "Actual % Comp."

FIGURE 1-17. PROJECT STATUS REPORT – DATA COLLECTION AND STATUS CALCULATION FIELDS

Taking the 250 man hours and multiplying by the Subtask Weight yields the Subtask Man hours subtotals. All of these fields, Total Man hours, Subtask Weight, and Subtask Man hours, would have been loaded at the beginning of the project by the Project Manager. Percentage Completed data are normally provided weekly by the design team lead for each subtask. From this, Man hours Earned can be calculated, by multiplying the estimated percentage completed and the Subtask Man hours. The PM will then load actual hours charged against the task. Finally, the design team will analyze the work remaining and provide an estimate as to the hours it will take to finish. Periodic updates should be requested by the project manager and scheduling team. The data to be updated are the Project Timeline dates and the man hours Estimate to Complete (ETC). The scheduling team plows the date information back into the schedule, and the project manager incorporates the man-hour data into the budget for analysis. Notice that the ETC for WBS-T01 shows 188 man hours to complete, with 250 available. From this, it might be inferred that the design staff is 188/250 = 25% finished. This is incorrect, as we will see in the next section.

B. Project Management Techniques

1. Assessing Project Status

The project status update data that were collected (Figure 1-17) can be used by the project manager to develop additional data that will help forecast the likelihood of success for the project (Figure 1-18). Some of the additional information needed includes:

- *Earned Hours:* The completion percentages are used to develop earned hours. This is done by multiplying the reported percentage complete by the total number of hours allocated to the task. For example, if a task has 100 hours allocated, and the design team reports 50% completion on the task, then the earned-hours value is 50.

- *Actual Hours*: Actual hours are the number of hours actually expended to date. This information is available from the timesheet system.

- *Estimate to Complete (ETC):* ETC data shown against the subtasks are data collected directly from the design team. The header row for each subtask has the total ETC hours for the category.

- *Estimate at Completion (EAC):* EAC data shown against the subtasks are calculated by summing Actual hours and ETC hours. The header row for each subtask has the total EAC hours for the category.

- *Apparent Percentage Completed:* This parameter looks at the Subtask Man hours, which was the original number of hours allocated to the task, and compares them to the number of hours spent to date (Actual). This gives a percentage that reflects status as compared to the original expectations. In the case of WBS-T01-1, the task appears to be 28% over budget, as Actual hours exceeds the Subtask hours allocated by 32 - 25 = 7 man hours.

- *Actual Percentage Completed:* This parameter compares the Actual man hours to the EAC man hours to derive an "actual" percentage completed value.

- *Efficiency Ratio:* This parameter compares the Actual Percentage Completed to the Apparent Percentage Completed, to come up with a value that shows the team's currently projected performance against the original budget.

Project Status Reporting Structure - by WBS Area

WBS Item#	Description	Total Mhrs	Subtask Weight	Subtask Manhrs	Est. %Comp	Manhrs Earned	Manhrs Actual	Manhrs ETC	Manhrs EAC	Apparent % Comp	Actual %Comp	Efficiency Ratio
T01	Railcar Unloading	250.0			42.4%	69.4	106.0	188.0	294.0	42%	36%	0.85
T01 -1	Feasibility Study		10.0%	25.0	100%	25.0	32.0	-	32.0	128%	100%	0.78
T01 -2	Definition		10.0%	25.0	95%	23.8	40.0	8.0	48.0	160%	83%	0.52
T01 -3	System Design		25.0%	62.5	25%	15.6	32.0	40.0	72.0	51%	44%	0.87
T01 -4	Software		20.0%	50.0	10%	5.0	2.0	60.0	62.0	4%	3%	0.81
T01 -5	Deployment		25.0%	62.5	0%	-	-	56.0	56.0	0%	0%	-
T01 -6	Support		10.0%	25.0	0%	-	-	24.0	24.0	0%	0%	-
T02	Bulk Storage	400.0			18.0%	111.0	72.0	260.0	332.0	18%	22%	1.20
T02 -1	Feasibility Study		10.0%	40.0	100%	40.0	16.0	-	16.0	40%	100%	2.50
T02 -2	Definition		10.0%	40.0	95%	38.0	32.0	32.0	64.0	80%	50%	0.63
T02 -3	System Design		25.0%	100.0	25%	25.0	16.0	16.0	32.0	16%	50%	3.13
T02 -4	Software		20.0%	80.0	10%	8.0	8.0	72.0	80.0	10%	10%	1.00
T02 -5	Deployment		25.0%	100.0	0%	-	-	100.0	100.0	0%	0%	-
T02 -6	Support		10.0%	40.0	0%	-	-	40.0	40.0	0%	0%	-
T03	River Water	80.0			0.0%	-	-	80.0	80.0	0%	0%	-
T04	Fire Protection	400.0			0.0%	4.0	-	376.0	376.0	0%	0%	-
T05	Material Handling	1,000.0			0.0%	-	-	980.0	980.0	0%	0%	-
T06	Crush & Slurry	1,500.0			0.0%	-	-	1,490.0	1,490.0	0%	0%	-
T07	Convey	400.0			0.0%	-	-	400.0	400.0	0%	0%	-
T08	Truck Loading	80.0			0.0%	-	-	80.0	80.0	0%	0%	-
T09	System Services	80.0			0.0%	-	-	80.0	80.0	0%	0%	-
T10	Miscellaneous	80.0			0.0%	-	-	80.0	80.0	0%	0%	-
	Project Summary	4,270.0			4.2%	184.4	178.0	4,014.0	4,192.0	4%	5%	1.04

FONT LEGEND: XXX-*Calculated Value;* **XXX**-*Data by Project Manager;* XXX *Data by Design Team Leads*
Note: Efficiency rating is calculated by dividing "Apparent % Comp" by "Actual % Comp".

FIGURE 1-18. PROJECT STATUS REPORT – ANALYSIS FIELDS

Note the differences between the Estimated, Apparent, and Actual completion percentages. In the case of WBS-T01-2, the design staff thought they were 95% complete (Estimated). But they had already exceeded the original budget by 60% (Apparent), and should have been done a long time ago. From the standpoint of the EAC, the one that counts, they are only at 83% (Actual).

2. STAFF MEETINGS

Communication is the grease in the design engine, facilitating proper operation of the machine. Regular meetings between design subgroups and among the project's leads are a necessary evil. Meetings are viewed by some as a waste of time—and most of them are—not because a meeting was not needed, but because the meeting was not managed properly. A meeting that is properly run will:

- Start promptly. If the meeting is supposed to start at 10AM, the facilitator should start the meeting at 10AM whether all the participants are present or not.

- Have an agenda. Topics should be specific and pertinent. Time should be allocated to major topics.

- Have a structure for collecting and classifying information gleaned during the meeting.

- End promptly.

Properly managing a meeting is something of an art form, and like a specification or document, should be considered to be its own sort of deliverable, an end in itself.

a. **The Meeting Facilitator**

Every meeting has a facilitator, whether that person knows that's what he is or not. The facilitator is the individual in charge of the meeting, and he or she has a responsibility to the project (to get some meaningful work done), and to the other participants (not to waste their time). This person can be the project lead, or possibly a person who is good at leading meetings. The facilitator's objective should be to facilitate meaningful communication among the meeting participants. To be meaningful, the topic should point to at least one of the project's primary controls: schedule, scope, or budget.

The facilitator should be focused and prepared. If the facilitator has a reputation for tightly controlling the meetings in terms of time and content, then the participants are more likely to attend and are more likely to arrive prepared for the discussion.

The facilitator should manage to the agenda. If the agenda gives a topic five minutes, then the facilitator should limit discussion to five minutes and then move to the next topic. If necessary, a timekeeper could be appointed whose function is to inform the facilitator of approaching time deadlines. Deviations from the agenda should be minimized and should be tightly controlled. If the discussion deviates from the topic at hand, the facilitator should step in and guide the discussion back towards the topic, or should ask the participants to shelve the discussion until after the meeting. Extending the meeting to handle emerging issues is always an option, but should be employed with caution. The best course to take would be to shelve the topic and schedule a second meeting, or if the new issue affects the majority of attendees, hold it until after the agenda has been fulfilled, and ask the attendees if they would mind extending the meeting to deal with the new issue. The facilitator should always respect the schedules of the other participants.

Finally, and most importantly, the facilitator should ensure that each participant understands his or her responsibility to the group as the meeting concludes. If an action was deemed necessary by the group, the facilitator should capture that action, the person who is supposed to address it, and the point in time at which it should be delivered. There are some tools to help in this endeavor, discussed next, after which the most important feature of a well-run meeting will be discussed: the agenda.

b. **The Facilitator's Toolbag**

How many of us have attended meetings that had no apparent purpose? Or maybe the meeting itself was edifying, but nothing resulted other than the participants feeling good about themselves for having contributed to an interesting discussion. The facilitator has a big job. He must arrive prepared, have the ability to steer a group towards a desired destination (a task similar, at times, to herding cats), and capture information for later use in the Notes of Conference. One way to do this is to categorize each topic, and list it by category. The mechanics of doing this can be improved by the following method:

- Collect Status Notes.
- Collect Emerging Issues Notes.
- Maintain a Needs List.
- Maintain an Action List.
- Maintain a Suggestion List.

Of these five forms, two are used only for making notes to be mentioned in the Notes of Conference, and three are formal lists to be maintained. Each of the five forms should be separate and easily accessible to the facilitator. Any issue being discussed will likely fall into one of those categories. The following is a brief discussion of each:

The Status Notes form (Figure 1-19) should be partially filled out prior to the meeting. It should follow the agenda topics, with an area for each department or group to provide information.

The Emerging Issues form (Figure 1-20) provides a means for collecting information on points of discussion that weren't anticipated. This would be a good place to note points of concern raised by meeting members or issues that may not need to be mentioned in the notes of conference but still should be captured as being points for consideration.

STATUS NOTES						
Project: _____			Meeting ID#: _____		Meeting Date: _____	
Agenda Item	Date	Issue	Member	Status	Comment	

FIGURE 1-19. MEETING STATUS NOTES FORM

EMERGING ISSUES NOTES						
Project: _____			Meeting ID#: _____		Meeting Date: _____	
Issue Item#	Date	Issue	Originator	Owner	Status	Comment

FIGURE 1-20. EMERGING ISSUES NOTES FORM

The Needs List (Figure 1-21) is a formal list that should be maintained and added to as issues arise that cannot be satisfied by the members of the project team. Each item is given a unique Needs Number and status reported against it at each subsequent meeting until a resolution is reached. This activity is usually directed toward outside organizations, such as the customer or a vendor.

NEEDS LIST							
Project: _____			Meeting ID#: _____			Meeting Date: _____	
Needs Item	Date	Issue	Originator	Owner	Date Due	Status	Comment

FIGURE 1-21. NEEDS LIST FORM

The Action List (Figure 1-22) is a formal list that should be maintained and added to as issues arise that can be satisfied by the members of the project team. Quite possibly, this is the most important tool available to the facilitator. Each item is given a unique Action Number and status reported against it at each subsequent meeting until a resolution is reached. An Action Item emerges if someone uses an action verb directed toward someone present, or some organization represented, at the meeting. Proper use of the Action List will ensure that people know their responsibilities and will provide a means to track their progress in resolving the issues.

ACTION LIST								
Project: _____			Meeting ID#: _____				Meeting Date: _____
Action Item	Date	Issue	Originator	Owner	Date Due	Status	Comment

FIGURE 1-22. ACTION LIST FORM

The Suggestion List (Figure 1-23) is a formal list that should be maintained and added to as opportunities arise to improve internal processes or provide cost savings to the customer or which constitute another improvement that someone should investigate for feasibility. This list can be an important piece of documentation to show a customer that the project team is thinking outside the box. Some suggestions may develop into major changes in project scope, if cost or time can be saved. Actions can emerge from suggestions. Each item is given a unique Suggestion Number, and status is reported against it at each subsequent meeting until a resolution is reached.

SUGGESTION LIST						
Project: _____			Meeting ID#: _____			Meeting Date: _____
Suggestion Item	Date	Issue	Originator	Owner	Status	Comment

FIGURE 1-23. SUGGESTION LIST FORM

For example, Fred is given a task to monitor the status of a pumping station fabrication supplier and report to the group. The item is assigned an Action Item number and is listed on the Action List. Each week Fred makes his report, and the Action List gets updated. Finally, the pumping station arrives on site, and the issue is deemed to be resolved. The action item is then closed and drops from the weekly open action items report.

Using these five tools will allow the facilitator to keep up with the conversation, minimizing the need to stop while notes are taken and minimizing the risk of losing information. Most importantly, proper use of the Needs

List and Action List will eliminate the risk of having the meeting result in nothing. These two lists should be deployed even if none of the other ones are used. They will help ensure that loops get closed and loose ends get tied off.

In fact, it is recommended that these two lists be maintained in a database, with one record per date of comment. For example, if Action Item #42 is discussed in three meetings, then there should be three records in the database, each with a different date, and each with its own set of comments.

c. **The Meeting Agenda**
The meeting agenda is the single most important tool available to the facilitator. But more than a tool for the facilitator, the agenda is a tool for each of the participants. Remember, the goal of a meeting is to communicate, to get some meaningful work done, and to not waste people's time. Therefore, the agenda needs to be prepared well ahead of time, and distributed to the participants in time for them to prepare for the topics being discussed.

A good agenda format is one in which news is broadcast and feedback collected. It needs to be short and to the point but with enough detail to allow the participants to prepare. If the tools described previously are used (e.g., Needs List, Action List, etc.), they should be incorporated as agenda items, with any non-resolved items distributed in advance as attachments to the agenda.

A basic agenda format[3] is shown in Figure 1-24.

3. MANAGEMENT OF CHANGE (MOC)

Every project, regardless of the type of contract and the type of project, has both implicit and explicit sets of expectations. This is because it is impossible to spell out every permutation of a project in a contract. Project parameters as noted in the contract (explicit expectations) can be met, and yet the project can still fail due to poor relationships, misunderstandings, bad perceptions, and the like (implicit expectations).

Thinking back to the Success Triangle (Figure 1-2), the buyer's focus is on quality and cost; the seller's focus is on quality and profit. Few projects get off the ground if either party knows that the triangle is either too small or non-existent. However, projects whose success triangle is satisfactory to both parties can still

---

3. This form, and others noted in this book, is available on the CD.

Meeting Agenda

Project: _____ **Project #:** _____

Date: _____ **Time:** _____ **Place:** _____ **Facilitator:** _____

Attendees:

Purpose:

Topics:

1. 8:00 - Attendance (5-min)
2. 8:05 - Greetings, News & Announcements (5-min)
3. 8:10 - Action List Review (5-min)
4. 8:15 - Needs List / Suggestion List Review (5-min)
5. 8:20 - Project Status Review (20-min)
6. 8:40 - Project Schedule & Budget Review (10-min)
7. 8:50 - Action, Needs, & Suggestion List Update (10-min)
8. 9:00 - Meeting Over

Notes:

FIGURE 1-24. SAMPLE PROJECT MEETING AGENDA FORM

fail. So in those cases there must have been a disconnect between what was expected at the beginning and what is known at the end.

The key to this mystery is frequently both parties' ability to manage change. Having a reasonable MOC plan in place that the entire team buys into makes for a less stressful work environment, as people are not carrying a lot of worries home in the evenings. While there is some risk of fostering a "cover yourself" mentality on the part of the designers, it is worth it, as this program helps spread the risks of making changes throughout the organization so that all carry their share of the responsibility. This not only makes the designers themselves more assured of their work, but also reduces some of the worry on the part of management that they may not be in full control of the design process.

The type and size of contract has a bearing on the amount of energy that is put into the MOC process, but MOC needs to be addressed in some fashion across the board. In fact, even in the most benign case of the Cost-Plus contract, in which price is not the prime consideration, the buyer still has an expectation of ultimate price based on his own budgetary estimate. If the scope of work is allowed to expand unchecked over a long period of time, no budget is safe, and the likelihood of the buyer's remaining happy with the seller, regardless of the seller's abilities, is remote.

Therefore, it is important that the Management of Change topic be discussed openly and frankly at the project's inception. A procedure for processing engineering change notices (ECN) should be approved by all parties and documented thoroughly—even to the point of amending the contract if necessary. In fact, it is best by far if the buyer takes the initiative and spells out the process formally in the RFP.

Regardless of whether or not an ECN process is spelled out by the buyer, the seller should have his own internal MOC program in place, even if it is called by some other name, such as Internal Change Request (ICR), which is what will be used in the following narrative and in Figure 1-25.

For the engineering organization, the change management philosophy should be one that encourages the generation of internal change requests by the rank and file. The purpose of doing this is not to drive up the project cost, but to keep personnel disciplined and focused on the PEP. If they find themselves deviating from the PEP, then an ICR should be generated. Some reasons to generate an internal change request might be:

- Any new customer request
- A better method
- A better idea

Refer to Figure 1-25 for the following discussion:

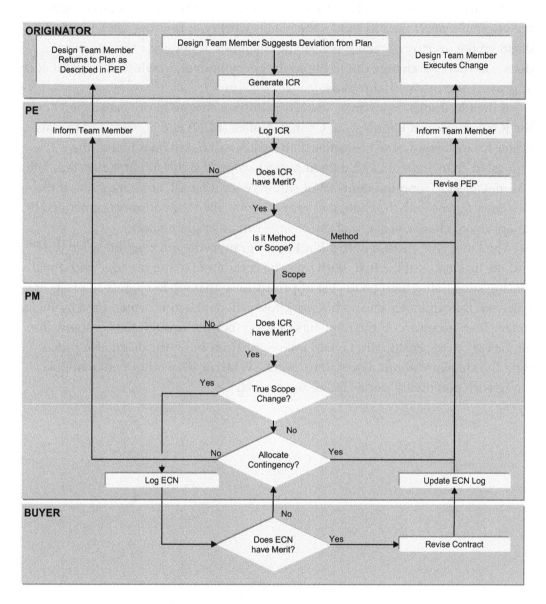

FIGURE 1-25. A MANAGEMENT OF CHANGE (MOC) PROCESS

After the ICR is submitted, the PE must evaluate it quickly in order to minimize any interruption of the workflow. If it has no merit, then he disapproves it at that moment, and counsels the originator to make sure that person understands why it's a bad idea. If the PE thinks it's a good idea, then another question should be asked: Will cost increase? If not, then the PE should immediately give the originator permission to proceed, and then update the PEP. If the project cost will go up, then the PM should be notified. The PM must then decide if the idea is good enough to warrant its impact on his budget. If not, then the item is dead, and the PE will either try to negotiate with the PM, or inform the originator to return to the original plan. If the PM thinks the idea is worthy of funding, then he must decide if it is worthy of taking before the customer. If not, then the PM may decide to fund the change out of his contingency budget, in which case the ICR is approved and the work proceeds.

If the PM thinks this worthy idea constitutes a change in project scope, then the ICR becomes an Engineering Change Notice (ECN) and is passed to the customer for his review. The customer then makes a decision on whether the requested change merits a change in project scope. If it doesn't, it is rejected. The PM may elect to fund the item out of contingency after all, or he may not. If the customer agrees with the change in project scope, then the contract is amended by an approved change order, and the change can be implemented.

The key to the success of this kind of program is the interface between the PE and the originator of the ICR. Both must be committed to the process, with the PE placing such issues high on his list of priorities so that when change requests are received, the originator knows his workflow will not be interrupted for a moment longer than necessary. If the issue has to wind its way through the company and all the way to the customer, it could take some time, but if the originator proceeds with the change without discussion, then he is taking on a lot of personal risk, along with potentially jeopardizing the project.

Part I – Summary

In summary, perspective is sometimes difficult to maintain, particularly when a person is involved in the fine details of getting the job done. But the loss of perspective can cause serious problems. Keeping an eye on the ball can be a challenge, particularly if a basic knowledge of project structure, participants, and management is lacking. Part I has been about providing perspective. It answers the questions "what," "who," "when," and "how" as they relate to a typical engineering project. What are the major project types? How should designers adapt their thinking based on the type of project?

Should interaction with the customer be minimized, or should a designer seek the customer's advice? Should the design effort only generate construction documents, or should it provide detail for maintenance as well? What are the real measures of success? Who is the designer's counterpart in the other organizations? What are some of the key concepts in setting up and managing a project? How do you measure your progress? It is hoped that you, the reader, now have a better grasp of these matters. Part II in this book delves into some of the key technical issues surrounding design engineering.

Part I - References

1. Cockrell, G. W. *Practical Project Management: Learning to Manage the Professional*. Research Triangle Park: ISA –International Society of Automation, 2001.

2. *ISA Comprehensive Dictionary of Measurement & Control*, 3d ed. Research Triangle Park: ISA –International Society of Automation, 1995.

3. Travathan, V. L., *A Guide to the Automation Body of Knowledge*, 2d ed., Research Triangle Park, ISA – International Society of Automation, 2006

4. Yeager, General C. and Janos, L. *Yeager: An Autobiography*. New York: Bantam Books, 1985

5. Courter, G. and Marquis, A. *Mastering Microsoft® Project 2000*. Alameda: Sybex Inc., 2000.

6. Kletz, T. Hazop and Hazan, 4th ed. Philadelphia: Taylor & Francis, 1999.

7. CSIA, *Guide to Control System Specification & System Integrator Selection*, Volume 1. CSIA: Exton, PA, 2000.

Part II
Table of Contents

List of Acronyms – Part II	109
List of Figures – Part II	111
Introduction – Part II	115
Part II – Chapter 4: Basic Design Concepts	**117**
A. Scaling and Unit Conversions	117
1. Definition of Key Terms	118
2. Accuracy and Repeatability	120
3. Resolution Effects on Accuracy	122
4. Instrument Range versus Scale	123
5. Instrument Calibration	123
6. Linearization and Unit Conversions	124
7. Practical Application	127
B. Introduction to Information Management	129
Part II – Chapter 5: Design Practice	**147**
A. Basic Wiring Practice	148
1. Inter-Cabinet Wiring	148
a. Generating a Cable Schedule	149
b. Generating a Conduit and Tray Schedule	149
c. Calculating Conduit Fill	150
d. Generating Physical Drawings (Location Plans)	154
2. Intra-Cabinet Wiring	155
a. Generating a Wiring Diagram	156
b. Relays, Switches, and Timers	156
c. Wire Numbering	164
B. Failsafe Wiring Practice	166
C. Hazardous Area Classification and Effects on Design	170
1. Hazardous Locations	170
a. Class I	172
b. Class II	173
c. Class III	174
2. Explosionproofing	174
3. Intrinsic Safety	175
4. Purging	175
a. Class X Purge	175

  b. Class Y Purge 175
  c. Class Z Purge 176
 D. Connecting to the Control System 177
  1. Discrete (Digital) Wiring 178
   a. Sinking and Sourcing 179
   b. Circuit Protection (Fusing) 181
   c. Digital Input (DI) Circuits 181
   d. Digital Output (DO) Circuits 182
  2. Analog Wiring 184
   a. Circuit Protection (Fusing) 185
   b. Noise Immunity 186
   c. Resistance Temperature Detector (RTD) 187
   d. Thermocouple 187
   e. 0–10 Millivolt (mV) Analog 188
   f. 4–20 Milliamp (mA) Analog 188
 E. Design Practice Summary 192

Part II – Chapter 6: The Control System 193

 A. Introduction 193
 B. The Cognitive Cycle 194
 C. Control System Overview 195
  1. A Historical Perspective 195
  2. PLC versus DCS 197
   a. The Distributed Control System (DCS) 197
   b. The Programmable Logic Controller (PLC) 199
  3. Major Control System Elements 200
   a. The Physical Plant 200
   b. The I/O Marshalling Area 201
   c. The Hardwired Control Board 201
   d. The PLC 201
   e. The HMI 202
  4. Control Modes and Operability 202
   a. Local/Remote (L/R) Mode Selector 204
   b. PLC/Benchboard Mode Selector 204
   c. Auto/Manual (A/M) Mode Selector 204
 D. The Human-Machine Interface 205
  1. The Graphic User Interface (GUI) 205
   a. Action Links 207
   b. Animation Links 207
  2. The HMI Database 210
   a. Tagnames 210
   b. Scan 210
  3. The HMI Alarm Manager Utility 211
  4. The Historian 212
  5. The Trend Utility 212
  6. Reports 213
 E. Programmable Logic Controller 214
  1. Major PLC Components 214
   a. The Rack Power Supply 215
   b. The Central Processing Unit (CPU) 215
   c. The Communications Module 215
   d. The I/O Module 216

Table of Contents – Part II

- 2. The PLC Program 216
  - a. I/O Map 216
  - b. Memory Map 217
  - c. Scaling 217
  - d. PLC Memory 217
  - e. Documentation and the PLC Database 217
  - f. Programming Languages 218
  - g. Recommended Program Structure 221
- 3. The I/O Interface 222
  - a. Physical (Hardware) Address 223
  - b. Software Address 224
  - c. I/O Map 225
  - d. Discrete (Sometimes called Digital or Binary) Signals 225
  - e. Analog Signals 226
  - f. Partitioning (I/O Mapping) 226

F. Networking 230
  1. Optimized/Proprietary Networks 230
  2. Optimized/Non-Proprietary Networks 232
     - a. Serial Communications (RS-232) 232
     - b. Optimized Local Area Networks 233
  3. Non-Optimized (Open) Local Area Networks 234
  4. Wireless Local Area Networks 234
     - a. The "Bluetooth" Standard 234
     - b. The "Wi-Fi" Standard (IEEE 802.11B) 235
  5. The Ethernet Client/Server Environment 235
     - a. "Thick" Client Architecture 236
     - b. "Thin" Client Architecture 236
     - c. Ethernet Hardware 236
  6. The Industrial Enterprise-Wide Network 238
     - a. The Remote I/O (RIO) LAN 239
     - b. The Hot-Standby PLC 239
     - c. Ethernet Subnet X 239
     - d. SCADA Server Pair 240
     - e. Ethernet Subnet Y 241
     - f. Thin Client (TC) Server Pair 241

G. Working with a Control Systems Integrator (CSI) 241
  1. Initial Search 241
  2. Writing a Control System Specification 242
     - a. Process Overview 242
     - b. Operability 242
     - c. Control 244

H. Selecting a Control System 246

References – Part II 249

List of Acronyms – Part II

A/M	Auto-Manual Selector Switch
CFC	Continuous Function Chart
COTS	Commercial, Off-The-Shelf
CPU	Central Processing Unit
CRT	Cathode Ray Tube
CSI	Control Systems Integrator
CV	Control Variable
DAS	Data Acquisition System
DCS	Distributed Control System
DFB	Derived Function Blocks
DIO	Distributed I/O
ENET	Ethernet
EU	Engineering Units
FJB	Field Junction Box
GUI	Graphic User Interface
HMI	Human-Machine Interface
HOA	Hand-Off-Auto Selector Switch
HSBY	Hot Standby
I/O	Input/Output
IEC	International Electrotechnical Commission
ISA	International Society of Automation
LAN	Local Area Network
LOR	Local-Off-Remote Selector Switch
mA	milliamp
MBP	Modbus-Plus
mV	millivolts
NEC	National Electric Code
NFPA	National Fire Protection Association
O&M	Operations & Maintenance
OEM	Original Equipment Manufacturer
PLC	Programmable Logic Controller
PPH	Pounds per Hour
PV	Process Variable
RIO	Remote I/O
RSCADA	Redundant SCADA (see SCADA)

RTD	Resistance Temperature Detectors
RTU	Remote Terminal Unit
SAMA	Scientific Apparatus Maker's Association (now defunct, absorbed by ISA)
SCADA	Supervisory Control and Data Acquisition
SFC	Sequential Function Chart
SP	Setpoint
TC	Thin Client
V&V	Validation & Verification
VAC	Volts, alternating current
VDC	Volts, direct current
WAN	Wide Area Network
WBS	Work Breakdown Structure
XMTR	Transmitter

List of Figures – Part II

Figure 2-1. Typical error pattern caused by deadband 123

Figure 2-2. Typical error pattern caused by hysteresis. 125

Figure 2-3. Conversion problems 127

Figure 2-4. Data translation process — from field device to HMI 128

Figure 2-5. Signal conversion at PLC input 129

Figure 2-6. Engineering unit calculation at the HMI 129

Figure 2-7. Spreadsheet versus database comparison 132

Figure 2-8. Typical relational database program structure 135

Figure 2-9. ICS-based project flow with database-intensive activities highlighted .. 137

Figure 2-10. The P&ID takeoff query 138

Figure 2-11. The I/O partitioning query 139

Figure 2-12. The software & logic assignment query 139

Figure 2-13. The cable and conduit schedule query (partially shown) 140

Figure 2-14. The instrument specification query (partially shown) 141

Figure 2-15. The construction checkout query... 141

Figure 2-16. The Validation & Verification (V&V) test queries. 142

Figure 2-17. The site acceptance test queries 143

Figure 2-18. Typical document handling process 146

Figure 2-19. Typical cabling scheme 150

Figure 2-20. Defining the cable route (wire W1, route C1/T1/T2/C2/C2a) 151

Figure 2-21. Sample cable schedule.. 152

Figure 2-22. Cable area fill 153

Figure 2-23. Cross-sectional views of cable orientation before, during, and after a conduit bend . 154

Figure 2-24. Conduit facts.. 154

Figure 2-25. Conduit sizing calculator 155

Figure 2-26. Sample conduit schedule 157

Figure 2-27. Sample instrument arrangement... 157

Figure 2-28. Interconnection wiring example 159

Figure 2-29. Form A contact set (SPST – NORMALLY OPEN) 160

Figure 2-30. Form B contact set (SPST – NORMALLY CLOSED) 161

Figure 2-31. Form-C contact set (SPDT) 161

Figure 2-32. 5-pole relay used as a motor starter (shown in shelf state, with interlocks, overloads, and PLC input) 163

Figure 2-33. Common types of switches and their diagrams 164

Figure 2-34. Types of contacts 164

Figure 2-35. Interval timer timing diagram 165

Figure 2-36. Time delay on de-energize (TDOD) timer timing diagram 165

Figure 2-37. Time delay on energize (TDOE) timer timing diagram 166

Figure 2-38. Sample ladder elementary format 167

Figure 2-39. Failsafe interlock chain (devices shown in shelf state) 171

Figure 2-40. Hazardous boundaries 173

Figure 2-41. Basic discrete (digital) circuit 180

Figure 2-42. Discrete (digital) circuit wiring technique 181

Figure 2-43. Simple switching 182

Figure 2-44. Sinking and sourcing digital input modules 184

Figure 2-45. Isolated digital output module 186

Figure 2-46. Sinking and sourcing digital output modules 187

Figure 2-47. Analog circuit wiring technique 188

Figure 2-48. Analog wiring methods: 2-wire vs. 4-wire 193

Figure 2-49. The cognitive cycle 197

Figure 2-50. Typical control system 205

Figure 2-51. The Human-Machine Interface (HMI) 208

Figure 2-52. Graphical User Interface with pushbutton configuration template 211

Figure 2-53. Trend screen 215

Figure 2-54. Typical PLC rack 217

Figure 2-55. Sequential function chart washing machine sequence control application 222

Figure 2-56. Continuous function chart washing machine temperature control application 223

Figure 2-57. Control detail sheet 225

Figure 2-58. Suggested program flow of control 226

List of Figures – Part II

Figure 2-59. I/O tally worksheet 229

Figure 2-60. Revised I/O tally worksheet reflecting new setup 231

Figure 2-61. I/O tally worksheet with split by I/O type 233

Figure 2-62. Remote I/O network 234

Figure 2-63. Industrial network 242

Introduction – Part II

In this book, Part I describes the engineering process and its participants, and discusses project management concepts. Part III takes a real-world scenario, defines a scope of work, and generates the design products necessary to meet that scope. To make the transition from managing the project to generating the product, some background information needs to be presented—hence, Part II.

In Part I, we discussed how a controls project is deterministic in structure—that the design process is generally repeatable from project to project and from supplier to supplier. It must be recognized, however, that project content, from one project to the next, can be anything but deterministic, with significant deviations even among copy-type projects, as the systems must be adapted to physical location or circumstance.

In addition, the controls field is inherently a dynamic one, as technology and technique adapt to a rapidly changing marketplace. For example, Distributed I/O (DIO), wireless technologies, commercial off-the-shelf (COTS) equipment, and other systems, subsystems, and equipment are becoming more prevalent as people gain experience and become more comfortable with cheaper, more open Ethernet-based communication protocols. The price of I/O (input/output) falls, while the number of options for its implementation rises.

A good design organization must be able to innovate—without reinventing. The aim of Part II is to present information that must be thoroughly understood by the entire design staff, thereby enabling them to innovate with confidence. Before developing a control system design package, it is important that the design staff understands the fundamentals of *good* design. What things should be considered when specifying a PLC (Programmable Logic Controller)? How should you adapt your design to a hazardous environment? What are ground loops, and how do you design to avoid them? When should you use shielded cable, and how is it deployed? What are some effective, time-saving techniques for generating a conduit schedule? What is meant by I/O modules that *sink* current, versus those that *source* current? Practical information is presented here that is, perhaps, difficult to find documented elsewhere.

Part II is broken into three sections:

- Chapter 4 provides some basic information about instrumentation scaling and unit conversions in order to prepare for some of the design practice topics that follow later. How is the output of a sensor processed into useful information for the operator? A scaling example is presented and a real-world scaling exercise is provided. Information and Data management is also addressed.

- Chapter 5 delves into design practice. What considerations should be included in the design? What is meant by the term *failsafe*? How is it possible to operate in explosive atmospheres? In a controls project, an important facet of design practice is documentation. A lot of consideration should be given to component identification. What is a good wire numbering scheme to facilitate maintenance? What about software tagnames, hardware tagnames, etc.? Is there a way to use tag naming conventions to help create a PLC program that is self-documenting?

- Chapter 6 dissects the control system. What is a DCS (Distributed Control System), and how has the DCS evolved? Where did the PLC originate, and what are the considerations for its specification? What is an HMI (Human-Machine Interface), and what is a good approach to screen design? What are some techniques for dealing with the I/O interface? What are some of the key considerations when laying out a network?

PART II – CHAPTER 4: BASIC DESIGN CONCEPTS

The topic of Instrumentation and Controls is all about managing information. Having the ability to take a sample of a physical process element, such as fluid temperature or gas pressure, is of little benefit unless someone can make use of the information. In today's high-speed, low-overhead environment, having an operator manually respond to the deflection of a needle is no longer an option. Today, process information must be prepackaged, timely, and accurate, so we in the controls profession must be adept at managing information, from instrument scaling and unit conversions to database management.

A. Scaling and Unit Conversions

In this context, the word *scaling* implies the scale of a ruler. A ruler is marked in inches and parts of inches, or in meters and parts of meters. The ruler, therefore, is said to be scaled for its units of interest (inches or meters). Process instrumentation scaling is similar, but it uses different units, such as pounds per square inch (PSI), pounds per hour (PPH), degrees Fahrenheit (degF), and so on. To "scale" an instrument is to calibrate it, making its output linear with respect to a given range or scale.

A process *instrument* generally consists of a sensing element and at a minimum, a needle and a graduated dial. As the measured process variable changes (for example, as temperature rises), the sensor detects and causes the needle to deflect and indicate a value. A process *transmitter* is an instrument whose output is an electrical signal instead of, or perhaps, in addition to a local indicator of some sort (like a needle).

From the physical measurement, to the numeric representation, to the electrical signal, to the displayed data in engineering units, the initial setup (calibration and scaling) and the unit conversions must be understood and properly implemented. Process control designers and troubleshooters, and PLC programmers in particular, must be able to maneuver through the various permutations of a data item as it transitions from sensor output to an electrical signal to a number in a computer.

Before we get into the specifics of signal scaling, a review of some instrumentation and controls basics is in order.

1. Definition of Key Terms

- *Accuracy*: Accuracy is expressed as a ratio of the percentage of error to the full-scale output. An instrument's accuracy reflects its ability to detect, transduce, and then report the value of a physical property. A highly accurate instrument is one that reports with a minimum of error introduced.

- *Calibrated Zero Point*: An instrument's lowest scale setting. An instrument that is scaled from 15 to 150 pounds per square inch (psi) has a calibrated zero point of 15 psi.

- *Calibrated Span Point*: An instrument's highest scale setting. An instrument that is scaled from 15 to 150 psi has a calibrated span point of 150 psi.

- *Calibrated Range (scale):* The difference between an instrument's calibrated span point and its calibrated zero point. An instrument's calibrated range is usually configured by the user. A transmitter that has been calibrated (scaled) from 15 to 150 psig has a calibrated range of 135 psig. This should not be confused with its design range, which can cover a much wider span than what has been calibrated and is the instrument's safe operating limit.

- *Design Zero Point*: An instrument's lowest factory scale setting. An instrument that is scaled from 0 to 500 pounds per square inch (psi) has a calibrated zero point of 0 psi.

- *Design Span Point*: An instrument's highest scale setting. An instrument that is scaled from 0 to 500 psi has a calibrated span point of 500 psi.

- *Design Range*: The maximum input span for which an instrument can provide a linear output signal. A transmitter that can provide a linear output signal for an input pressure range from 0 to 500 psig has a full range of 500 psig. This is also referred to as the span. This should not be confused with its calibrated range, or *scale*, which can cover all or just a portion of the unit's full range.

- *Drift:* The shifting of an instrument's calibration over time due to aging components, temperature shifts, and other external forces.

- *Floating Point*: A numeric designator that indicates a value with a decimal point. The decimal point may "float," depending on the level of precision of the number. In PLC programming, it implies the use of certain mathematical functions.

- *Linearity*: For signals that are supposed to be linear, linearity is the measure of deviation, expressed in the percentage of deviation between the slope of the expected line and the slope of the actual line.

- *Live Zero Offset*: A "live" zero is one in which an undershoot is possible, thus allowing a negative number. For example, a 3mA signal present on a circuit calibrated for a span of 4-20mA would yield a scaled value that would be less than zero, if the engineering unit scale was 0-100.

- *PID*: Refers to a control algorithm that takes a Process Variable (PV), compares it to a Setpoint (SP), and generates a Control Variable (CV) that automatically reacts to the error as PV deviates from SP. If SP = PV, then the CV remains where it is. If the PV increases above the SP, then the CV falls if the PID is configured as direct-acting; or it rises if the PID is configured as reverse-acting.

- *Offset*: The percentage-of-scale error between a stable output value and its desired value. Offset is usually measured during control loop tuning, and can sometimes be reduced by incrementing the integral aspect of the PID equation.

- *Rangeability (turndown)*: The ratio of an instrument's maximum range to its minimum range. If an instrument's output signal will remain linear for input ranges that are between 0–50 and 0–100 psig, then its rangeability ratio is 100:50, or 2:1.

- *Repeatability*: The capacity of an instrument to report the same output value each time a specific input value is detected over a short time span.

- *Resolution*: The degree to which a change in an instrument's input can be detected. If the sensing element of a load cell can only detect weight changes in 0.1 lb increments, then its resolution is 1.6 ounces. A computer's resolution is defined by the size of the data word that can be accommodated. A PLC's analog input might have 12-bit resolution, which equates to 4096 discrete values that can be used to represent an input voltage or current. For example, if an input signal's range is 0–10 VDC, a change of 10 V/4096 =.00244 V will be detected.

- *Response Time*: A measure of the amount of time it takes an output to respond to a given shift in input value.

- *Scale*: The portion of an instrument's design range for which it is calibrated to provide a full scale linear output. It is the calibrated range of the instrument. An instrument's scalability is the ratio by which the scale can be different from the range.

- *Set point*: The point at which a process is supposed to operate (control set point), or the point at which an alarm should occur (alarm set point).

- *Span*: See *Design Range*.

- *Transduce*: To transduce is to convert from one type of energy to another. A relay is a transducer, converting electrical energy to mechanical, as the coil magnetizes and closes its contacts. A pressure transducer converts pressure, as measured mechanically, to electrical energy. An I/P transducer converts electric current to a pneumatic pressure.

- *Turndown*: See *Rangeability*.

- *Units, Engineering*: Units of measure useful to the operator. Examples of engineering units are pounds per hour, feet per second, and degrees Fahrenheit.

- *Units, Raw*: Units of measure at the transmitter. Examples of raw units are millivolts (mV) and milliamps (mA). These values are generally of little direct use to the operator since they are difficult to interpret until converted into engineering units.

- *Variable, Process (PV)*: A process variable is the measured value fed to a control equation, like a PID algorithm.

- *Variable, Control (CV)*: A control variable is the output of a control equation, like a PID algorithm.

There are several good books available that provide a more detailed look at instrumentation concepts and principles. Those discussed here are only the most basic. Refer to the ISA Press online library at www.isa.org/books.

2. Accuracy and Repeatability

Measurement accuracy is critical to proper system performance, but to be useful, it must be repeatable. A sensor that is inaccurate but repeatable with respect to the error's magnitude is preferable to an instrument that can be calibrated but then is not repeatable. Such an instrument is usually deemed to be inaccurate. The following are some of the factors that may limit measurement accuracy and/or repeatability:

- Mechanical factors

- Leaks in the sensor lines (causes a shift in the span-point toward the zero-point.)
- Mechanical loading (causes a shift in the zero-point toward the span-point)
- Impurities in sensor lines (causes nonlinearity)
- Deadband (Figure 2-1): an attribute of an instrument that describes how well it responds to a change of direction in its input. For a pressure transmitter scaled 0–100 psig with a 1% deadband, it will take 1 psig of downward movement of the pressure to overcome inertia and/or mechanical slop that was in place due to previous upward movement of pressure. Deadband affects repeatability because, for a given measured input, the output of the device will report a different value, depending on whether the input value is increasing or decreasing.
- Hysteresis (Figure 2-2): caused by internal friction, hysteresis is usually evidenced when there is a different output value if the input is at 50% of scale and rising versus when the input is at 50% of scale and falling. Hysteresis, like deadband, affects repeatability.

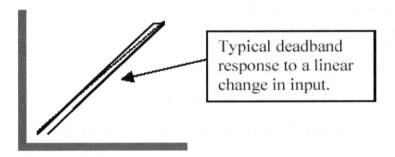

FIGURE 2-1. TYPICAL ERROR PATTERN CAUSED BY DEADBAND

- Repeatability: the ability of an instrument to precisely (see "precision" just below) duplicate its output in response to input signals.
- Environmental factors
    - External heating and cooling (causes span shift)
    - Humidity
    - Vibration

- Electrical factors
  - Output linearity
  - Component age (causes drift)
  - Resolution
- Human factors
  - Calibration
  - Installation
  - Manipulation

3. Resolution Effects on Accuracy

An instrument's accuracy is limited by its precision. An instrument's precision is determined by making minute variations in the measured process variable (e.g., pressure, temperature) and then monitoring the instrument's output for a response. The smallest magnitude of change an instrument can detect and then reflect at its output is its precision. This precision rating can be listed in percentage of span or in engineering units. For example, a hypothetical ultrasonic tank level transmitter might have a detection range of 0–12 ft with its output scaled 4–20 mA, precise to ±0.1 inch. Precision can be related to the instrument's resolution. Rather than a smooth movement of the output as the tank level changes, the output will step to a new milliamp value as the level changes in 0.1 inch increments. The transmitter's resolution is

12 ft x 12 in/ft x 10 steps/in = 1440 steps across the range.

Therefore, the 4–20 mA signal will change in increments of 16 mA/1440 steps = .01111 mA/step.

In the digital world, *bits* are units of resolution. The term *12 bit resolution* tells us the smallest change of signal magnitude that can be interpreted by the computer or device. This resolution value relates to the margin of error. Since each device in a system has its own associated resolution (hence, its own margin of error), the overall accuracy of a system is decreased every time a signal is converted or retransmitted. In most cases, the best move is to digitize the signal at the sensor and then transmit the value as a digitized data stream rather than as a pure electrical signal, thus eliminating several layers of conversion.

Most PLCs currently have 12-bit resolution on their analog inputs. This means the analog signal is converted to a 12-bit binary integer. Once in integer format, the value can be converted to a decimal integer, or hexadecimal, or whatever, for

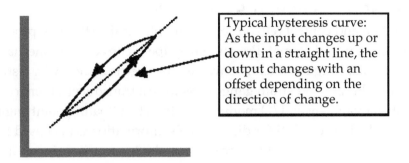

FIGURE 2-2. TYPICAL ERROR PATTERN CAUSED BY HYSTERESIS

display. A 4–20 mA signal has a span of 16 mA (20 – 4 =16). This 16 mA span might represent the 0–100% output of an ultrasonic transmitter or other device.

For an ultrasonic meter with a calibrated range of 12 ft, how much error will the PLC introduce to the system?

12-bit resolution: 4096 divisions

Resolution error: 144 inches/4096 = .0352 inches

4. INSTRUMENT RANGE VERSUS SCALE

There is a difference between an instrument's design range and its scale. When an instrument is received, it comes with an inherent range. This range value defines the maximum input span for which the device can provide a linear output. For example, an ultrasonic level transmitter might have a range of 12 ft. If we place that transmitter in a 6-ft tank and do not recalibrate it, we will lose resolution. Its output will only change 50% for a 100% change in tank level. To rectify this, the unit needs to be recalibrated to provide a 0–100% change in output for a 0–50% change in its input, provided the device is scalable to that degree. Sometimes a device is not scalable beyond, say, 10% of its range. This limit is called *rangeability*.

5. INSTRUMENT CALIBRATION

An instrument's *scale* and its *calibrated range* are synonymous. Once a scale is decided on, the instrument must be calibrated. If its range is 0–550 psig and its scale is 10–350 psig, its *calibrated span* is 350 - 10 = 340 psig. The calibration end points are the *zero point* and the *span point*. The point at which the process is expected to operate is called the *set point*, though this is not generally part of instrument calibration other than to confirm the anticipated set point is near the center of the calibrated span.

6. Linearization and Unit Conversions

Instrumentation and control rely on converting physical or thermal process variables into a more useful format. Pressure in a pipe is converted to mechanical deflection of a diaphragm, which is converted to electrical energy by a strain gauge (the diaphragm and strain gauge constitute a transducer), then to a numeric integer value by an I/O module, and then to a floating point engineering unit value by the PLC or HMI for display. This information is also used to help generate output commands, which are converted into electrical signals and then to mechanical action. The trick is to understand the I/O relationships of the various converters.

For example, a flow orifice will cause a predictable pressure drop as fluids flow across it. A pressure transmitter can measure this pressure drop by comparing the upstream pressure to the downstream pressure. Though this pressure differential is not linear with flow rate, it has a repeatable relationship to it. This relationship is best approximated as a square-root function. Taking the square root of the differential pressure signal effectively linearizes it with the flow rate.

After a linear relationship has been established, the entire conversion sequence from transmitter to computer display can be deduced from one measurement.

To make use of this phenomenon, the technician must be able to move readily from one set of units to another and from one numbering system to another. Computers can communicate using binary, octal, hexadecimal (hex), decimal, BCD, or even more arcane numbering schemes. It is important to know when one is more appropriate than the others.

For fun, here are a few conversion problems. Work the problems using the charts in Figure 2-3 (no calculators needed!).

Figure 2-4 depicts two typical temperature measurement circuits as follows:

The top configuration uses the internal power supply of the transmitter to power the signal loop. This configuration is referred to as a four-wire loop. The bottom configuration uses an external power supply to power the loop. This configuration is referred to as a two-wire loop.

The following discussion about unit conversions applies to both circuit types.

Focus on the top circuit. A thermocouple is the sensing element. Thermocouples are devices that use the principle of bimetallic contact to generate a small millivolt signal. Note that the temperature-voltage curve presented in the chart is relatively linear throughout the temperature interval of interest. Outside of that temperature interval, the signal can become less linear (a characteristic of a thermocouple), but that is of no importance here.

Instrument scaling must always begin at the process measurement. The designer consults the heat and material balance (HMB) sheet for our imaginary

PART II – CHAPTER 4: BASIC DESIGN CONCEPTS   125

Binary Value															
Dec	32768	16384	8192	4096	2048	1024	512	256	128	64	32	16	8	4	2	1
Bin	2^5	2^4	2^3	2^12	2^11	2^10	2^9	2^8	2^7	2^6	2^5	2^4	2^3	2^2	2^1	2^0
Bit	16	15	14	13	12	11	10	9	8	7	6	5	4	3	2	1

Table 1. Decimal-to-Binary conversion

Table 2. Decimal-to-Hexidecimal conversion

Dec	Hex
0	0
1	1
2	2
3	3
4	4
5	5
6	6
7	7
8	8
9	9
10	A
11	B
12	C
13	D
14	E
15	F

Problems

1. Convert 001101100110 (bin) to decimal:

2. Convert 101101000111 (bin) to decimal:

3. Convert 101101000111 (bin) to decimal:

4. Convert 2020 (dec) to binary:

5. Convert 6060 (dec) to binary:

6. Convert B5A2 (hex) to binary:

7. Convert 1EC2 (hex) to decimal:

ans: 1) 870, 2) 2887, 3) 0B47, 4) 011111100100, 5) 001010010100, 6) 1011010110100010, 7) 7874 (hint: hex-to-binary-to-decimal)

FIGURE 2-3. CONVERSION PROBLEMS

system and finds the expected temperature at the measurement point is approximately 105°C. The upstream heater is capable of heating the system to approximately 130°C before it shuts down due to its over-temperature interlock. The design engineer knows a properly calibrated span would place the normal operating point at about the middle of the curve. The upper end would need to be above 130°C. After some thought, the engineer decides on a calibrated span of 15 to 150°C and chooses a type K thermocouple, which provides an output of 0.597 to 6.138 mV over that temperature interval.[1]

The temperature transmitter, then, must be bench calibrated to provide a 4–20 mA output signal that is proportional to the 0.597 to 6.138 mV input signal expected from the thermocouple. The transmitter, being a current source (as opposed to a voltage source), varies its power output as necessary to maintain a steady milliamp output that is proportional to the millivolts on its input.

126 SUCCESSFUL INSTRUMENTATION AND CONTROL SYSTEMS DESIGN

(Note: A voltage source, such as a battery, tries to maintain a constant voltage regardless of load, while a current source tries to maintain a constant current regardless of load).

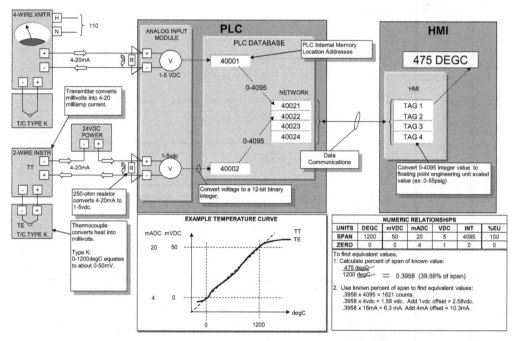

FIGURE 2-4. DATA TRANSLATION PROCESS — FROM FIELD DEVICE TO HMI

The temperature transmitter then converts this signal into a 4–20 mA signal that has been scaled, in this case for a span of 15–150°C (Figure 2-5).

The PLC has an analog input module that detects the output of the temperature transmitter. Virtually all analog input modules are voltmeters, even though they are listed as milliamp inputs. Sometimes the resistor is external on the terminal strip, and sometimes it is internal on the PLC I/O module (Figure 2-4). In either case, the 4–20 mA signal will be converted to a voltage. Typically, this voltage is 1–5 VDC because the resistor used is 250 ohms. This analog value must then be converted to a binary value.

In our example, the PLC specification lists this particular PLC I/O module as having 12-bit resolution. To find the resolution of the module in terms of the process variable, perform a binary conversion: $2^{12} = 4095$. So, for an input span of 1–5 VDC, the PLC I/O module provides an integer value to the PLC program that ranges from 0 to 4095.

The PLC program may fetch this data to use as needed. One of the possible actions of the PLC program is to move this data value into a network interface buffer (a series of contiguous locations in PLC memory) for transmittal upstream to the HMI. The raw-count integer value is then made available for data transmittal across the network.

The HMI receives this transmitted data stream, which is then stored in an input data buffer. The HMI computer has a tag-file database, which contains instructions about how to manipulate each data item for presentation to the operator. Many of the tags in the tag file are linked to data items in the input data buffer. One such tag is linked to this particular location. The 0–4095 raw value is extracted and converted to engineering units by use of the formula embedded in either the tag-file database or the graphic screen software that uses the information. The formula in our sample case is shown in Figure 2-6.

The value produced (85.88) would be the value displayed to the operator in °C as follows in Figures 2-5 and 2-6:

Signal Conversion at the PLC Input
E = IR : Vots = Amps x Resistance

Zero: E = .004 Amps x 250-ohms = 1 - volt at 15 degC
Span: E= .020 Amps x 250-ohms = 5 -volts @ 150 degC

FIGURE 2-5. SIGNAL CONVERSION AT PLC INPUT

Engineering Unit Calculation at the HMI
For a scaled range from 15-150 degC,

EU = ((Raw Counts / 4095) x (150-15)) + 15
if Raw Counts = 2150, then EU = 85.88 degC

FIGURE 2-6. ENGINEERING UNIT CALCULATION AT THE HMI

7. PRACTICAL APPLICATION

Let's have some fun with this. Billy, an instrument technician at Crazy Al's RubberWorks, a manufacturing facility, is on his break. He is just sitting down to eat some linguini when he gets a call that solvent flow into a mixing kettle is not reading right, and the rubber material looks like bad oatmeal. He drops his fork and makes his way to the drawing file, where he locates the proper drawing and uses it to wipe the noodles off his shirt. He then goes to the kettle and finds the solvent line and locates the flowmeter. He finds this differential pressure (dp) flowmeter is calibrated 0–10 inches of water column (inwc) pressure across its diaphragm, which equates to a flow rate of 0–100 pounds per hour (lb/hr) of solvent.

After getting approval from Operations and confirming the throttling valve is in manual mode and is closed, Billy disconnects the signal output cable from the transmitter and hooks the cable to a current source (a portable test device) to simulate the transmitter output signal. He then tries to inject a signal into the system. The test device's batteries are dead, so he has to go get more. With new batteries, he measures the injected signal with an ammeter, which reads 14.2 mA on a 4–20 mA scale. He knows the resolution of the PLC input module is 12 bits, and that the operator reads the flow in PPH on a scale of 0–100.

- What is the simulated pressure drop across the orifice in inches of water column? (Square root extraction is already accounted for in the transmitter.)
- What is the decimal integer value in the computer?
- What should the operator be seeing in pounds per hour?

The solutions follow, but first, try to solve it yourself.

SOLUTION

First, Billy adjusts the analog value to account for the live-zero offset: 14.2 mA reading at the meter, minus the 4.0 mA offset, equals 10.2 mA. He knows he has a live span of 16 mA (20-4=16), and this span equates to 0–10 inwc at the dp transmitter and 0–100 PPH at the HMI.

Once the relative ranges are known, percentage of scale can be used to equate the values:

1. The inwc value is (10.2 mA/16 mA) x 10 inwc = 6.375 inwc.

   Then, Billy decides to check the PLC to make sure it is operating correctly. He knows the PLC has 12-bit analog input resolution, making the valid range of the PLC integer 0 to 4095. He uses a percentage-of-scale relationship to determine the proper integer value:

2. The integer value should be: (10.2 mA/16 mA) x 4095 counts = 2611 counts.

   Billy confirms the PLC is reading 2611 by checking the integer value in the PLC memory. He then checks the PLC program to make sure the integer is being moved to the proper memory location so it can be uploaded to the HMI. He finds it is being moved to address 40021, which is in a data array that is being placed on the network. Further, the blinking light on the interface module confirms that the network is functioning.

So, he reasons, if a problem exists, it must be in the HMI. Billy goes to the control room and views the screen. He calculates the value he expects to see using the percentage-of-scale technique:

3. The PPH value is (10.2mA/16mA) x 100 PPH = 63.75 PPH, or

4. The PPH value is (2611 counts/4095 counts) x 100 PPH = 63.75 PPH

But the screen is reading 54.2 PPH. Billy takes the HMI system into development mode and checks the tag file for the configuration of that tag. He finds the value animation being displayed for FT1001 = (13$40021/4095) x 85, where 13$40021 is the memory location of the PLC-stored integer value. He remembers the flow data array location in the PLC was address 40021, and the PLC was at network node 13. But the proper scaling of the displayed value should be 0–100, not 0–85. Billy mentions his finding to the operator, who informs him that the transmitter, which was broken yesterday, was changed out last shift. Billy surmises the technician changed everything but forgot to modify the HMI screen. He changes the 85 to 100, recompiles the HMI program, and re-launches the application. He confirms the displayed value equals the expected 63.75 PPH value. He removes his test equipment from the circuit and reconnects the wiring to the transmitter, but forgets to turn off the current source (which ensures the batteries will be dead next time), informs Operations that the problem is resolved, and returns to his linguini.

B. Introduction to Information Management

The proper management of information has always been an integral part of the controls profession. Those who do it well, and make proper use of the information as the design progresses, are generally more successful than those who view information management as a chore. In the past, I/O lists, instrument lists, and equipment lists were always required by the customer, but usually were provided as an afterthought. The design manager would obtain a list of information items the client desired for the list, and he would engage a designer to extract that information from an already completed drawing package. The designer would copy the information down and hand it to a data entry clerk to "pretty it up." In fact, that method is still used all too often.

In today's marketplace, profit margins are small, customer expectations are high, and the design team must make full use of the tools available. The single easiest way to reduce or eliminate the duplication of effort is to properly manage data collection and retrieval. Design teams who can demonstrate that their product has the highest quality at the lowest price will generally win in the end—no matter what the other considerations may be. So it is important to come to a full understanding of the productivity gains that can be realized by effective management of a living database, a database that is used in real time, while the design process is in full swing. The database can be the glue that binds the design package together as the design moves from Electrical/Instrumentation & Controls (E/I&C) to the Control Systems Integrator (CSI) to the Control Panel Fabricator (CPF) to the end user.

The database is usually a deliverable in its own right, but more than that, it can be an effective productivity tool for design, a quality management tool, and an indispensable construction and startup aid. Given a database's multiple uses, designers should not limit their databases to the information elements being requested by the client or to the basic fields generated from any of the various automatic design tools available today that extract information from the P&IDs to create other documents. Those fields should be merely the starting point. A large database can always be pared down later to conform to whichever formats the client desires.

The first question we must ask is, "Which kind of program should we use?" There are basically two ways to store and retrieve data: (1) a spreadsheet and (2) a database. There are several viable options on the market for each type, but what is the difference between a spreadsheet and a database? In very broad terms, a spreadsheet treats each data item independently in its own *cell*, while a database inherently links a whole set of data items to a unique *record*. The difference is subtle, but important (Figure 2-7).

Spreadsheet Vs. Database Comparison			
Attribute	Spreadsheet	Database	Comment
Data Quantity	Smaller	Larger*	*Better Data Management Tools
Data Integrity	Less	More*	*Data Stored as it is entered.
Data Entry	Easier	Harder	Cell Manipulation vs. Record Manipulation
Sorting	Easy*	Easy	*Risk of data corruption if done wrong
Filtering	Easy	Easy	Differences are key, however.
Single-Function	Better	Good	Big edge to Spreadsheet
Multi-Function	Fair	Great*	*Enter data once, use many times
Report Gen	Poor	Great	Big edge to Database…
Multi-User	Poor	Great	Big edge to Database…
Indirect Links	Great	Poor	Big edge to Spreadsheet
Calculations	Great	Poor	Big edge to Spreadsheet

FIGURE 2-7. SPREADSHEET VERSUS DATABASE COMPARISON

Time spent in managing data is generally time well spent, provided the data being entered can be applied to multiple purposes and is easy to retrieve. To be effective, a particular data item should only be keyed once, yet be available for many purposes. The following is a more detailed examination of each approach:

1. The Spreadsheet

    Microsoft® Excel is probably the most recognized spreadsheet program on the market today, though there are others such as Lotus Symphony, KSpread, etc. A spreadsheet is a two-dimensional array of individual data cells. The data cell is the base element of the spreadsheet. Each cell is independent, with its own individual characteristics. Each cell can contain a constant, a calculation, an indirect address that allows data displayed in other cells to be fetched and re-displayed—or some combination of the three.

    The greatest benefit of the spreadsheet is its simplicity. It lends itself well to data entry, and with respect to format, what you see is pretty much what you get. Data entry utilities are optimized for entering data cell-by-cell, giving you plenty of flexibility. Data is immediately available, provided the spreadsheet is properly designed for the purpose intended. This feature is particularly valuable on small projects with only a few elements to manage.

    Spreadsheets have a huge edge on data operations that are dynamic, requiring calculations or the presentation of data using indirect addressing. They are also very useful for real-time data collection and display operations and for charting. For example, a workbook may contain many worksheets, each having detailed information relating to a particular process area or building. A summary worksheet can then be used to extract totals from each of the detailed worksheets for presentation or inclusion in a report. Or a chart may be generated from data generated by multiple worksheets within the workbook, or even in remote workbooks. This makes the spreadsheet perfect for the generation of a project estimate or a bill of materials.

    Spreadsheets are at a disadvantage in a number of ways when managing large amounts of static data, however. For example, if data is collected in an I/O List format, it can be difficult to reorganize the data into a report that presents the data sorted by equipment number. The tendency would then be to create a separate worksheet with the data reorganized in the format desired. Doing this creates the need for time consuming layers of edits if changes are made and increases the likelihood of having inconsistencies in reports. Also, spreadsheet data is saved only periodically, not automatically, as data is entered, so it is possible to lose data.

For this reason, it is wise to resist using spreadsheets as your master data management tool for I/O lists and other such data repository applications. For these types of operations, the spreadsheet guarantees a limited return on the data entry time investment.

2. The Database

    Microsoft® Access is probably the most readily available database program on the market today, as it is included in the Microsoft Office Professional software suite, along with Excel and several other programs. It is generally adequate for most project database applications, though it can become limited as the number of records and/or complexity of the application increases, or if long-term database maintenance operations are envisioned. This discussion will center around Access because it is a representative database in structure and utility.

    A database is a two-dimensional column of data records, each record containing a number of individual data cells called *fields*. The data records exist in a data table. There can be multiple data tables in a relational database, each table being linked through key fields. Entering data in a key field causes that data to be automatically distributed to the other data tables, which themselves contain data that is unique to that table's purpose. For the purposes of this discussion, and this comparison to the spreadsheet, we will ignore the relational aspects of the tables, and concentrate on a single data table.

    The following are a few of the key elements of a database:

    - *Table*: A two-dimensional array of information consisting of *records* (rows) and *fields* (columns). Each record describes a single element in the system, as stored in the fields that are associated with that record. For example, in a drawing database, data related to a specific drawing are found in that drawing's record. The drawing's number is in the DWG NO. field, its title is in the DWG TITLE field, and so on.

    - *Record:* A collection of data items related to a single entity, such as an instrument or drawing. In the table, data records are organized into rows made up of data fields.

    - *Field:* A column in a data table that defines a specific type of information found in all records. In the drawing list, for example, DWG NO. would be a field in the data table.

    - *Query:* A preconfigured filter/sort that allows the user to optimize data read/write activities. A Select Query displays subsets of records in a tabular format. An Update Query allows the user to perform global data entry operations and other activities.

- *Form:* Also a preconfigured filter/sort that allows the user to optimize data read/write activities. Generally a form will display items related to one record.

- *Report:* A preconfigured filter/sort that allows the user to optimize data presentation.

The data record is the base element of the database. Each record is independent, with its own individual characteristics. Each record can be displayed and edited in a number of ways, through the use of forms, queries, or reports.

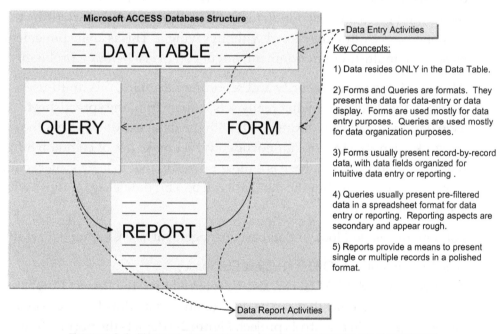

FIGURE 2-8. TYPICAL RELATIONAL DATABASE PROGRAM STRUCTURE

Subsets of the data fields for each record can be fetched for either display or edit in queries, which are used to retrieve specific data or data sets, and forms, which allow the user to optimize data read/write activities. Reports are for output only; no editing can be done from a report.

Compared to the spreadsheet, the database is more restrictive and requires more effort (overhead) to get started. A structure has to be defined, and the data is a bit more difficult to enter. For example, since spreadsheet data is stored by cell, there are handy utilities for entering data into one cell and then dragging the data from that cell to others, thus quickly and easily copying data. A database program, organized by independent record, has no way to do this other than by an update query or a macro, though there are shortcuts and workarounds that minimize the disadvantage.

One of the advantages of a database that is immediately apparent is increased security. When the user finishes editing a record and clicks on another, the data is stored. The program does not wait for a Save keystroke or icon selection. This allows multiple persons to be editing the database at the same time from different locations with minimal risk of overlap, as the likelihood of both parties being on the same data record at the same time is remote. If that does occur, the database informs the second person to leave the record; an edit has occurred since he opened the record, and gives him the option to continue with his edit, overwriting the first one, or of exiting the record without saving it. This is an inconvenient aspect of a multi-user environment, but not an overwhelming problem.

Along with the added security and reporting flexibility, this multi-user aspect of the database is what really sets it apart from the spreadsheet. Two designers can be working, both with the database open across the network, each able to work independently, yet make use of common information. In short, the database is a more powerful tool to use for inert data retrieval and storage. But the database can be a bit unwieldy for those who are unfamiliar with it.

For a more detailed look at database management, see Chapter 9, Part III.

3. The Instrument & I/O List (Project Database)

    The Instrument & I/O List can also be referred to as the Project Database, as it is the one product that can be used to unify data flow between organizations engaged in a controls project. Figure 2-9 depicts the normal flow of a project as major design products are generated by the different participants. Activities that can and should be heavily influenced by a properly designed Instrument & I/O List database are highlighted.

    Far from being a mere deliverable, this database can be the bedrock of the design, and can become a valuable management and quality assurance tool. In order to extract the most benefit from the data-entry investment—an investment that will need to be made anyway—the process should begin early, as part of the Front End Loading (FEL) process (for more about FEL, see Part I). The design team lead should own the database him-

PART II – CHAPTER 4: BASIC DESIGN CONCEPTS

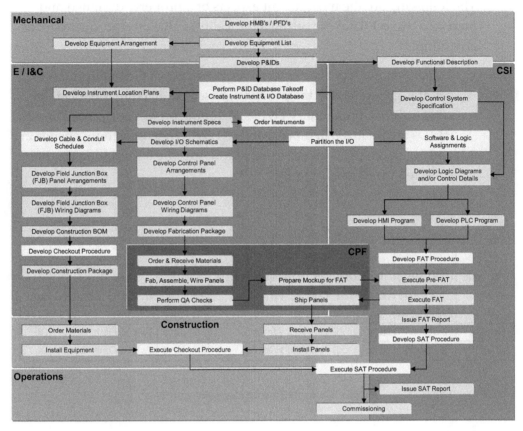

FIGURE 2-9. ICS-BASED PROJECT FLOW WITH DATABASE-INTENSIVE ACTIVITIES HIGHLIGHTED

self, taking responsibility for it, and using it to direct workflow. He or she should make assignments of a particular instrument to a floor plan drawing, for example, and assign that instrument's I/O points to the various I/O schematics as a part of the partitioning process. If the database is done in a multi-user format, the Instrument & I/O List can become a living element of the design process that allows data, entered once, to be used over and over by multiple parties as the project develops. Changes made are available real-time to the users, reducing lag times and the likelihood of having multiple hardcopy revisions floating around.

To realize the full potential of such a database requires a certain level of commitment by the design staff, as the most efficient and powerful database is one that is developed, updated, and modified as an integral part of the design process. If the design staff can be trained to populate the database on the front end, then maintain it throughout the design process, it will be less of a drain on resources.

The following are the most salient aspects of a properly designed and managed project database, using a typical fill valve on a product tank as an example:

- *P&ID (Piping & Instrumentation Diagram) Takeoff:* Generally, an I/O count is generated for the proposal, based on an equipment list, or sometimes on preliminary P&IDs. After the project is awarded and the P&IDs are finalized, a more detailed P&ID Takeoff activity ensues in which, for a typical fill valve on a tank, specific information might be collected via a query (Figure 2-10):

P&ID Takeoff Query - NOTE: (g) Data Gathered; (e) Data Existing; (de) Data Entered							
Instr. Tagname	P&ID Dwg#	Related Equip.	Related Process	Service Descr.	Item Descr.		I/O Type
HV-042	229-02-040	TK-10 Prod. Tank	DI Water	Fill Valve	Vlv, On/Off, 3", CS		N/A
SV-042	229-02-040	TK-10 Prod. Tank	DI Water	Fill Valve	Solenoid		DO
ZSC-042	229-02-040	TK-10 Prod. Tank	DI Water	Fill Valve	Switch, Limit, Clsd		DI
ZSO-042	229-02-040	TK-10 Prod. Tank	DI Water	Fill Valve	Switch, Limit, Opnd		DI
(g)	(g)	(g)	(g)	(g)	(g)		(de)

FIGURE 2-10. THE P&ID TAKEOFF QUERY

- *I/O Partitioning:* Before it is possible to assign I/O to particular I/O modules in the control system, physical location and process function need to be understood. Redundant equipment that provides damage recovery or maintenance capacity should be wired to separate I/O modules. For most installations, I/O racks are mounted remotely, in the field, in order to reduce cable lengths. In those cases, it is important to be able to partition instrument I/O assignments by location. During the I/O partitioning process, the following information, if known, is added to the database (Figure 2-11).

- *Software & Logic Assignments:* After the I/O has been partitioned to fit the physical configuration of the facility, the CSI team can get involved, adding key information to the database (Figure 2-12). Data added includes the software address, which is specific to the control system, a device control detail sheet number and/or logic diagram drawing number linking the item to its source logic documentation and its software tagname, which can be either a derivative of its instrument tagname or its functional description (as was used in this example). After the software tagname assignments are made, the

PART II – CHAPTER 4: BASIC DESIGN CONCEPTS

I/O Partitioning Query - NOTE: (g) Data Gathered; (e) Data Existing; (de) Data Entered							
Instr. Tagname	Piping Plan Dwg #	Instr. Plan Dwg #	RIO Cabinet Dwg #	Remote I/O Drop	Remote I/O Rack	Remote I/O Slot	Remote I/O Point
HV-042	229-06-029	229-09-001	229-10-001	N/A	N/A	N/A	N/A
SV-042	229-06-029	229-09-001	229-10-001	D03	R02	S04	P01
ZSC-042	229-06-029	229-09-001	229-10-001	D03	R02	S05	P01
ZSO-042	229-06-029	229-09-001	229-10-001	D03	R02	S05	P02
(e)	(g)	(de)	(de)	(de)	(de)	(de)	(de)

FIGURE 2-11. THE I/O PARTITIONING QUERY

programming team can create a file that forms the genesis of the respective databases used by their development software packages. NOTE: The HW Address field shown below can be created, using an Update Query, by combining the information entered during the partitioning stage discussed above.

SW & Logic Query - NOTE: (g) Data Gathered; (e) Data Existing; (de) Data Entered						
Instr. Tagname	P&ID Dwg#	HW Address	SW Address	DCDS Sheet#	Software Tagname	
HV-042	229-02-040	N/A	N/A	N/A	N/A	
SV-042	229-02-040	D03-R02-S04-P01	%O00042	VSD01	HV042_YOPN	
ZSC-042	229-02-040	D03-R02-S05-P01	%I0012	VSD01	HV042_XCLSD	
ZSO-042	229-02-040	D03-R02-S05-P02	%I0013	VSD01	HV042_XOPND	
(e)	(e)	(e)	(de)	(de)	(de)	

FIGURE 2-12. THE SOFTWARE & LOGIC ASSIGNMENT QUERY

- *Cable and Conduit Schedule Generation:* As the physical drawings develop, cabling designations must be made to extend signal and power cables and conduit to each instrument. The project database can be used as a simple way of managing those cable and conduit assignments. In the Figure 2-13 example, entries are made into the cable schedule, as the signal is tracked from the instrument floor plan to the Field Junction Box (FJB) via signal cables HV042-01 through -03. A homerun cable, RIO09-01-1, extends the signals from the FJB to the Remote I/O (RIO) panel using its pairs 7, 8, and 9. Additional fields that could be added to this query include PowerCable#, FieldConduit# and HomerunConduit#. Using relational links, separate tables could be generated called Cable Data and Conduit Data that would tabulate each cable or conduit as it is added to the Master Instrument and I/O data table. Data could later be added to those tables that describe the properties of the specific cable or conduit segment. (For more detailed information on field wiring, refer to Part III.)

Cable Schedule Query - NOTE: (g) Data Gathered; (e) Data Existing; (de) Data Entered							
Instr. Tagname	Instr. Plan Dwg #	FieldSig Cable#	Field JBOX#	Homerun Cable# / Pair#	RIO CAB#	RIO Cabinet Dwg #
HV-042	229-09-001	N/A	N/A	N/A	N/A	N/A
SV-042	229-09-001	HV042-01	FJB-09-01	RIO09-01-1/07	RIO-01	229-10-001
ZSC-042	229-09-001	HV042-02	FJB-09-01	RIO09-01-1/08	RIO-01	229-10-001
ZSO-042	229-09-001	HV042-03	FJB-09-01	RIO09-01-1/09	RIO-01	229-10-001
(e)	(e)	(de)	(de)	(de)	(de)	(de)

FIGURE 2-13. THE CABLE AND CONDUIT SCHEDULE QUERY (PARTIALLY SHOWN)

- *Order Instruments* (Figure 2-14): As instrument specifications are developed, the database provides a convenient place to log specification numbers, part numbers, instrument type, etc. These fields are useful to the design staff, but are also useful for procurement, allowing the information to accumulate as the specifications are developed, and then allowing report(s) to be generated for procurement purposes later. Note for the on/off valve, there is no Engineering Units or Scale field. If the item were an analog transmitter, those fields would have information in them. Additional fields that could be added to this query include Alarm Setting Fields, Normal Control Setpoint, Interlocks, Permissives, Vendor Phone Number, and other information of interest that could be collected for future use.

- *Construction Checkout* (Figure 2-15): Construction checkout occurs prior to power-up, as construction managers audit the installation and wiring of the instrumentation and equipment. The Construction Checkout Query provides key information to the construction team, helping them navigate through the design package. In addition, it provides a place for sign-off as each device is checked for physical installation and wiring. In addition to the fields shown in the charts below (Figures 2-15, 2-16, 2-17), there should be a comment field. For the sign-off data, if initials and date are in the field, then the installation is complete. If a Punch List Item Number is in the field, then there was a

PART II – CHAPTER 4: BASIC DESIGN CONCEPTS

InstrSpec Query - NOTE: (g) Data Gathered; (e) Data Existing; (de) Data Entered							
Instr. Tagname	Item Descr.	Spec#	Manuf.	Vendor	P/N	Engrg Units	Scale
HV-042	Vlv, On/Off,	VLV001-01	Xomox	Joe's Instr.Shop	8001-1112-02-1	N/A	N/A
SV-042	Solenoid	VLV001-01	Xomox	Joe's Instr.Shop	Part of HV-042	N/A	N/A
ZSC-042	Switch, Limi	VLV001-01	Xomox	Joe's Instr.Shop	Part of HV-042	N/A	N/A
ZSO-042	Switch, Limi	VLV001-01	Xomox	Joe's Instr.Shop	Part of HV-042	N/A	N/A
(e)	(e)	(de)	(de)	(de)	(de)	(de)	(de)

FIGURE 2-14. THE INSTRUMENT SPECIFICATION QUERY (PARTIALLY SHOWN)

problem, and the construction punch list (kept separately) should be consulted as to the resolution of the problem. A construction audit report, consisting of the database and punch list with resolution notes, can be provided to the customer as a part of the project documentation package upon completion of the project.

Constr. Checkout Query - NOTE: (g) Data Gathered; (e) Data Existing; (de) Data Entered							
Instr. Tagname	FieldSig Cable#	Field JBOX#	Homerun Cable# / Pair#	RIO CAB#	Hardware Installed?	Wiring Terminated?	
HV-042	N/A	N/A	N/A	N/A	*DJ 081112*	*N/A*	
SV-042	HV042-01	FJB-09-01	RIO09-01-1/07	RIO-01	*PL Item 27*	*DJ 081112*	
ZSC-042	HV042-02	FJB-09-01	RIO09-01-1/08	RIO-01	*DJ 081112*	*DJ 081112*	
ZSO-042	HV042-03	FJB-09-01	RIO09-01-1/09	RIO-01	*DJ 081112*	*DJ 081112*	
(e)	(e)	(e)	(e)	(e)	(de)	(de)	

FIGURE 2-15. THE CONSTRUCTION CHECKOUT QUERY

- *Validation & Verification (V&V):* The V&V test typically occurs prior to shipment of the control system core. The core consists of the control processors and HMI- or Supervisory Control and Data Acquisition (SCADA)-level equipment, along with network devices such as Ethernet switches, fiber patch panels, etc. In most cases, the I/O racks and remote I/O panels are not required, as process simulators or loop-back software is used to simulate field devices. The RIO panels are tested in a separate process at the assembly point. The purpose of the V&V test is to ensure the PLC or DCS programmer has completed the control logic programming for every field device, and the control logic is properly integrated with the HMI package. The HMI animation should be verified for every device to minimize the problems likely to exist when the system goes to the field. The V&V Queries provide a place for sign-off as each device is checked. In addition to the fields shown in Figure 2-16, there should be a Comments field. If initials and date are in the field, then the test was successful. If a Punch List Item Number is in the field, then there was a problem, and the punch list (kept separately) should be consulted regarding the resolution of the problem. The V&V report is usually printed as hardcopy, with empty fields where the V&V manager can make pencil entries as the test progresses. These entries can be put into the database as a separate data entry activity later, with the manually written hardcopy being retained as the official record of the V&V.

V&V HMI Query - NOTE: (g) Data Gathered; (e) Data Existing; (de) Data Entered						
Instrument Tagname	SW Address	Software Tagname	DCDS Sheet#	HMI Screen	FAT HMI Config'd ?	FAT HMI Tested ?
HV-042	N/A	N/A	N/A	*N/A*	*N/A*	*N/A*
SV-042	%O00042	HV042_YOPN	VSD01	*TK10 Ctrl*	*MW 081211*	*MW 081211*
ZSC-042	%I0012	HV042_XCLSD	VSD01	*TK10 Ctrl*	*PL Item 22*	*PL Item 22*
ZSO-042	%I0013	HV042_XOPND	VSD01	*TK10 Ctrl*	*MW 081211*	*MW 081211*
V&V PLC Query - NOTE: (g) Data Gathered; (e) Data Existing; (de) Data Entered						
Instrument Tagname	SW Address	Software Tagname	DCDS Sheet#	PLC Logic Section	FAT PLC Config'd ?	FAT PLC Tested ?
HV-042	N/A	N/A	N/A	*N/A*	*N/A*	*N/A*
SV-042	%O00042	HV042_YOPN	VSD01	*TK10 Ctrl*	*MW 081211*	*MW 081211*
ZSC-042	%I0012	HV042_XCLSD	VSD01	*TK10 Ctrl*	*PL Item 22*	*PL Item 22*
ZSO-042	%I0013	HV042_XOPND	VSD01	*TK10 Ctrl*	*MW 081211*	*MW 081211*

FIGURE 2-16. THE VALIDATION & VERIFICATION (V&V) TEST QUERIES

- *Site Acceptance Test (SAT):* The SAT typically occurs at the end of the construction process as a precondition to Mechanical Completion. Mechanical Completion is a major milestone in a construction project where all of the mechanical equipment is installed, all cables pulled and terminated. At that point, equipment can begin to be checked out. A SAT procedure is written to exercise every end device (motor, valve, actuator) in each of its operating modes prior to commissioning. This is an activity typically driven by the construction organization, but attended and supported by the CSI team, who should maintain the database as a way of tracking progress. Software problems uncovered during the SAT need to be addressed and retested. The SAT is sometimes called "dry commissioning" because materials and product are generally not handled at this time. Equipment is operated "dry" and "cold" (no process materials, and at ambient temperature) to the fullest extent possible in order to bring the control system to a high state of readiness prior to dedicating resources and materials to the process, and prior to full-scale participation by operations personnel, as will happen during system commissioning. The SAT report is usually printed as hardcopy, with empty fields where the SAT manager can make pencil entries as the test progresses. These entries can be put into the database as a separate data entry activity later, with the pencil-whipped hardcopy being retained as the official record of the SAT.

SAT HMI Query - NOTE: (g) Data Gathered; (e) Data Existing; (de) Data Entered						
Instrument Tagname	SW Address	Software Tagname	DCDS Sheet#	HMI Screen	SAT HMI Config'd ?	SAT HMI Tested ?
HV-042	N/A	N/A	N/A	N/A	*N/A*	*N/A*
SV-042	%O00042	HV042_YOPN	VSD01	TK10 Ctrl	*MW 090202*	*MW 090202*
ZSC-042	%I0012	HV042_XCLSD	VSD01	TK10 Ctrl	*MW 090202*	*MW 090202*
ZSO-042	%I0013	HV042_XOPND	VSD01	TK10 Ctrl	*MW 090202*	*MW 090202*
SAT PLC Query - NOTE: (g) Data Gathered; (e) Data Existing; (de) Data Entered						
Instrument Tagname	SW Address	Software Tagname	DCDS Sheet#	PLC Logic Section	SAT PLC Config'd ?	SAT PLC Tested ?
HV-042	N/A	N/A	N/A	N/A	*N/A*	*N/A*
SV-042	%O00042	HV042_YOPN	VSD01	TK10 Ctrl	*MW 090202*	*MW 090202*
ZSC-042	%I0012	HV042_XCLSD	VSD01	TK10 Ctrl	*MW 090202*	*MW 090202*
ZSO-042	%I0013	HV042_XOPND	VSD01	TK10 Ctrl	*MW 090202*	*MW 090202*

FIGURE 2-17. THE SITE ACCEPTANCE TEST QUERIES

As can be seen from the examples above, the project database can be far more valuable for management and quality assurance purposes if it is set up properly in the beginning, and then maintained. If these things are done, the customer can have a database that is a useful cross-reference, but the real value of it is to the service provider, as data, once collected, is used for multiple purposes: for quick reference during design, for auditing during design checks, and for validation during acceptance tests. Its potential uses cross organizational boundaries, provided each organization has real-time access to the data and is disciplined enough to maintain the data as the design moves toward completion.

4. The Document List

    Documents are the stock in trade of the design engineering organization. As such, engineering organizations should give their documents the same consideration a car manufacturer gives his automobiles. Sometimes the transfer of documents is the main point of contact between the customer's organization and the service provider's. Service providers who cannot meet the customer's specifications regarding the content of the drawing title block, or who have problems being able to track the progress of documents as they are being developed, will have customer relations problems, regardless of the technical quality of their work.

    All too often, however, this kind of attention is lacking. In the first place, it is easy to dismiss the document list as a mere tabulation of documents. In reality, the document list should be much more dynamic, reflecting the current state of the document package, and thereby becoming a valuable management tool. Any documents to be delivered to the customer should be given a document number and should be listed in this table. This includes specifications, calculations, memos and sketches. The document list can be managed in its own database, or it can be a separate data table in a master project database that also includes the instrument list & I/O table.

    The Document List table should be designed around the document control sequence, since formal document control procedures probably exist to some degree. Figure 2-18 shows a fairly typical drawing handling process, with nine stages. The Document List table should have a date field for each of these nine stages, where the designer can record when each stage is reached.

    In addition to the dates, in most projects, every document created is assigned to a work package that is tied to the project Work Breakdown Structure (WBS). It is important to be able to sort the document list by WBS package, so the WBS Package ID field should be included.

    The sequence of events depicted in Figure 2-18 can change slightly, depending on the customer, but each step should be logged as it is reached. Date fields with data imply completion of the step, while blank date fields indicate remaining milestones. It is possible, using this process, to develop a quick-and-dirty approach for calculating completion percentage by dividing the number of filled dates by the total number of date fields. This approach is crude, but it is sometimes good enough.

The following is a commentary on a typical drawing handling cycle for a large capital project, as depicted in Figure 2-18:

- *Order Prints from the Customer:* A design project typically begins with a tentative drawing count. This count must be developed as a part of the project estimate. If the project is a retrofit, with new systems being added to existing ones, the existing drawings that are pertinent to the design need to be obtained. Usually that is done by ordering print hardcopies from the owner, or by requesting read-only CADD files. These drawings need to be combed to identify additional drawings that will need to be ordered for revision by the design team. Drawings ordered as hardcopy prints should be logged—if for no other reason than to note that a print has been received and is available to the design team for reference.

- *Identify Drawings for Revision/Creation:* After the prints/CADD files are in hand, the design team reviews them to see which ones need to be revised as part of the project. When a drawing is designated as a possible deliverable, it should be so designated in the drawing list. In addition to existing drawings, any new drawings that need to be created by the team need to be listed so drawing numbers can be requested. Usually, the customer has an ordering process that entails filling out a form or providing information in a specific format. The process area and suggested title are usually included in the request.

- *Order Drawing Files from the Customer:* After a list of needed drawings has been compiled, a request needs to be sent to the customer for transfer of the drawing files, which exist at the customer's location. At this point, the design team is requesting ownership of the drawings. Until the drawings or drawing files have been officially received from the customer, they are subject to change by others outside the purview of the project, which poses some risk that the needed drawings will either not be available, or will have changed in a way that will negatively impact the project.

    Another point to consider is the effect on the customer's perception of the engineering provider. Many times, the drawing ordering process is the engineer's first contact with the customer. If the engineering provider shows himself to be poorly organized in this area, then it is likely that he will be poorly organized in other areas. The customer may

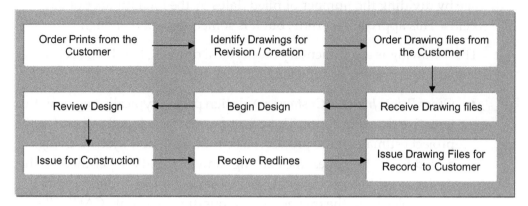

FIGURE 2-18. TYPICAL DOCUMENT HANDLING PROCESS

begin to lose confidence in the service provider before the project is fairly started.

- *Receive Drawing Files:* When the drawing files are officially received from the customer, the event should be logged, along with the revision number assigned for this project.

- *Design Process:* When a drawing is completed and checked, that event should be logged in the database. Other intermediate milestones can be logged as well if desired. This is a project management tool for determining project status.

- *Issue for Construction:* The Issue for Construction milestone signifies the end of the design phase ("Phase 2" in Part I) and the beginning of the construction (deployment) phase ("Phase 3" in Part I). This does not, however, imply that drawing revision activities are over. Sometimes a drawing might be revised and reissued several times during the construction process. Each time the drawing is reactivated for design changes, a new record should be created for the drawing with an incremented revision number to indicate the drawing is back in the design arena. Each of the field copies should be located and destroyed or marked up with a "Hold" note to indicate the area under revision. When the redesign is completed, then the drawing review process should be followed prior to the drawings being reissued.

- *Receive Construction Redlines:* The constructor is sometimes empowered to make modifications "on the fly." For example, a common on-the-fly change is that of color designations. Many times, the designer will assume a certain cable with a particular color code will be purchased. However, the purchasing agent may find a deal on another cable of a different color equal in all other respects. The constructor will then mark up the print to show the correct color combinations. Redline drawings of this type are called *as-built* drawings since the changes are required to bring the drawings up to the as-built conditions.

Incidentally, the typical color code for as-built markups uses *green* to indicate items that should be deleted, and *red* to indicate items to be added. Other colors are for information only.

- *Issue for Record:* The final step in the life of a drawing revision is the issue-for-record step. At this point, all drawings need to be scanned to make sure they comply with the customer's standards for design content and file structure. Most engineering service providers have design content and file structure guidelines that may not be enforced until it is time to submit the drawing to the customer for inclusion in the drawing control system. The design team should contact the customer about drawing guidelines well in advance to make sure there are no surprises.

Part II – Chapter 5: Design Practice

What does a customer have a right to expect as a given in a design package? The National Electrical Code (NEC) is an example of a set of standards that should be carefully followed, as these define universal safety issues and approaches that should be followed across the board. One of the more useful standards to the control panel designer is National Fire Protection Association (NFPA) 79. This standard defines many of the basic tenets that should be followed as a design package is developed. ISA—The International Society of Automation—promulgates standards specific to instrumentation and process controls.

In addition to national standards, many industries have developed standards that are specific to the needs of that industry. The pharmaceuticals industry, for example, requires that design products be developed in a way that integrates into their overall validation program, which is unique to that industry. The next step down in the standards hierarchy is plant standards, and after that, project specifications. From the control design team's perspective, this hierarchy should be followed bottom to top (local to global) in order of priority. For example, if a plant's control specification requires that an approach be followed that conflicts with a national standard, then the service provider should either follow the plant specification, or take exception to it in his proposal, citing the conflict. If this is not done, once the project is awarded, the customer will have the expectation that the stipulation in his project specification will be followed, and a misunderstanding could result.

Beyond published standards, there are also guidelines and best *practices*, which are design approaches commonly understood to be acceptable for a particular industry. A best practice may align perfectly with a stipulation spelled out in a standard, but it may also be ignored by the standard. A customer has a right to expect that his supplier will be conversant with the industry's best practices, but it also behooves that customer to use the project specification to clearly spell out the best practices that are most important to him.

This section describes basic wiring practice, failsafe wiring, hazardous area classifications and their effects on design, control system wiring, PLC programming setup and architecture, and other topics, with references to governing standards and best practices.

A. Basic Wiring Practice

Generating a document that properly depicts interconnection wiring is something of an art form. There are two basic types of wiring diagram: the inter-cabinet wiring diagram, in which cables connect cabinets, and the intra-cabinet wiring diagram in which point-to-point wiring is described inside the cabinet. The former is relatively simple to produce, while the latter is not.

One of the best guidelines for this topic for both types of wiring is the NFPA 79 standard. While this section deals with more than is contained in the standard, nothing in it conflicts with the standard.

Wiring activities comprise a high percentage of the electrical portion of the construction project. Figure 2-19 depicts a typical cabling scheme that collects and distributes signals to and from the field devices. An FJB is used to concentrate the signals into multi-conductor homerun cables. The homerun cables then terminate in a remote cabinet where the signals are marshalled (reorganized) as necessary to efficiently terminate at the I/O interface of the DCS or PLC system.

Homerun cables serve multiple end devices. They generally consist of multi-conductor or multi-pair cables that must have their insulation jackets removed (stripped) to get at the individual conductors within. It is desirable to reduce the amount of stripping an installer must do; thus, it is convenient to mass terminate these cables on terminal strips in the FJB, then distribute the signals to the field devices using individual field cables.

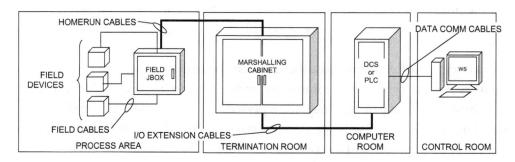

Figure 2-19. Typical cabling scheme

1. Inter-Cabinet Wiring

Managing information about wiring from cabinet to cabinet or from device to cabinet is very important to the constructor. It is also important to maintain this information as a permanent record for the purpose of disaster recovery in case of fire or other catastrophe.

Since this type of information presents well in a tabular format, a series of tables, called schedules, has evolved. The schedules are most meaningful when

PART II – CHAPTER 5: DESIGN PRACTICE

accompanied by another tool called a location plan drawing. The following is a brief discussion of each of these tools:

a. Generating a Cable Schedule

Point-to-point wiring information is provided in a table called a cable schedule. Each cable is a line item in this list and is given a unique identifier. Source and destination locations are specified (preferably in a cabinet/terminal strip/terminal number format), along with the more salient aspects of the cable specification, such as conductor size and quantity, voltage ratings, and jacket type. Routing information is also provided, in which each conduit and/or cable tray used is listed.

Figure 2-20 depicts a cable (Cable W1) routed between a control panel (CP1) and a junction box (JB2). The cable passes through conduit to a cable tray system. It then transits back to conduit, and on to JB2.

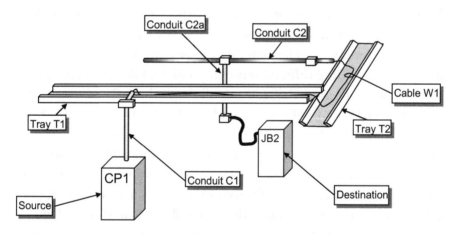

FIGURE 2-20. DEFINING THE CABLE ROUTE (WIRE W1, ROUTE C1/T1/T2/C2/C2A)

Figure 2-21 shows a record in the cable schedule for cable W1 shown in Figure 2-20. Drawing numbers are omitted here but should be provided on the source and destination. Note the routing code (C1/T1/T2/C2/C2a) includes the entire path of the cable through each conduit and tray leg.

b. Generating a Conduit and Tray Schedule

The conduit and tray schedule is a chart that lists each conduit and tray section and provides a cross-reference to the cable schedule by listing each cable that it contains. The results of the fill calculation (see below) should be listed here, along with any associated drawing numbers. Physical location could also be added to this schedule.

Figure 2-22 shows two of the conduits in Figure 2-20. Conduit C2 contains three cables at a 25% fill, while conduit C2a contains only a single cable (W1).

CABLE SCHEDULE								
Cable Number	Type	AWG	OD	Source	Destina.	Route	Est. Len (ft)	Purpose
W1	19/cond	14	.531"	CP-1	JB-2	C1/T1/T2/C2/C2a	50	120VAC Signals
W2	3/cond	12	.12"	CP-1	JB-4			

FIGURE 2-21. SAMPLE CABLE SCHEDULE

c. **Calculating Conduit Fill**

To make a plan drawing or conduit schedule useful to construction, it must provide conduit fill capacities that allow successful pulling to occur. Clearly, a conduit sized just slightly larger than the cables it contains would make pulling difficult if not impossible. Likewise, certain cable combinations are more difficult to pull than others. The National Fire Protection Association (NFPA) has conducted tests, and has published the results for rigid metal conduit in Article 344 of the NEC.[2] Surprisingly, three cables are easier to pull than two, as evidenced by the area fill percentages permitted. But three cables have an inherent tendency to bind when coming out of a bend if the conduit size is in a certain proportion to cable area. Four or more cables do not exhibit this tendency. Therefore, an additional step is necessary when calculating the area fill of a conduit carrying three cables.

The NEC clearly defines the maximum percentage of area fill for conduits in Chapter 9, Table 1, *Percent of Cross Section of Conduit and Tubing for Conductors*.[3] This table describes maximums that should not be exceeded and gives an assumption that the pull is based on common conditions of proper cabling and alignment of conductors where the length of the pull and the number of bends are within reasonable limits.

This table recommends, in general, a 40% area fill if more than two cables are in the conduit (Figure 2-22). If one cable is in the conduit, a fill of 53% is permitted, and if two cables, 31%. Ampacity issues can be ignored because these are low-voltage applications in which heating due to high current is not an issue.

Qty	Max % Fill
1	53
2	31
> 2	40

FIGURE 2-22. CABLE AREA FILL

There are a couple of exceptions to the 53/31/40 rules. There is a cautionary note in *Fine Print Note #2* about the possibility of jams (binding) occurring when pulling three cables, if the ratio of conduit inside diameter (i.d.) to cable outside diameter (o.d.) is between 2.8 and 3.2.[4] This jamming could occur if the conduit is deformed due to bending, thus allowing the three cables room to line up in cross-section, as opposed to forming a triangle, while within the bend (Figure 2-23). In that case, the cables jam when the conduit resumes its round shape after the bend is negotiated. So if pulling three cables, it is best to avoid a jam ratio of between 2.8 and 3.2, as calculated by:

Jam Ratio = ID of raceway/OD of conductor.

To calculate the area fill of a conduit, the size of the conduit must be known. This is not the trade size (nominal diameter), but the inner area. NEC Chapter 9, Table 4, Article 344 lists the data for rigid metallic conduit (RMC).[5] Based on the total inner area values provided in that chart, Figure 2-24 lists the particulars for most of the conduit sizes E/I&C designers are likely to use.

So the NEC lists the key statistics for each type of conduit as they relate to the fill calculation. The cable manufacturer provides statistics relating to the outer dimensions of the cable. For the purposes of this discussion, we are using RMC-type conduit, and we will pull cables having identical cross-sectional areas. The cable type is a 16 AWG (American Wire Gauge), 16-pair cable with an overall shield and an outside diameter (OD) of 0.866 inch, and the number of cables in the pull is three.

- Find the cross-sectional area of the cable using one of the formulas provided:

$$\text{Area} = \pi r^2 = 3.14159(.866/2)^2 = 0.589 \text{ in}^2$$

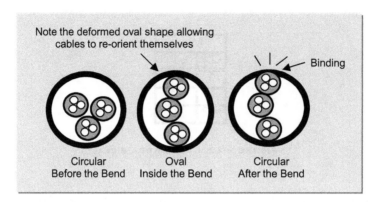

FIGURE 2-23. CROSS-SECTIONAL VIEWS OF CABLE ORIENTATION BEFORE, DURING, AND AFTER A CONDUIT BEND

BASIC CONDUIT DATA				
Conduit	I.D.	Area	40% Fill	31% Fill
1/2"	0.622	0.314	0.1256	0.0973
3/4"	0.824	0.549	0.2196	0.1702
1"	1.049	0.887	0.3548	0.275
1-1/4"	1.38	1.526	0.6104	0.4731
1-1/2"	1.61	2.071	0.8284	0.642
2"	2.067	3.408	1.3632	1.0565
3"	3.068	7.499	2.957	1.8483

FIGURE 2-24. CONDUIT FACTS

- Find the total area of the cables: 3 cables × (0.589 in^2) = 1.767 in^2.

- Refer to the basic conduit data table in Figure 2-25 to find the proper conduit size to meet the 40% fill specification. Find the RMC section of the table. In the 40% fill column, the smallest conduit size that can accommodate a 1.767 in^2 cable area would be a 1½ inch conduit with an inside area of 2.071 in^2.

- Next, compare conduit inside area with cable area to avoid the problematic 2.8 to 3.2 binding ratio noted above:

    2.071/1.767 = 1.172, a satisfactory result.

If there is a mix of cables in a particular conduit, this complicates things a bit. In that case, the cable area calculation has to be done for each cable type, and the results must be summed before consulting NEC Chapter 9, Table 4.[6]

Figure 2-25 presents a method for automating this process.

CONDUIT SIZING CALCULATOR

CONDUIT CONTENTS		CABLE SPECIFICATION DATA								
CABLE QTY	Calc'd Area	Cable	Awg	Shld?	Manuf.	Rated Volts	Part No.	Svc	COST/FT	O.D.
	0.000	RG62A/U	na	NA	Anixter	na	B9268	na	$0.39	0.260
	0.000	1-cond	18	NA	Belden	600	8918	Cond.	$0.01	0.110
	0.000	1-cond	16	NA	Belden	600	8917	Cond.	$0.01	0.123
	0.000	1-cond	14	NA	var	600	?	Cond.	$0.01	0.133
	0.000	3-cond	18	US	Alpha	300	1898	Tray	$0.20	0.225
2	0.174	4-cond	16	US	Anixter	600	2A-1604	Tray	$0.70	0.333
	0.000	4-cond	16	US	Alpha	600	45174	Tray		0.450
	0.000	6-cond	16	US	Belden	600	27615	Tray		0.390
	0.000	1-tupr	16	US	Belden	600	27337	Tray		0.290
	0.000	1-tupr	16	US	Belden	600	9487	Tray		0.301
	0.000	1-tupr	18	US	Belden	300	8461	Cond.	$0.25	0.234
	0.000	1-tupr	18	IS	Belden	600	1120A	Tray		0.289
	0.000	1-tupr	16	IS	Alpha	600	5616B1801	Tray		0.276
	0.000	1-tupr	18	IS	Belden	300	8790	Cond.	$0.25	0.241
	0.000	1-tupr	22	IS	Belden		9182	Cond.	$0.30	0.350
	0.000	1-triad	16	IS	Belden	300	8618	Cond.	$0.40	0.327
	0.000	1-triad	18	IS	Alpha	300	2258/3	Cond.	$0.34	0.230
	0.000	4-triad	18	IS/OS	Anixter	600	2AS-1804-S	Tray	$0.57	0.587
	0.000	3-tupr	16	IS	Anixter	600	2AS-1603-S	Tray	$0.39	0.486
	0.000	4-tupr	18	IS/OS	Belden	300	9388	Cond.	$1.39	0.486
	0.000	12-tupr	18	IS	Belden	600	1477	Tray		0.784
	0.000	16-tupr	16	IS	Anixter	600	323-221-16	Tray		0.860
	0.000	16-tupr	16	OS	Anixter	600	2AS-1816P	Tray		0.905
	0.000	24-tupr	18	OS	Anixter	600	2AS-1824P	Tray		1.040
	0.000	30-cond	16	US	Belden	600	27616	Tray		0.770
	0.000	16-cond	16	US	Anixter	600	2A-1616	Tray	$0.98	0.602
2	0.174	<<< CAUTION - USE 31% MAXIMUM FILL ! >>>								

Complex Fill for Rigid Metal Conduit (Contents Mixed)						
SIZE	1/2"	3/4"	1"	1-1/4"	1-1/2"	2"	3"
AREA	0.314	0.549	0.887	1.526	2.071	3.408	7.499
FILL	55.5%	31.7%	19.6%	11.4%	8.4%	5.1%	2.3%

FIGURE 2-25. CONDUIT SIZING CALCULATOR

CAUTION: The Excel-based calculator depicted in Figure 2-25 is provided in the CD[1] that accompanies the book. Cable data shown in the table could be dated and is provided here for exercise purposes only. Actual values should be gathered from cable and conduit manufacturers and plugged into the calculator. Fill calculations are based on Rigid Metal Conduit (RMC) values listed in the NEC.[2]

---

1   For an in-depth look on how to size conduit using the Microsoft® Excel-based tool shown in Figure 2-25, a CD-ROM titled "Software Tools for Instrumentation and Control Systems Design" is included with the book.

Notice that two four-conductor, 16 AWG, 600 volt, unshielded cables were selected for this conduit. The calculator checks the number of cables being pulled and issues the cautionary 31% fill guideline that is appropriate for two cables. Then, the fill is calculated for each of the most-used conduit sizes. In this case, a 1-inch conduit will suffice.

d. Generating Physical Drawings (Location Plans)

Cable and conduit routing information is typically provided in a drawing as well as in a schedule. In some cases, if properly designed, the drawing can incorporate enough information to make separate schedules unnecessary (see the example in Part III). It should be noted that these drawings are rarely intended to be perfectly accurate. Rather, they are usually diagrammatic, providing a reasonable approach to routing the cable and conduit. The construction team will sometimes find better paths to follow, but it should be possible for them to at least group the cables as shown.

Whatever the format, these drawings are typically called location plans or conduit arrangements. At minimum, these drawings provide a physical view of an area from the standpoint of the cable and conduit and raceways. These drawings, first and foremost, show the location of process equipment, instrumentation, junction boxes, and other major artifacts relative to the control system. Next, they depict the conduit connections to these devices and possible routing paths, including tray and raceways.

It is possible to make the conduit numbers meaningful by embedding the plan drawing number into the routing number. For example, Figure 2-26 has a fragment of a conduit schedule in which a conduit number C2 is a 2" galvanized steel conduit section that runs from Tray T2 to conduit section C2a. Contents in this section of conduit are cable types W1, W5 and W8, which represent a 25% Area Fill. The cable types Wx references will point to a Cable Table somewhere else in the drawing package, either in a table on the drawing itself, or referenced in a note on that drawing. Conduit numbers are shown without a prefix, which could either be a portion of the floor plan drawing number, or a physical location designator of some sort.

In Figure 2-27, the process area instrument plan drawing depicts the top view of a vessel, with various instruments depicted as numbered boxes. Conduit fittings (bodies) are shown as circled letters. A component schedule is used to provide detail about each instrument. Conduit routings are depicted graphically, with additional detail available in the component schedule.

CONDUIT SCHEDULE

Conduit Number	Size (in)	Type	Len (ft)	Source	Destina.	Contents	Fill (%)
C2	2	GS	8	Tray T2	Cond C2a	W1/W5/W8	25
C2a	1.5	GS	4	Cond C2	JB-2	W1	40
C3							

FIGURE 2-26. SAMPLE CONDUIT SCHEDULE

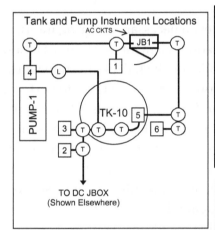

COMPONENT SCHEDULE				
STA#	TAG#	DESCRIPTION	ELEVA.	CABLE
1	LSL-47	LO LEVEL SW	6'-0"	4/C US #14
2	PT-48	PRESS XMTR	4'-0"	TWPR OS #18
3	PCV-48	CTRL VALVE	8'-0"	TWPR OS #18
4	HS-15B	HOA HANDSW	4'-0"	4/C US #14
4	ST-15	SPEED XMTR	4'-0"	TWPR OS #18
4	TT-15	BRG TEMP RTD	3'-0"	TWPR OS #18
5	LT-10	LEVEL XMTR	15'-0"	TWPR OS #18
5	LSH-10	OIL PRESS XMTR	15'-0"	TWPR OS #18
6	HV-10	FILL VALVE	18'-0"	2/C US #14
6	ZSC-10	CLOSED LIMIT SW	18'-0"	2/C US #14

NOTES:
1) Numbered boxes depict instruments or groups of instruments. Conduit bodies are represented by the "circle" symbol with the body type letter designator.

FIGURE 2-27. SAMPLE INSTRUMENT ARRANGEMENT

Each process area plan drawing should have associated cable and conduit schedules that provide detailed information about each conduit routing that originates from that drawing. These schedules may be shown on the drawing itself, or they may be separate documents such as those previously described. In either case, information presented includes routing number, contents of the routing (i.e., number of cables and types), size of the conduit, destination cabinet, and destination plan drawing.

2. INTRA-CABINET WIRING

When external (inter-cabinet) multi-conductor cables arrive at a panel, the cable insulation is stripped revealing individual conductors that must be landed on terminals. Interior wiring is then done using single-conductor wires to effect point-to-point connections from these terminals to either internal devices or other terminals. This kind of wiring is inter-cabinet wiring. This type of interior wiring is frequently done using hookup wire, sometimes referred to as "SIS" wire. SIS wire is a specific type of single-conductor wire that is most commonly used for switchboards in the power industry and other places. It is characterized by being extremely flexible and easy to work with inside panels. The specification for SIS wire is UL3173.

The intra-cabinet wiring activity can be extremely time-consuming for the constructor. But it can be made easier by a design product called a wiring or intra-connection diagram. The intra-cabinet wiring diagram presents wiring intra-connection information in a graphic format that is very useful to the installer. Its usefulness diminishes somewhat after construction, so it is sometimes considered to be merely a construction document. As such, this drawing is sometimes not very well maintained after startup. This makes it a favorite target for elimination for those trying to reduce engineering cost.

So, given the cost, are such wiring diagrams even necessary if wiring information is sufficiently detailed on the loop sheet or ladder elementary? In most cases, the answer is yes. Unless it is a very simple installation, installers forced to work solely from the loop sheet must gather and interpret many loop sheets to perform this task. They will most likely distill information from these sheets into a rough sketch that can be used as a reference during termination. Doing this makes the task much more difficult, fosters mistakes, and requires more construction supervision than would otherwise be the case. Negative effects from this practice may be noted at startup. The office environment, with its availability of database tools and other aids, is the best environment for packaging this information.

In any case, the designer should remember that the ultimate purpose of the drawing is to communicate wiring information to an installer, and possibly to a maintenance technician downstream. Selecting a format that makes the design process simpler might not translate very well to the field. For example, it is possible to generate wiring information using a coordinate system. Each wired device terminal has the drawing coordinate of the other end of the wire. The installer must interpret the coordinate to terminate the wire. This method is somewhat easier to draw but is not very user friendly.

a. **Generating a Wiring Diagram**

    The most effective approach for laying out the wiring diagram is to start from a physical cabinet mounting arrangement that shows the location of each terminal strip, switch, relay, transmitter, and so on, mounted on it. The devices should be depicted as viewed from the back of the panel by the installer (see Figure 2-28). Refer to Chapter 13 for practical examples.

b. **Relays, Switches, and Timers**

    For a controls designer to be versatile and to turn out quality work, a basic understanding of the terminology is important. Relay circuits, in particu-

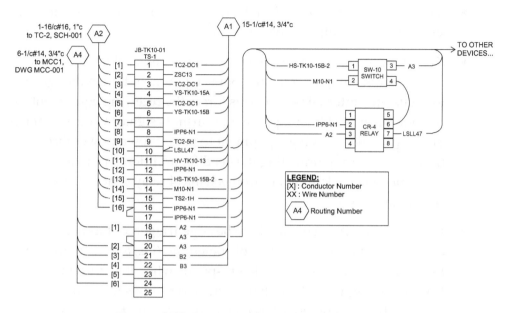

FIGURE 2-28. INTERCONNECTION WIRING EXAMPLE

lar, are becoming anachronistic, except in voltage or current isolation circumstances. They are still used, however, and some of the terminology that evolved during that era is still pertinent in today's computerized environment. Therefore, many of the definitions provided in this section deal with relays and switching devices. Other terminology is included as a means of clarification.

- *Bit:* A binary element that can have the value of either 1 or 0 and is usually associated with computer programming. A bit can be used as a flag to denote status, or as a PLC signal to energize a relay or solenoid. There are eight bits in a byte and either 16 or 32 bits in a word.

- *Coil:* A coil is a component of a relay. Relay coils provide the switching action. They consist of an electromagnet and a hinged ferrous plate to which electrical contacts are mounted. As current passes through the coil, the plate is drawn toward the electromagnet, which pulls the contact wiper to a new position. Coils also are used on occasion to reference discrete outputs in a PLC program – virtual coils.

- *Contacts:* The parts of a switch or relay that alternately close to pass electrical current and open to stop its flow. A set of contacts consists of one common and either one normally open contact or, more typically, one normally open and one normally closed. The latter configuration is sometimes referred to as "form-C" configuration.

- *DPDT (double-pole, double-throw):* Switching terminology that implies a switch or relay with two form-C contact sets.

- *Dry Contacts:* This term describes an unpowered contact closure that is isolated from any power source. Some equipment transmits signals using its own internal power. This forces the receiving equipment to provide isolation. Other equipment transmits signals via dry contacts. This allows the receiving equipment to pass its own power through the contacts, thus eliminating the need for additional isolation. The term *dry contact* relates to electromechanical relay contacts, not solid-state relays.

- *Edge-Triggered:* A circuit is said to be edge-triggered if its action depends on the change of state of something (i.e., a transition) rather than a steady state. A circuit that responds when its input transitions from low to high, as when a light switch transitions from "off" to "on," is said to be "leading-edge" triggered. A circuit that responds when its input transitions from high to low is said to be "trailing-edge" triggered.

- *Flags:* A generally accepted programming term to describe a binary bit used to communicate status. By monitoring a flag, you can learn something about the program or system. Flags are raised and lowered, as opposed to bits, which are set and cleared. Raised and Set are the same logically, but if you set a bit, then you expect an action, if you raise a flag, then you are communicating status.

- *Form-A:* A generally accepted term to describe a set of contacts with a common (COM) terminal switched to a normally open (N.O.) contact (Figure 2-29). This is classified as a single-pole, single-throw (SPST) configuration.

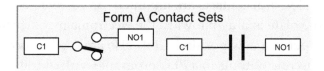

FIGURE 2-29. FORM A CONTACT SET (SPST – NORMALLY OPEN)

- *Form-B:* A generally accepted term to describe a set of contacts with a COM terminal switched to a normally closed (N.C.) contact (Figure 2-30). This is also classified as an SPST configuration.

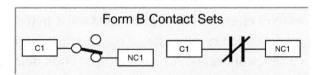

FIGURE 2-30. FORM B CONTACT SET (SPST – NORMALLY CLOSED)

- *Form-C:* A generally accepted term to describe a set of contacts with a COM terminal switched to two destination terminals, one N.O. and one N.C. (Figure 2-31). This is classified as a single-pole, double-throw (SPDT) configuration.

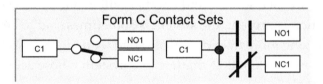

FIGURE 2-31. FORM-C CONTACT SET (SPDT)

- *Ice-Cube Relay:* This term refers to the physical appearance of a class of low-cost relays used in most low-current (<10 amp) logic applications. An ice-cube relay is enclosed in a clear plastic cube. The contacts are usually rated for 10 amps, and come in either a DPDT or 3PDT (three pole, double throw) configuration. These relays require a separate plug-in base, which provides termination screws for connecting external wiring.

- *Interposing Relay:* A term describing the function of a relay when its only purpose is to convert one signal source type into another. If a digital signal is available as a 110 VAC (Volts AC) form, but can only be read if it is 24 VDC (Volts DC), then an interposing relay is used. The coil of this example relay would be operated by the 110 VAC signal, and its contacts would switch 24 VDC.

- *Normally Closed:* The set of contacts on a switch or relay that make electrical contact when in the shelf state (i.e., non-activated).

- *Normally Open:* The set of contacts on a switch or relay that do not make electrical contact when in the shelf state.

- *Poles:* See Switch.

- *Propagation Delay*: In terms of a device, it is the time it takes the device to respond to a change in its commanded state. In terms of a system, it is the total delay that results when a signal must propagate, or "ripple" through a series of events. A relay's propagation delay is the time it takes to switch its contacts from one position to the next. With electromagnetic relays, this time is less when energizing than when de-energizing due to the residual magnetism that must be overcome to de-energize.

- *Race Condition*: A situation that occurs when two devices are driven to function at the same time logically, but due to real-world conditions, when one attains the desired state, it prevents the other from attaining its desired state. Frequently, this is not a repeatable condition, but one will "win" in one case, and another will in another case. When a race condition is present, the system is sometimes said to be metastable, i.e., having more than one stable state for a given condition. The result could be a system that chatters or a system whose actions are not repeatable.

- *Relay:* In its most common form, a relay is an electromechanical device made up of a coil wound around an iron core to form an electromagnet, a hinged plate called an armature, and up to several sets of contacts (poles) (Figure 2-32). The armature is mechanically linked from one to several mechanical arms, each with a contact pad at its tip. When the coil is de-energized, the spring-loaded arms pull the armature away from the coil. This is called the relay's shelf state, allowing current to pass from the common terminal through the spring arm to the N.C. terminal. The magnet, when energized, attracts the armature, which in turn causes the mechanical arms to move up, thus placing them in their energized state. Current then flows from the COM terminal to the N.O. terminal. Similarly, N.C. terminals open, interrupting the current flow.

- *Relay, Timing:* Logic circuits frequently require some form of timing function. This timing is achieved by the use of relays that have some onboard intelligence. An input signal (trigger) is monitored to start the timing function. Once a trigger is detected, one of several actions can result, based on the type of timer. Most timers today can be configured to provide any of the timing functions. A timer can delay its response to the arrival of the trigger signal (see Timer, On-Delay), delay its response to loss of the trigger signal (see Timer, Off-Delay), or respond to the trigger being present for a preset duration (see Timer, Interval).

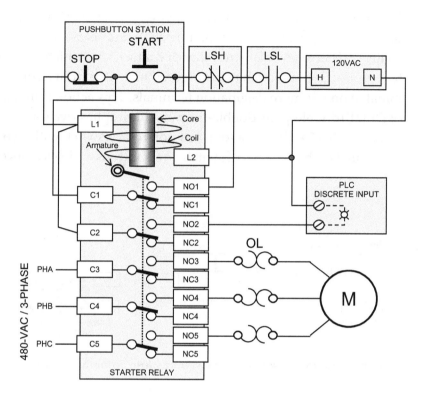

FIGURE 2-32. 5-POLE RELAY USED AS A MOTOR STARTER (SHOWN IN SHELF STATE, WITH INTERLOCKS, OVERLOADS, AND PLC INPUT)

Most timing relays are manually adjustable (knob or potentiometer with several specified ranges from which to choose. Contact configurations vary.

- *Ride-Through Timer:* See Timer, Ride-Through.

- *Shelf State:* The state of a device as it would be if it were on the shelf (i.e., unwired). A relay's shelf state is de-energized, with its N.C. contacts closed and ready to pass current.

- *Solid-State Relay:* An electronic device made up of a transistor or, for higher current applications, an SCR (Silicon Controlled Rectifier). There is generally a single output, which causes this device to operate as an SPST relay. One advantage is its small size and low power consumption. A disadvantage is the fact that a solid-state relay always passes some current, even when in the off (open) state. This off-state leakage current is sometimes high enough to prevent a load from releasing when the relay moves from the on state to the off state. Sometimes a shunt resistor is needed to reduce the amount of current forced through the load.

- *Switch:* In its most common form, a switch is a mechanical device made up of a lever, pushbutton, or rotating shaft and its associated sets of contacts. There can be any number of contact sets, which are referred to as poles. Each pole can be configured to switch from a common terminal to one or more destination terminals. This action is in the form of DPDT (double-pole, double-throw). This implies two sets of form-C contacts. DP4T indicates a four-position rotary switch with two poles. There are two basic types of switch: Pushbuttons and Selectors (Figure 2-33):

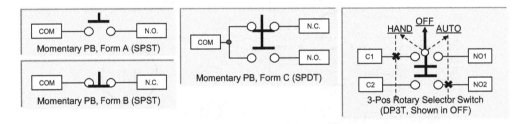

FIGURE 2-33. COMMON TYPES OF SWITCHES AND THEIR DIAGRAMS

- Switching Action:
  - *Make-Before-Break (Figure 2-34):* A type of switch contact that *does not* interrupt current flow when being switched from one "on" position to another.
  - *Break-Before-Make (Figure 2-34):* A type of switch contact that *does* interrupt current flow when being switched from one "on" position to another.

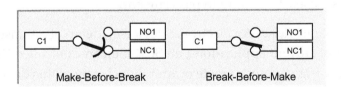

FIGURE 2-34. TYPES OF CONTACTS

- *Timer, Interval (Figure 2-35):* An interval timer is powered at all times, allowing it to continually monitor a signal input. If the input goes from low to high (logically true), the timer is activated, causing its output contacts to change state. The timer's output remains activated for a calibrated interval. After that interval, the timer's output is de-activated, regardless of the state of the input signal. Only when the input signal is off, and the timing interval is past, does the timer arm itself again, waiting for another low to high transition on its input.

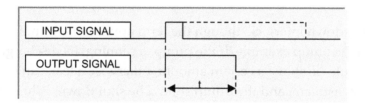

FIGURE 2-35. INTERVAL TIMER TIMING DIAGRAM

- *Timer, Off-Delay, or Time Delay on De-Energize (TDOD) Relay:* A timing relay that energizes when a signal is applied and remains active after the signal is removed. Since the TDOD timer's electronics must be powered at all times, this relay coil requires at least three wires. Figure 2-36 shows the timing diagram of a typical TDOD relay.

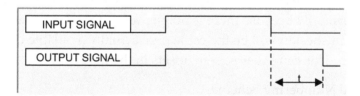

FIGURE 2-36. TIME DELAY ON DE-ENERGIZE (TDOD) TIMER TIMING DIAGRAM

- *Timer, On-Delay, or Time Delay on Energize (TDOE) Relay (Figure 2-37):* A timing relay that begins timing when a signal is applied and energizes only if the signal remains present throughout the timing period and beyond. This relay immediately de-energizes after the signal is removed. Since the TDOE timer's electronics are only active (timing) when the signal is present, a third wire such as that used on the TDOD relay is not required.

- *Timer, Ride-Through:* A timing relay that is used to mask, (cover up) a condition for a period of time. This type of relay is most commonly used when starting up a pump that uses low discharge pressure as a

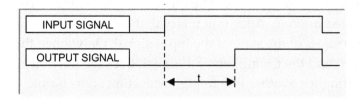

FIGURE 2-37. TIME DELAY ON ENERGIZE (TDOE) TIMER TIMING DIAGRAM

shutdown interlock, though the term applies in all similar situations. In the pump example, if the pump is running, its discharge pressure should be above a certain amount, otherwise a leak can be assumed downstream, and the pump should be shut down. When starting that pump, however, the pressure will always be low, preventing it from running. Therefore, a ride-through timer is used to mask the low pressure long enough for the pump to get up to speed.

c. **Wire Numbering**

A wire is an electrical conductor that can either be a single-conductor standing alone, or a single conductor as part of a multi-conductor cable. Each wire in an installation should have a unique designation, either with a unique wire ID number (can be alpha-numeric), or with a color code mated to a cable ID. Wire numbering is more prevalent than designing by color, primarily because there is no guarantee that the cable type being used in the design will be the cable type actually available during construction. This can cause severe upsets in the field.

1. Wire Numbering Schemes
   Numbering schemes can vary widely and still be effective. The only truly important things are that the wire number be unique and that the scheme be consistent throughout the design package. There is some added value if the wire number conveys some useful bit of information, such as an encoded drawing number reference or other vector to cabinet or instrument.

   One method, if the drawing package is in a loop-sheet format, is to use the loop or instrument number as the wire number prefix. Then, the designer may simply add a sequential suffix to each wire. If the loop number happens to be embedded in the drawing number, then the field electrician is able to quickly ascertain the proper drawing to pull when troubleshooting.

Another method, if the drawing package is in a ladder elementary (schematic) format, is to use some element of the schematic drawing number and add two digits to it. For example, if the drawing number is 29A-4296-1, then a good wire number range might be 4296-1-XX where XX is a number between 00 and 99. Ladder elementaries typically have such numbers listed in columns outside the wiring area of the drawing (Figure 2-38). In this ladder scheme, wires drawn near a particular ladder number assume that numbering sequence. As devices are crossed (see below) and numbers change, a letter suffix may be added to make each wire uniquely numbered. If the circuit crosses a boundary between two ladder numbers, the wire number sequence assumes the new number and begins again with the A suffix.

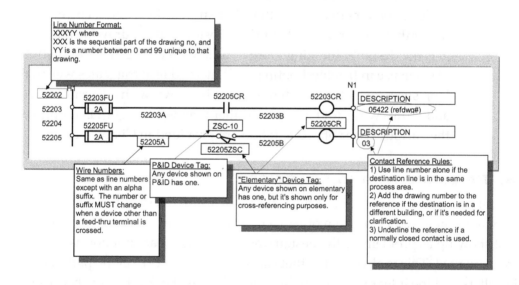

FIGURE 2-38. SAMPLE LADDER ELEMENTARY FORMAT

2. Wire Numbering Rules of Thumb

   When must a wire number be changed? A wire number must be changed only when it crosses a device of some sort. Fuses, relays, resistors, and switches are some examples of devices. Feed-through terminals are *not* considered devices because they do not change the characteristics of the signal.

   When must a wire number *not* be changed? A wire number must not be changed when the wire exits as a node. An electrical *node* is an electrical junction with any number of wires attached, all with the same voltage present. A node is typically represented by a black dot. Two electrical lines crossing does not necessarily imply they are the same node unless a black dot is present at their junction. In such cases, wires leaving a node will have the same wire numbers as those entering the node. Exceptions to this rule are generally noted on the drawings. A dot on a drawing may or may not imply a feed-through terminal, but in either case, the number should not change.

What is a good practice to employ when assigning wire numbers? Sometimes designers use the left-to-right method. In this method, letter suffixes change as the circuit moves from left to right across the page. This is a reasonable approach but is somewhat limited. A better method is to increase the letter suffixes in the order of current flow. In some cases, the designer might find the information presents better with current flowing from right to left across part of the drawing. This designer prefers to increase the letter suffixes in order of current flow rather than the circuit's left-to-right position on the drawing page. This gets the designer thinking more in tune with the circuit and might be of some use in troubleshooting. This method also translates well to the loop sheet format in which the circuit is shown with adjacent source and return wires (out and back rather than left to right). Examples of the various styles of presentation are provided in Chapter 14.

B. Failsafe Wiring Practice

Failsafe wiring practice is one of those topics that separates control system designers and electricians from other technical specialties. This is one of the areas that show up as problems if the design/installation team are not normally controls oriented. This is also an area that causes a lot of rework on the part of the installers and the integrators when they meet during checkout onsite because it requires a lot of crosstalk in order to get in sync.

To enter a discussion of the merits of failsafe wiring, we need to come to an understanding of some of the basics terms:

- The term failsafe implies fault-tolerant, as opposed to fault-free, operation. In other words, a device or system is allowed to fail, but only to a known safe state. An example of a failsafe signal is one that is wired to generate an alarm if power flow is interrupted to an alarm detection device, such as a relay that drives a horn or a "system OK" light.

- The phrase failsafe wiring refers to a design practice that causes an interruption of current flow when the sensing device is in any condition other than its normal operating condition. For a wiring scheme to be failsafe, the device in question needs to be energized when process conditions are normal.

- As was introduced earlier, the term *shelf state* refers to the state of a device's output switches as they would appear on the shelf, with the device being unwired or unpowered. Depiction of a relay's shelf state would show its contact sets in the de-energized state, with N.C. contacts in the closed position, ready to pass power if connected to a source, and N.O. contacts shown in the open position, blocking current flow. This is the default condition typically shown in schematic and connection drawings.

The term *normal operating condition* should not be confused with the terms *normally open (N.O.)* or *normally closed (N.C.)*. Normal operating conditions are those in effect with the equipment running normally, and the process variable being measured is within tolerance. Turning the equipment off, having the process variable go out of tolerance, or having any other component in the system fail will cause a loss of voltage (logical zero) at the annunciator or PLC, causing the alarm to be generated. Note that such an alarm does not necessarily indicate that an alarm condition exists in the process (e.g., tank level too high), but that either the alarm condition exists or the alarm condition is no longer being monitored.

Figure 2-39 depicts the same simplified motor control circuit previously shown in Figure 2-31 but drawn in a more schematic format. In this circuit, a motor will start or stop based on an operator pressing the spring-loaded start or stop pushbuttons. The operator presses the start button, the relay energizes, and then the operator can release the button as the relay has sealed a set of contacts around the pushbutton. If, however, the tank level is not in range, the motor starter coil will not energize because the level switches will not permit current to flow to the starter coil, and the motor will not start. If, after the motor starts, the level subsequently changes out of range, the relay will de-energize. The motor will not restart, even if the level returns to normal, until the operator presses the button.

This circuit shown in Figure 2-39 possesses all the key elements of a failsafe circuit. The end device operates only under prescribed process conditions <u>and</u> prescribed electrical conditions. If anything happens to the power supply or any other part of the circuit, rendering it inoperable, the relay de-energizes, and an alarm is generated. The only circumstance that would cause this circuit to fail in its function are mechanical problems, with either the relay contacts fusing together (which is rare now that most relays are encased and better protected from moisture) or the level switches failing to respond to changes in head pressure (level) as they are designed to do.

Most of the time, failsafe circuits use normally open contacts for interlock chains. In the case discussed above, however, a set of normally closed contacts was properly used. This was proper because the level switches used here are dumb—unpowered, non-electronic switches that switch the state of their outputs strictly based on pressure. As the level in the tank rises, so does the pressure at the sensing point, which is fed to each switch via tubing. The increased pressure causes a bellows to inflate inside the switch body. Eventually, the bellows inflates to a point where it exerts enough force on a contact set to overcome its mechanical reluctance (a mechanical setting that can be adjusted, or tuned, to a particular pressure), and the switch activates. So using one switch as a low level switch and another as a high level switch depends simply on where you mount the switches and how you adjust their response to pressure changes. The fact that the switch has a Form-C contact set (Figure 2-30) allows the same model switch to be configured for failsafe operation, using the N.C. set for high levels and the N.O. set for low levels. With the tank empty, the low-level switch's N.O. contacts are open, removing the interlock for the motor. As the level rises, the low-level switch operates its N.O. contacts, closing them and enabling the circuit. The next time the operator presses the start button, the motor will start and run until the tank empties or until the level reaches the high setting, at which time the relay de-energizes. The operator cannot restart the motor until the level falls below the high-level point.

Most modern electronic level switches give the installer an option for how a switch's output should behave, so those can and should be configured to always use N.O. contacts since the outputs will only stay energized if the unit has power and the process conditions being monitored are within tolerance. In all cases, the PLC or annunciator looks for a loss of signal to signify an alarm condition.

Whenever possible, failsafe wiring practice should be employed on feedback signals (digital inputs), non-mission-critical control relays, and annunciator systems. This gives the plant operators knowledge that the sensing or alarm system is in fact monitoring the process and is standing ready to inform them of upset conditions. Judgment does need to be exercised, however. On some control circuits that are mission-critical, it might be better to let the circuit fail unnoticed than to bring down the plant due to a faulty relay. But the default should be to make all circuits failsafe. This causes an increase in power consumption because the load is always energized. Nonetheless, the personnel and process safety considerations usually outweigh the relatively minor economic ones.

To summarize, the following are some rules of thumb for failsafe wiring practice:

- If the sensing device is a dumb switch (such as a float switch) employed as a high-process alarm (e.g., high temperature, high level), then its normally closed contacts should be used to support failsafe operation. Why? The switch will not change from its shelf state until it detects an alarm condition. So it needs to pass power when in its shelf state and when the process is in its normal state.

- If the sensing device is a dumb switch, employed as a low-process alarm (e.g., low temperature, low level), then its normally open contacts should be used to support failsafe operation. Why? The switch will change from its shelf state as soon as the process variable (e.g., temperature, level) reaches its normal operating condition. If the process variable falls below the alarm point, the device will return to its normally open shelf state, and the circuit will de-energize.

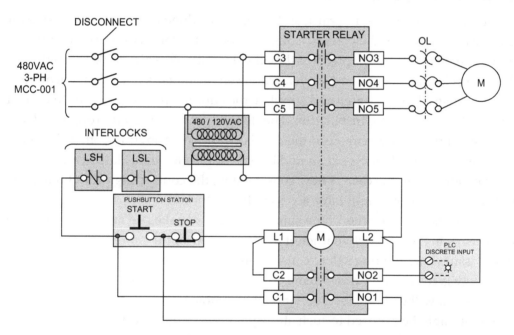

FIGURE 2-39. FAILSAFE INTERLOCK CHAIN (DEVICES SHOWN IN SHELF STATE)

- If the sensing device is electronic, its normally open contacts will generally support failsafe operation because it will probably be configurable. Most sensing devices today are electronic, and most of them provide a user-configurable setting that allows the device to be configured for failsafe operation. Whenever possible, the normally open contact should be made to close during normal operating conditions. It is wise to place a note on the loop sheet to that effect to remind the installer of the need to make that field adjustment.

C. Hazardous Area Classification and Effects on Design

Those who work in industry, particularly in certain sectors of industry, will invariably work in hazardous areas at some point in their careers. It is important to realize processes have been developed, through trial and error in many cases, that bring the risk to personnel and equipment to within acceptable limits. No system or process can guarantee a risk-free environment when it includes dusts, fibers, gases, or volatile fluids. But if the guidelines are followed, as published by the NFPA and other agencies, the risk is minimized.

1. Hazardous Locations

What exactly constitutes a hazard, and what can be done to render an area safe for the use of electrical and electronic components? A thorough description of the NEC[2] can be found in Chapter 5 of the NEC Handbook, articles 500 through 517. These articles classify the various hazardous environments one might encounter in the industrial process area and stipulate the equipment and practices that must be used to avoid unsafe installation. The designer must have a thorough understanding of these requirements if the area is classified as anything other than General Purpose. The plant owner should be able to provide area classification information. Or, in the case of new construction, the contracting project manager or process engineer should have access to that information.

Due to the complexities of the subject, this book makes no attempt to be comprehensive in its approach. Rather, it presents a few of the more common real-world issues surrounding safety considerations for design.

Safety is certainly a primary concern when designing control systems or subsystems that will be immersed in flammable atmospheres (NEC Class 1, Division 1) or that might be exposed to such atmospheres on a periodic basis (NEC Class 1, Division 2). Such designs must take into account not only normal operating conditions but also abnormal events that could occur.

Regulatory agencies, such as the Occupational Safety and Health Administration (OSHA), strive to prevent catastrophic events by promulgating regulations geared toward eliminating hazards. The design team should carefully observe these regulations during the design process.

The NEC defines and classifies the hazard boundaries in place in the process area.[5] The NEC also makes suggestions for applying remedies, or at least provides acceptable options. Insurance carriers, such as Factory Mutual, factor the myriad regulations and interpret the NEC to assist, if asked, in selecting the right remedy on a case-by-case basis. If a brand new design issue arises at a site, it is a good idea to discuss the issue with the customer's insurance carrier before finalizing the design. Insurance carriers are excellent resources, as they have a vested interest in maintaining a safe plant (to reduce their risk) and in their client's being able to make a reasonable profit (to maintain their contract). As a result, their opinion is often workable and affordable.

The NEC rating system is broken into three sections: class, division, and group. For example, a typical hazardous location may be described as a Class I, Division 2, Group D area. In this method of classification, class refers to the type of hazardous location (i.e., gases and vapors, dusts, fibers), division refers to the likelihood of the area being hazardous at any given moment, and group refers to the ignition temperature of the material, its explosion pressure, and other flammable characteristics.

Hazardous locations are classified by the NEC as normal or abnormal in terms of the likelihood of hazardous conditions being present. For example, most flammable vapors encountered in industrial processes are heavier than air. Such a vapor, if released, will collect in the lowest area. Therefore, any trench or ditch transiting from the containment area would also be classified as a hazardous zone (Figure 2-40).

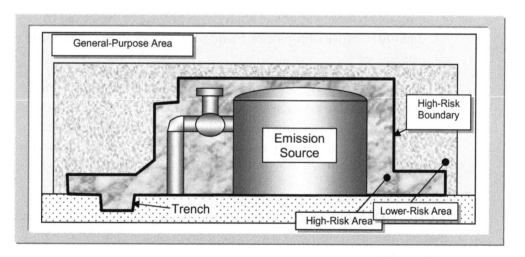

FIGURE 2-40. HAZARDOUS BOUNDARIES

Figure 2-40 depicts a fairly typical situation in which a vessel and its support equipment, such as a valve and transfer pipe, are considered emission sources. One or more hazardous area envelopes surrounds this equipment, the dimensions of which depend on the degree of hazard and the likelihood of a release. The hazardous area classification usually decreases with greater distance from the source. Any electrical equipment mounted within this boundary needs to be specially modified to eliminate the risk of ignition. Conduit seals are required for conduits that cross from one level of hazard into another in order to prevent transmission of hazardous gas, or vapor, to a remote location.

Refer to the current NEC articles for the latest definitions. The following is a basic overview of the hazardous area classifications, according to NEC Article 500.5:

a. **Class I**

   Class I locations are those in which flammable gases or vapors are or may be present in the air in quantities sufficient to produce explosive or ignitable mixtures.

   1. Class I, Division 1
      A Class I, Division 1, location is one in which it may be assumed that hazardous conditions always exist. Some of the areas listed in the NEC that would fall into this category are areas in which:

      - Ignitable concentrations of flammable gases or vapors can exist under normal operating conditions,

      - Ignitable concentrations of such gases or vapors may exist frequently because of repair or maintenance operations or because of leakage, or

      - Breakdown or faulty operation of equipment or processes might release ignitable concentrations of flammable gases or vapors and might also cause simultaneous failure of electrical equipment in such a way as to directly cause the electrical equipment to become a source of ignition.

   2. Class I, Division 2
      A Class I, Division 2 location is a location in which it may be assumed that hazardous conditions could occur on occasion. The NEC details these as areas:

      - In which volatile flammable liquids or gases are handled, processed, or used, but in which the liquids, vapors, or gases will normally be confined within closed containers or systems in case of abnormal operation of equipment;

      - In which ignitable concentrations of gases or vapors are normally prevented by positive mechanical ventilation and which might become hazardous through failure or abnormal operation of the ventilating equipment; or

- That are adjacent to a Class I, Division 1, location and to which ignitable concentrations of gases or vapors might occasionally be communicated unless such communication is prevented by adequate positive-pressure ventilation from a source of clean air and effective safeguards against ventilation failure are provided.

3. Class I, Division x, Group A, B, C, or D
   The group designation provides some idea of the flammability or explosivity of the material in question in descending order of its minimum igniting current ratio, or its maximum experimental safe gap rating. Materials listed in groups A through D are all gases, vapors, or vapor-producing liquids, which will always be used as Class I category modifiers.

b. **Class II**

   Class II locations are those that are hazardous because of the presence of combustible dust.

   1. Class II, Division 1
      A Class II, Division 1, location is a location in which:

      - Combustible dust is in the air under normal operating conditions in quantities sufficient to produce explosive or ignitable mixtures;

      - Mechanical failure or abnormal operation of machinery or equipment might cause such explosive or ignitable mixtures to be produced and might also provide a source of ignition through simultaneous failure of electrical equipment, through operation of protection devices, or from other causes; or

      - Combustible dusts of an electrically conductive nature may be present in hazardous quantities.

   2. Class II, Division 2
      A Class II, Division 2, location is a location in which:

      - Combustible dust is not normally in the air in quantities sufficient to produce explosive or ignitable mixtures, and dust accumulations are normally insufficient to interfere with the normal operation of electrical equipment or other apparatus, but combustible dust may be in suspension in the air as a result of infrequent malfunctioning of handling or processing equipment, and

- Combustible dust accumulations on, in, or in the vicinity of the electrical equipment may be sufficient to interfere with the safe dissipation of heat from electrical equipment or may be ignitable by abnormal operation or failure of electrical equipment.

3. Class II, Division x, Group E, F, or G
   Materials that are listed in groups E, F, and G are all dusts. Group E is metal dusts, Group F is carbonaceous dusts, and Group G is miscellaneous dusts.

c. **Class III**

Class III locations are those that are hazardous because of the presence of easily ignitable fibers or flyings, but in which such fibers or flyings are not likely to be in suspension in the air in quantities sufficient to produce ignitable mixtures.

1. Class III, Division 1
   A Class III, Division 1, location is a location in which easily ignitable fibers or materials producing combustible flyings are handled, manufactured, or used.

2. Class III, Division 2
   A Class III, Division 2, location is a location in which easily ignitable fibers are stored or handled other than in the process of manufacture.

2. EXPLOSIONPROOFING

If a field location is classified as hazardous (e.g., Class I, Division 2), it is possible that the electrical and electronic enclosures and equipment in that area will need to be rated explosionproof, or purged with nitrogen, air, or some other nonflammable gas. Purging is discussed in a later section.

The term explosionproof is a bit misleading. A more accurate characterization for such an enclosure would be explosion-containing, for that is exactly what it does. Semantics aside, an explosion-proof enclosure (as defined in the NEC) is built to contain any explosions that might originate within it. Small baffles in the form of bolt threads prevent the hot combustion gases from exiting the enclosure until they have cooled sufficiently to prevent propagation of the explosion to the surrounding atmosphere. Thus, the enclosure is very bulky. Its metal walls are thick enough to withstand an explosion and provide enough baffle surface area to cool the gases as they expand outward.

Such an explosion, however localized, would tend to be harmful to the contents of the enclosure and must be avoided. To accomplish this, the various components inside the enclosure should be properly rated on an individual basis whenever possible. In some cases a "non-incendive" rating would be acceptable. In others, only an "explosion-proof" rating will do. Consult the NEC for clarification.

3. Intrinsic Safety

Intrinsically safe circuits are those in which electrical energy is limited to levels insufficient to initiate combustion. Per the NEC (Article 504.2), this condition occurs in a "device that will neither generate nor store more than 1.2 volts, 0.1 ampere, 25 milliwatts, or 20 microjoules." This specification can be met by the inherent characteristics of a device, such as a thermocouple (i.e., a "simple apparatus") or by virtue of some energy-limiting device, such as an intrinsic safety barrier (i.e., an "associated apparatus").

Intrinsically safe equipment may be installed in hazardous areas as if the areas were general purpose, provided the proper approvals have been acquired. Gaining approval means independent validation of the circuitry, either by a committee set up by the owner or by Factory Mutual or some other body. Consult NEC Article 504 for additional information on intrinsic safety.

4. Purging

If a field location is classified Class I hazardous, it is possible that field equipment, such as field junction boxes, analyzers, or other equipment, will need to be purged with nitrogen or some other inert gas. "Purging" simply refers to the practice of maintaining a slightly higher pressure inside the box than the surrounding atmosphere. This greatly reduces the likelihood that flammable or ignitable gases will enter the box, where a spark-producing element like a relay can precipitate an explosion. NFPA7 496 defines three purge categories:

a. **Class X Purge**

   A Class X purge reduces the classification within a protected enclosure from Division 1 to unclassified.

b. **Class Y Purge**

   A Class Y purge reduces the classification within a protected enclosure from Division 1 to Division 2.

c. **Class Z Purge**

A Class Z purge is perhaps the most commonly used purging scheme. This purge reduces the classification within a protected enclosure from Division 2 to unclassified.

Probably the greatest benefit of Class X or Class Z purging is neither the enclosure nor its contents need to be rated for hazardous atmospheres. An enclosure located in a Class I area that is purged to those specifications can have components inside it that are only rated for general-purpose use. This is of great benefit when compared to the explosion-proof enclosure method, in which all the components inside the enclosure must be individually rated as explosionproof. *Caution:* In either the purging case or the explosion-proof case, components mounted on the door of the enclosure, and thus outside the protective envelope of inert gas or an explosionproof container, must be properly rated for the outside conditions.

Refer to the NFPA specification for particulars regarding purging. The following list paraphrases some of the key elements of the Class Z purge specification:

- The protected enclosure shall be constantly maintained at a positive pressure of at least 0.25 Pa (Pascal) (0.1 inwc) above the surrounding atmosphere during operation of the protected equipment.

- Measures shall be taken to protect the enclosure from excessive pressure of the protective gas supply. (NOTE: This implies a relief valve of some sort, and a regulator.)

- If positive pressure is not maintained, a suitable device for alarm or indication must warn of the failure or automatically de-energize power to the ignition-capable equipment.

- An alarm or indicator shall be provided to indicate failure of the protective gas supply. (The NFPA specification goes on to explain the types of alarms that are appropriate.) One key stipulation is that the alarm must be observable at a location that is "constantly attended."

- Adequate instructions shall be provided on the enclosure to ensure that the system can be used properly. Several labels are specified that must be mounted on or near the enclosure.

- Opening a purged enclosure allows the protective gas blanket to dissipate, as will any protection that it provided. So before opening a purged enclosure, one of two things must be done: (1) Remove power from the enclosure or (2) use a combustible gas analyzer to test the area.

- If the enclosure is opened under power, the combustible gas analyzer must remain in the area and in operation during the time the enclosure is open. If the enclosure is unpowered when opened, power cannot be restored to it until either a combustible gas analyzer is used to confirm safe conditions or the door is closed and several volumes of purge gas are forced through the enclosure. NOTE: This last requirement forced the advent of the "fast purge" valve, which may be opened during the pre-powerup purge period and then closed after the proper number of purge gas volumes have flowed through the enclosure.

If all this seems too complicated or risky, there are purge packages available commercially that will do the trick. Enclosures can be purchased pre-fitted with the appropriate purge fittings, so it is only necessary to mount the package on the enclosure and hook up the purge gas.

The comments made here give only a brief overview of purging and purge systems, so be sure to refer to the NFPA guidelines before attempting to design a purge system.

In closing: Operating within a hazardous area is sometimes unavoidable in process industries. If a HazOp or other form of recognized hazard analysis process is used to analyze the hazard risks, and proper steps are taken to accommodate those risks, then it is possible to work in hazardous areas with confidence.

D. Connecting to the Control System

I/O modules come in a variety of configurations called *form factors*. They consist of an electronics section and an I/O interface section. In a typical rack-mount configuration, the electronics section is powered from the backplane (motherboard) of the I/O rack. This power may or may not be distributed to the field devices. In most cases, field device power (also called wetted voltage) is from a separate source. In the case of most PLC installations, this source is a power supply that is entirely separate from the PLC itself. In the case of many DCS installations, the wetted voltage is internally distributed but is still separate from the electronics power distribution system.

From the manufacturer's standpoint, a major objective is to pack as many I/O points into a given amount of space as possible (achieving "high point density"). Another objective might be to be able to switch large loads. Since the two objectives are mutually exclusive (due to the heat dissipation requirements in the second case), the PLC or DCS manufacturer must provide his customer with several alternatives.

In the drive to increase point density, probably the single largest limit in low-power applications is the physical space needed for terminating field wiring. By internalizing as much wiring as possible, the space needed for field wiring terminations can be reduced. So internally bussed power structures evolved. Today, most manufacturers provide modules with a density of up to 32 points. This is accomplished by providing an internal signal "common" that gives each point either a path to power (sourcing) or a path to ground (sinking). By this means, it is possible to terminate only one field device wire to the module itself, rather than both wires that return from the field device. However, this practice increases the risk of having power distribution conflicts between wetted and unwetted power. Other wiring concerns should be considered, given the many configurations possible. This section describes some of the more important wiring basics.

1. Discrete (Digital) Wiring

The terms discrete, binary, and digital are interchangeable when used to describe on/off type circuits. Turning on a light by flipping a switch is an example of a digital circuit, whereas using a dimmer to modulate the light intensity is an example of an analog circuit. Digital circuits are composed of at least three elements: a power source, a signal source, and a load.

Figure 2-41 shows a typical lighting circuit, as you might find in your home. Current flows only when the switch is flipped. Since the circuit is either on or off, it is considered a digital circuit.

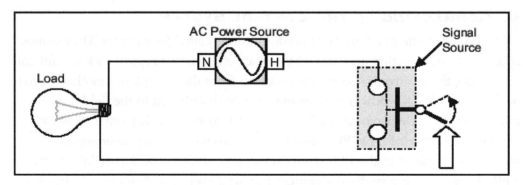

FIGURE 2-41. BASIC DISCRETE (DIGITAL) CIRCUIT

Today's on/off control is done through the PLC or DCS. Figure 2-42 is an overview of one discrete (on/off) circuit, showing the entire process from the power supply through the sensor and on to the PLC.

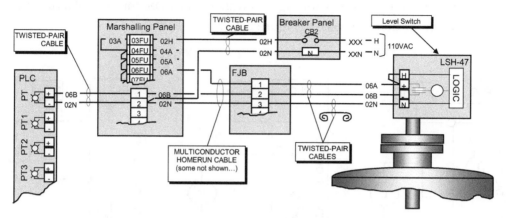

FIGURE 2-42. DISCRETE (DIGITAL) CIRCUIT-WIRING TECHNIQUE

In Figure 2-42, a level switch is mounted to a vessel. The switch is monitored by a PLC digital input module. The circuit is powered through a circuit breaker (CB2) in an instrument power panel. The main power feed is brought to a marshalling panel, where the power is split, feeding multiple fused circuits. Fuse 03FU is the main disconnect fuse, while the remaining fuses are distribution fuses. Fuse 06FU feeds our circuit. Hot (electrically live) wire 06A is passed to the field junction box (FJB) as one wire in a multi-conductor cable. This cable, sometimes called a homerun cable, is broken out at the field junction box (FJB), where, in this example, two unshielded twisted pair cables are fed to the end device, LSH-47. This leaves one spare conductor. The hot wire 06A hits the + terminal of the form-A contact and jumpers to the H terminal to power up the electronics of the switch.

The wire number changes across the relay contact to 06B. This wire feeds the signal back to the FJB, where the signal is passed back to the termination cabinet via the multi-conductor homerun cable. There the signal and neutral are paired and passed to the PLC module. Note that the return neutral wire, labeled 02N (since it is the return wire for CB2), is split to the PLC and the level switch. NOTE: It is always advisable to use twisted-pair wire when connecting to a PLC system. Twisted-pair cables exhibit excellent noise immunity, which is particularly useful when connecting to high-impedance loads, such as those found on PLC I/O modules. A high-impedance load can be particularly sensitive to noise since the attendant current is so low, and the amount of actual work being done is minimal.

That is it, in a nutshell. The following is a commentary on connectivity issues related to PLC signal wiring.

a. **Sinking and Sourcing**

The terms sinking and sourcing are used to describe the way a particular component in the circuit relates to power flow. These terms actually stem from the days of transistor logic. A transistor can be thought of as a simple switch for this discussion (Figure 2-43).

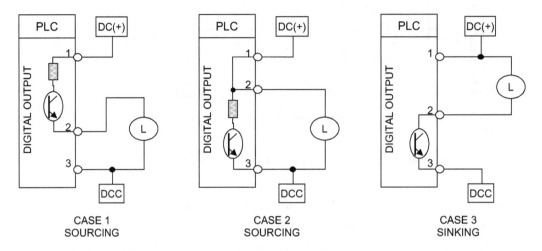

FIGURE 2-43. SIMPLE SWITCHING

This type of transistor requires a small resistance on its collector (the upper side) for current limiting. In the Case 1 example, the resistor is in place, with the load shown in series with the transistor's emitter. When the transistor conducts, current flows through the resistor, the transistor, and then through the load. This circuit was not used very much because the current divides across the internal circuitry, leaving less power available to drive the load and driving temperatures up in the I/O module.

Case 2 provides a more typical sourcing circuit, where the PLC output, by turning off, switches the full-load current to drive the load. When the output turns on, the transistor conducts, causing most of the current to be shunted through it, starving the load, and thereby de-energizing it. The downside of this configuration is that a small leakage current will continue to be present across the load, as a certain amount of current will continue to be directed across the load, though not enough, typically, to cause the load to stay energized. When troubleshooting, however, a small voltage will be detected across a de-energized load.

In the Case 3 example, the load is the collector resistor. When the transistor conducts, the load energizes. From the standpoint of the board electronics, this is a better configuration because most of the heat is dissipated by the load. The downside of this configuration is that "switching the neutral" is counter-intuitive and can be unsafe, as full voltage is present at both the positive and negative terminals of the load when it is de-energized.

For these reasons, Case 2 has evolved to become the most common output configuration. This sinking/sourcing concept can be extended to any circuit.

b. Circuit Protection (Fusing)

Most I/O modules are internally fused. However, that does not mean all that much to the user. While the internal fuse does limit damage to the module itself, in most cases the module still must be sent to the factory to be repaired. So the end result is the same to the user—a broken module.

As a result, it is good practice to add external fuses to each I/O point, with a rating just below the fuse rating on the module circuit board. While this limits the size of the load that can be driven directly by the module, the internal fuse and module are protected.

Caution: If internally fused discrete outputs are embedded in interlock chains, or if they are in circuits that depend on normally closed contacts to initiate safety actions, then another type of module that is unfused should be used. Or interposing relays could be deployed. It is possible to have the I/O point function normally (e.g., close its contacts, and report to the program that it has closed them) but still not pass power due to a blown internal fuse.

c. Digital Input (DI) Circuits

Digital input (DI) modules continually scan their input points for the presence or absence of voltage. If voltage is present, a 1 is written to a memory location. If voltage is absent, a 0 is written there. The required voltage type and magnitude are two of the factors that distinguish one DI module from another. Most DI points have high impedance, thus minimizing the amount of current absorbed, and so have a relatively minor effect on the power distribution system.

Each digital input point can be thought of as a lamp, one that is either on or off. DI modules can be electrically isolated point to point, or they can be grouped by internally bussing the I/O common. Most modules today are grouped, as grouping allows for higher density. As we have seen, point densities of up to 32 points per module are common in the grouped configuration.

Figure 2-44 shows two different DI modules. The first module internally busses the DC(+) side of the circuit. The I/O point then passes power to the field device. This type of module is called a sourcing module. This configuration is unusual. Switching the common side in the field is typically not done.

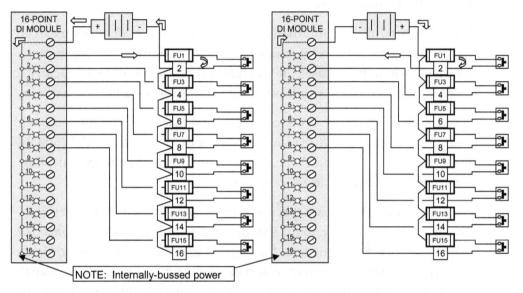

Current-sourcing digital input module   Current-sinking digital input module

FIGURE 2-44. SINKING AND SOURCING DIGITAL INPUT MODULES

If the module internally busses the DC common side of the circuit, then the module is considered a sinking module. The I/O point completes the path to DC common. This configuration is used in the vast majority of cases because it allows each I/O point to be individually fused near the power supply before power is distributed to the field device. In either case, the current flows in the same direction through the field-mounted switch.

d. **Digital Output (DO) Circuits**

Relay contacts are considered output devices because they force other devices to react when they change state. PLC digital outputs can be thought of as relay contacts. In many cases, that is just what they are. In others, the switching element may be a solid-state device of some sort. Even in that case, the relay analogy works as long as the designer remembers to consider leakage current. (Review the definition of Relay and Solid-State Relay in the Definition of Terms section if necessary.)

DO modules switch voltage on and off to cause an external device to change state. These modules are either "isolated" or "non-isolated." If a module is non-isolated, then it is either sinking or sourcing.

1. Isolated DO Circuits

   An isolated DO circuit (Figure 2-45) is one in which the power source can be isolated between I/O points. The source is not internally bussed. The cost is two terminals per point, so it is expensive in terms of real estate. In Figure 2-45, there are three sources of wetted power, with points 1, 2, 4, 5, and 6 being isolated from point 3 and points 7 and 8. In this example, AC is being fed to Point 3, while DC signals are on the remaining points. Doing this demonstrates the possibilities. In practice, it is a good idea to separate AC and DC signals if at all possible.

2. Non-isolated DO Circuits

   As with the DI PLC module, point density is an important feature of DO modules. As can be seen in the isolated module in Figure 2-42, isolation comes with a price. A 16-terminal module has a point density of only eight since two terminals are needed per point. By internally bussing a common, the point density can be improved dramatically. However, the result is a non-isolated module that places limits on the designer. Power sources must be managed. In most cases this is not a problem since extending PLC I/O power to the field device is feasible. However, if a field device must source its own signal, then an interposing relay must be added to the circuit to provide isolation.

Figure 2-46 shows two different digital output modules. The first internally busses the DC(+) side of the circuit. The I/O point then provides a path to power, making it a sourcing module.

If the module busses the DC common side of the circuit, as shown in Figure 2-46, Example 1, then the module is considered a sinking module. The I/O point completes the path to common. This type of module is rarely used today due to the common-side switching. Example 2 is far more common, as it puts the switching action ahead of the load in terms of current flow.

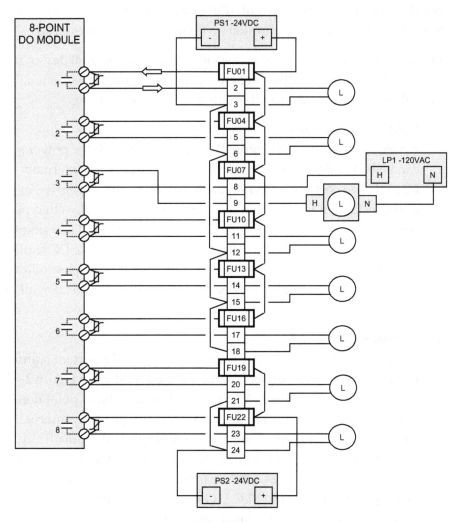

FIGURE 2-45. ISOLATED DIGITAL OUTPUT MODULE

2. ANALOG WIRING

Unlike the discrete (on/off) circuit, analog signals vary across a range of voltage or current. Taking the same vessel described previously in Figure 2-42, how would the wiring change if we replaced the switch with a level transmitter?

Figure 2-47 has the same circuit-breaker panel as in Figure 2-42, but now it is feeding a DC power supply. The power supply could be in its own cabinet, or it could be in the marshalling panel. In any case, DC power is distributed in the marshalling panel. A single fuse could power several circuits, or each circuit could be fused. The transmitter is fed +24 VDC at its positive terminal. The 4–20 mA current signal is sourced from the (-) terminal of the transmitter to the PLC. Cabling is twisted pair and shielded. The signal cable is numbered with the transmitter number, and the wires inside are numbered to provide power source infor-

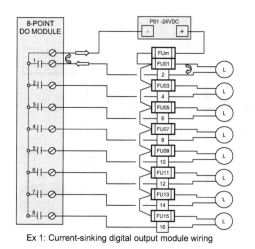
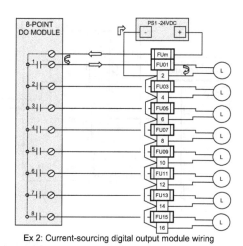

Ex 1: Current-sinking digital output module wiring

Ex 2: Current-sourcing digital output module wiring

FIGURE 2-46. SINKING AND SOURCING DIGITAL OUTPUT MODULES

mation. The shield is terminated in the marshalling panel, where all shields are gathered and terminated to a ground lug that is isolated from the cabinet.

Note: Care should be used to ensure that the shield is only grounded at one spot. Shields that are grounded in more than one spot may inject large noise spikes onto the signal. This condition is called a ground loop and can be a very difficult problem to isolate, as the problem is intermittent. A "quiet" ground should be used to ground all the shields at one point. A quiet ground is one that is either tied to a dedicated ground triad, or one that is tied to the center-tap of an isolation transformer. A noisy ground would be one that is physically located far from the transformer, and one that services motors, lights, or other noisy items.

That is the basic two-wire analog input circuit. The following is some specific information regarding the various analog possibilities:

a. **Circuit Protection (Fusing)**

   Analog circuits are always low voltage, usually 24 VDC. As a result, fusing individual analog circuits is not required for personnel safety.

   Also, most analog I/O modules have current-limiting circuits onboard. So fusing is generally not required to protect the modules. If these two conditions are true—and the designer should confirm this with the manufacturer—then per-point fusing can be avoided if desired. If a designer wishes to save money by not fusing every point, then grouping the circuits into damage control zones should be considered. For example, if there is a pump pair, a primary and a backup, instruments for the two should be in separate fuse groups to prevent a single blown fuse from taking them both out. For more information, see *I/O Partitioning* in the index.

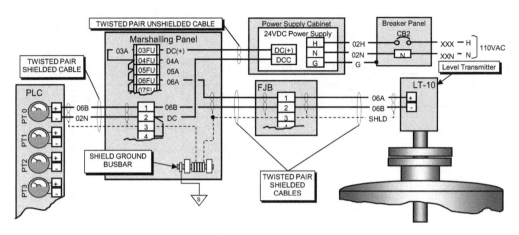

FIGURE 2-47. ANALOG CIRCUIT-WIRING TECHNIQUE

b. Noise Immunity

Analog circuits are susceptible to electronic noise. If, for example, an analog cable lies adjacent to a motor's high-voltage cable, then the analog signal cable will act as an antenna, picking up the magnetically coupled noise generated by the motor. Other sources of noise exist, such as radio frequency (RF) radiation from a walkie-talkie. Noise on an analog signal cable can cause errors in reading the value of the signal, which in turn can cause a multitude of problems in the control system. Some ways to mitigate noise include:

- *Twisted-Pair Cables:* Electronic noise may be greatly reduced by the use of twisted-pair cabling. Most instruments use two wires to transmit their signals. Current flows out to the device in one wire and back from the device in the other. If these wires are twisted, then the noise induced will be very nearly the same in each wire. The magnitude of induced current flow is identical in each conductor, but it travels in opposite directions, thus canceling out most of the noise.

- *Shielding*: A further refinement in noise rejection is shielding, i.e., the use of a grounded braid or foil shield around the conductors. As previously mentioned, the shield should never be grounded in more than one place to avoid ground loops. Most instrument manufacturers recommend grounding the shield at the field instrument. However, a better place to do it is in the marshalling panel. It is easier to verify and manage the grounds if they are in one place. Also, it is possible to ensure a good ground at that point.

- *Conduit:* A final refinement in noise rejection is grounded metallic conduit. This is rarely required, except for data communications cables and for particularly critical circuits.

c. Resistance Temperature Detector (RTD)

An RTD is made of a special piece of wire whose electrical resistance changes in a predictable way when the wire is exposed to varying temperatures. The material of choice today is 100 ohm platinum, though other types, such as 10 ohm copper, are sometimes used. For the platinum RTD, the rating is for 100 ohms at 0°C. Resistance changes with temperature are very small, causing voltage variations in the millivolt range.

RTDs are connected to a Wheatstone bridge circuit that is tuned to the RTD. But this tuning occurs on the bench. What about the field environment? We have already discussed the line attenuation difficulties inherent in millivolt signals (Chapter 4). This problem is overcome in the RTD circuit by the use of one or two sense inputs. These inputs help negate the effects of copper losses due to long lines and temperature variations along them and are additional wires that must be included in the RTD cable, hence the terms three-wire and four-wire RTDs.

d. Thermocouple

As we have discussed, a thermocouple exploits the electromotive force (EMF) that arises from changes in temperature affecting two dissimilar metals that have been laminated together. This EMF manifests itself as a millivolt (DC) signal. When certain combinations of these dissimilar metals are joined, a predictable curve of temperature to voltage results as temperature at the junction changes. The signal is measured at the open end of the two wires, and a millivolt-per-degree scale is used to convert the voltage to engineering units.

The thermocouple is thus a two-wire device. It is susceptible to radiated and induced noise and so is usually housed in a shielded cable if extended for a very long distance. The thermocouple signal is also susceptible to degradation due to line loss, so minimizing the cable length is desirable.

Also, it is important to use the proper extension wire. A thermocouple usually comes with a short pigtail connection to which extension wire must be attached. If a different wire material, such as copper, is used to extend the signal to the PLC, a spurious "cold junction" is created that causes a reverse EMF that partially cancels out the signal. Therefore, the proper extension wire should be used, or a device called a cold-junction compensator or ice-point reference needs to be installed between the copper wiring and the thermocouple wiring. Thermocouple I/O modules already have the cold-junction compensation onboard, so using the proper thermocouple extension wire is required.

Specific types of thermocouples exhibit different temperature characteristics. A type J thermocouple is formed by joining an iron wire with a constantan wire. This configuration provides a curve relatively linear between 0 and 750°C.[8] A type K thermocouple has a nickel-chromium wire mated to a nickel-aluminum wire, sometimes called chromel/alumel. The type K thermocouple spans a useful temperature range of -200 to 1250°C. Other combinations yield different response curves.

e. **0–10 Millivolt (mV) Analog**

Analog signals were first generated by voltage modulation. In the old days, a transmitter would generate a weak signal that had to be captured and then filtered and amplified so it could be used to move a pen on a recorder, or a needle on a gauge. The Achilles' heel of the millivolt signal is its susceptibility to electrical noise. This signal-to-noise ratio problem increases as a function of cable length. So the transmitter needed to be in close proximity to the indicator or recorder.

Millivolt signals today are, by and large, fed to transducers that convert the small signal to a current or to other media (like digital data values) less susceptible to noise and decibel (dB) loss before leaving the vicinity of the sensing element. However, some recorders and data acquisition systems still operate on the millivolt signal.

f. **4–20 Milliamp (mA) Analog**

The drive to overcome the line attenuation shortcomings of the millivolt signal resulted in the development of the 4–20 mA current loop. As a result of its greatly increased performance, this method of transmitting analog signals quickly became the industry standard.

Most field instruments on the market have a sensing element (sensor) and a transmitting element. The transmitter is tuned to the sensor, which may provide any type of signal from frequency-modulated analog to millivolts DC. Whatever the form of the signal, the transmitter interprets it and converts it to an output current between 4 and 20 mA and within that span is proportional in magnitude to the input. The process of tuning the output to the input is called *scaling*.

Thus, the transmitter becomes what is referred to as a variable-current source. Just as a battery, as a voltage source, tries to maintain a constant voltage, regardless of the amount of load applied to it, the current source tries to maintain a constant current (for a given input signal), regardless of load. Since current is common at all points of a series circuit, the problem of cable length—as noted as a problem with the millivolt signal—is nullified.

Of course, the ability of the device to force a constant current through a circuit can be overcome if enough load is applied. Therefore, the designer must know how much energy the current source is capable of producing. Generally, today's instruments are able to maintain 20 mA at a circuit resistance of 1000 ohms. Since a typical instrument has no more than 250 ohms of input resistance, it is possible to power several instruments from a single current source without needing an isolator. For example, a single transmitter should be able to feed its signal to a PLC, a chart recorder, and a totalizer at a cost of 750 ohms, plus the line resistance. This should still be within the comfort zone of a typical transmitter. Note: There are still instruments with 600 ohm ratings on the market, so the designer should always check whenever a complex circuit is contemplated.

To determine the energy available to the circuit, the designer must be able to identify the provider of that energy. That task is sometimes not as straightforward as it might appear, and the answer to the question will greatly impact the wiring of the circuit. There are two primary types of analog circuit, as described from the point of view of the transmitter. Transmitters with two wires are considered to be passive devices that sink current, while transmitters with four wires are active devices that source current.

Figure 2-48 depicts three temperature transmitters, each connected to different I/O points on the same PLC module. One transmitter is directly powered (i.e., four-wire), while the others are indirectly powered (i.e., two-wire). Each transmitter is connected to a control device—in this case, a PLC input.

From the PLC's perspective, all 4–20 mA current inputs are really voltage inputs. Resistors, either user-provided external ones, as shown here, or internal ones, are used to convert the current to a voltage. The computer points themselves are actually high-resistance voltmeters, which give them excellent isolation from the field devices and minimize additional loading on the input circuit.

The I/O points on the PLC are shown with internal power available for each point, so the module is capable of being the voltage source for the loop. The following is a detailed commentary on the differences between two-wire and four-wire devices:

1. Four-Wire Circuit
   As seen below, a four-wire transmitter is one that provides the energy to power the loop and generate the current-modulated signal. Most level transmitters, for example, are four-wire devices. Four-wire devices always have power connections in addition to the signal connections. Yet not all such powered transmitters are four-wire. If a powered transmitter's output is noted as passive, then the device may be treated as a two-wire unit from the standpoint of the signal circuit.

   Most recording devices are externally powered, but are passive on the circuit. In these cases, the external power is for the internal electronics of the unit only. The signal circuit is isolated from this power source. Note that the recorder shown on the bottom circuit is a powered, passive device.

2. Two-Wire Circuit
   A two-wire device is said to be loop powered. This means the device functions by absorbing the energy it needs to generate the signal from the current loop. This is also referred to as "current sinking." This nomenclature can be a bit confusing because a transmitter that is current sinking is still the signal source for the circuit. Power for the current loop is supplied elsewhere.

   A transmitter classified as two-wire must typically be the first device in the circuit with respect to current flow. In other words, the positive terminal of the transmitter must be directly connected to the positive terminal of the voltage source. The voltage source is usually a 24 VDC power supply.

   (a) Two-Wire Circuits with Stand-Alone Power Supply
   Referring to Figure 2-48, PLC I/O point 2 depicts a two-wire circuit with an external DC power supply. Notice the wires must be rolled (polarity-wise) at the PLC for the proper polarity to be present across the I/O point. That is because current flow is now reversed with respect to the previous example because the transmitter must become

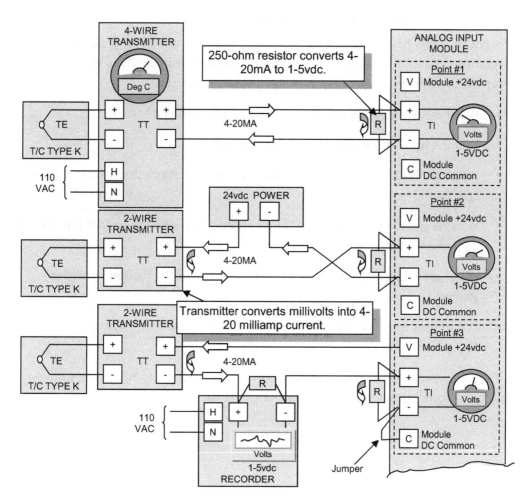

FIGURE 2-48. ANALOG WIRING METHODS: 2-WIRE VS. 4-WIRE

the first load in the loop as opposed to being the energy source for the loop.

(b) Two-Wire Circuits with PLC Internal Power Supply
Most PLC systems today are able to source the loop current themselves by simply connecting the positive terminal of the transmitter to a different terminal at the PLC. The negative terminal of the transmitter is then tied to the positive side of the I/O point, and the negative side of the I/O point is jumpered to the PLC system's DC common. That is depicted in the I/O point 3 example. In that example, a recorder has been added to the loop.

E. Design Practice Summary

To be comfortable in a design role, one needs to be knowledgeable about the expectations of customers and of those in the design business. This chapter has attempted to address some of the more important questions:

- What is an instrument database, and what good is it?
- What is involved in document management?
- What are I/O lists and instrument lists, and what is a good way to maintain them?
- What is good wiring practice? How should wires be tagged? What should be presented on a wiring diagram?
- What is a hazardous area, and what are some of the design considerations?
- What does intrinsically safe mean, and how is that different from explosionproof?

The discussion surrounding these questions should provide a basic understanding of some of the topics dealt with in Part III.

PART II – CHAPTER 6: THE CONTROL SYSTEM

A. INTRODUCTION

The control systems world is wide and deep. It is a world of acronyms and keywords that represent arcane systems and concepts, interfaces and networks, and hardware and software. Like an iceberg, much of its mass is below the surface. To most plant engineering professionals, the plant's control system consists of a set of enclosures bolted to a raised floor behind the control room, or a set of monitors. To the control systems professional—the systems integrator—it is a lush world unto itself.

An industrial plant exists for one purpose: to convert raw material into something useful. This can hopefully be done in such a way as to make the end product more valuable than the sum of the uncombined source materials, the energy expended to effect the conversion, and the cost expended in bringing the end product to market. Otherwise, why bother?

This chapter is written for the project manager who needs some perspective and the new engineer or technician who needs some basic understanding. It is written around no particular control system, though more time is spent on the PLC than on the DCS. I/O interface concepts, for example, are valid regardless of the flavor of the control system. This chapter explores some of the different types of control systems in wide use today and delves into some of the major questions.

- *Process Controller:* How did the process controller evolve? What are the origins of the two most common controller styles, the DCS and the PLC? How are they different, and how are they used today?

- *Human-Machine Interface (HMI):* How did the HMI evolve, and what are some of the issues related to the specification and design of the basic HMI?

- *Data Communications Interface:* What are some of the major communications protocols in use today, and what are some of their features?

- *I/O Interface:* How is the control system mated to the production facility? What are some of the basic concepts? How is an I/O map derived? How is an I/O count derived? What I/O module types are available, and what are some of the considerations for proper wiring practice?

In much of the discussion that follows, as in industry, the term "process" is used frequently and in various applications. For example, the act of energizing a washing machine is one step of the clothes-cleaning *process*. This is the more common use of the word. However, in the controls industry, the term *process* is used as a generic term to describe the production method and the related equipment. To expand on the washing machine analogy, by selecting "permanent press," the operator has selected a specific method, or *process*, that changes the way the agitator functions, among other things. The agitator itself is then referred to as *process equipment*.

B. THE COGNITIVE CYCLE

The physical plant is an entity capable of processing raw materials into a desired product. It is inanimate. Yet by mating this inanimate object with an operator (or more and more a computer), it is transformed into something else entirely. By adding a properly designed control system, the plant approximates a central nervous system. Like the nervous system of any creature, this one detects the condition of its environment, encodes the information, and transmits it to a "brain," where the information is decoded and processed. During this processing stage, the information is used as the basis for decision making. The end result is a set of commands encoded and transmitted to the "muscles," where action is initiated. Action occurs, causing changes that are sensed and analyzed, and which result in additional actions. In the background, some of the raw and resulting processed information is stored in historical memory for later retrieval.

This *"cognitive cycle"* (Figure 2-49) is a natural phenomenon, as basic as the motion of a wave. Sentient creatures everywhere follow the same pattern as they traverse their environment. They sense their surroundings to gather information, they analyze the information, and then they react to it. At its heart it is rather elementary, as are most things that work well. And this is a process that works, as evidenced by the fact that we are here! However, few creatures are able to affect the physical world and change it to fit their needs however. That is left mostly to humans. The machines we have built to manage our world follow this same cycle.

In modern industry, process sensors detect the condition of the material being processed and of the conditions surrounding that material. This class of sensors, made up of detectors and transmitters, convert the thermal, electrical, and physical properties of the material and the process into electrical signals. These signals are sent to a central processing site, where they are converted to numeric values and analyzed. Whether human or electronic, the central processor compares the actual process and material conditions to the desired conditions and reacts by developing a set of corrections that are expected to result in the elimination of errors—differences between the measured condition and the set point. These numeric correction commands are then converted to electrical signals fed to final control elements (most often, actuators), where conversion from electrical energy to mechanical energy takes place.

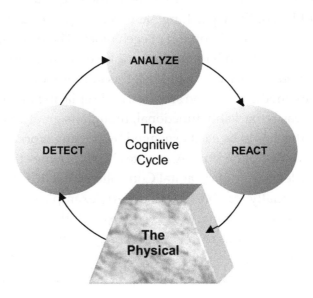

FIGURE 2-49. THE COGNITIVE CYCLE

C. CONTROL SYSTEM OVERVIEW

Processing facilities today are, for the most part, computer controlled. Computer systems used in control applications can be broken into several basic components as shown in Figure 2-50.

1. A HISTORICAL PERSPECTIVE

Originally, process control was accomplished using pneumatic controllers, for analog systems (such as continuous processes) and relay logic for discrete systems (such as automated assembly lines). Pneumatics was a better fit for analog than were relays for discrete circuits. The big limitation with relays was the difficulty in modifying the complex circuits. Also, relays, being electromechanical devices, required regular maintenance. Few early relays were sealed, and corrosion was a real problem. Troubleshooting problems, and even getting through startups, was always an adventure.

In 1958, the Square D Company launched NORPAK, the first commercially successful, completely encapsulated solid-state logic control system.[3] This transis-

---

3   See Square-D Website: http://www.squared.com/us/squared/corporate_info.nsf/unid/
    8E39107DD6826EE585256CE90072D73D/$file/timeline.htm

tor-based system relied exclusively on NOR gates and logic inverters (NOTs). The NOR gate could be used as a building block in the construction of almost any other Boolean element. NOR gates could be patched together to form flip-flops, inverters, and AND gates, among others. Each modular block of NOR gates varied in size from the dimensions of an ice cube to the size of a brick. The Boolean (binary) logic could be built by patching NOR gates together using jumper wires that plugged into the patch strip on top of each block. The resulting controller could be the size of a wall and had the appearance of a furry hide because of the thousands of exposed jumper wires. In 1985, this author worked on a large NOR-PAK-based system, modifying some sections, and replacing others with a new PLC. The system was remarkably functional, though its great weakness was the exposed jumpers that could be easily snagged by passing workers.

In 1968, Dick Morley of Bedford Associates in Massachusetts went to market with something called a Modular Digital Controller (MODICON). This was the world's first commercially feasible PLC.[4] An early example of a DCS was produced by the Australian company, Midac, in 1981.[5] Until the mid-1990s, the PLC was thought of as a relay replacer, handling mostly discrete circuits, while the DCS was thought of mostly as a process controller, managing analog-heavy applications. Together, the two could be made to function as a hybrid process control automation system.

At first, the PLC and DCS functioned behind the scenes. Most discrete (Boolean, or on/off) operator action still took place at the control board with hardwired relay logic or a PLC, and the DCS merely monitored the temperatures and pressures and reported any alarms via control-room annunciators. Then, DCS manufacturers began bundling cathode ray tube (CRT) workstations with their systems. Eventually called Human-Machine Interfaces (HMI), these initial workstations were text-based, which required the operator to spend time keying in information and manipulating a specially designed keyboard. This was a tall order in those days and created a need for a high degree of operator training. Even so, this innovation provided a means for entering recipes, tracking analog data, creating reports, and managing alarms.

As memory and computing power increased, this text-based interface quickly gave way to animated color graphics. The graphics packages, called Graphic User Interfaces (GUI—pronounced "gooey") greatly increased the designer's ability to make the controls intuitive to the operator. In the example in Figure 2-51, the GUI provides cartoon-like pictures to represent the process facility. It is animated, using (depending on the project's approved color scheme) green to show a pump

---

4   Jim Pinto, INTECH, 8 February 2006, http://www.isa.org/Content/ContentGroups/News/2006/February24/History_of_the_PLC.htm
5   Wikipedia http://en.wikipedia.org/wiki/Distributed_Control_System#The_Network_Centric_Era_of_the_1980s

running and red for a pump stopped. Later versions of these interfaces have touch screens and are very interactive with the operator and with the plant process equipment. Continued improvements to the HMI have included the absorption of chart recorders into a software application called a data historian, and the incorporation of the annunciator into an alarm management utility.

Thus, the control board, with its switches, lights, annunciators, and paper trend charts, began to be supplanted by the DCS operator workstation. As computing power increased, and better software became available, the HMI came into its own as a standalone software package outside of the DCS umbrella, running on a standard PC.

The modern control system, through its HMI, gives specific information the operator needs to oversee the production process, while providing *managed* access to the system controls. The operator is able to work within the confines of the pre-configured environment, which reduces the variations induced by shift change and attention span. Today's operator interface is extremely intuitive, allowing the information to be packaged for display and providing operator access to the specific controls needed for the situation at hand. This has resulted in greatly enhanced process reliability and repeatability.

2. PLC VERSUS DCS

Today, for all practical purposes, there are two flavors to choose from when selecting a process controller—the DCS or the PLC. Other options are emerging (PC-based controls, for example), but as of this book's publishing, the PLC and the DCS still enjoy the lion's share of the market. For several years now, the PLC has been making inroads into the DCS market. DCS manufacturers have retaliated by downsizing their systems, making them more competitive on a smaller scale.

What exactly is a DCS? How does it differ from the PLC in today's controls environment? When should one be selected over the other, and when is a hybrid mix of the two appropriate? The following is a discussion of both types.

a. The Distributed Control System (DCS)

Until recently, the term *distributed control system* has been something of a misnomer. In fact, rather than "distributed control," as the name implies, the practical effect has been "centralized control" since all signals are collected and marshalled to a central location. But whatever the name, for many years the DCS has been the center of gravity for process controls.

The DCS began as a DAS, or data acquisition system. In the early days, nobody trusted computers for control, depending instead on the tried-and-true standalone analog panel-mounted controllers and switch-and-lamp control boards to manage their processes. If a computer-based management system existed, it was used for data acquisition with analog sensors being wired to both elements (the DAS and the panel-mounted controller), allowing the local controller(s) to control the plant and the DAS to collect the data.

Confidence grew as computing systems became more robust, and minor control tasks were eventually attempted. Another milestone was reached when the DAS started handling the entire spectrum of analog I/O, including the control outputs, thus morphing into today's DCS. Then, over a very short time span, the DCS system evolved again to include discrete I/O as well as analog I/O.

Perhaps the biggest advantage of the DCS over other options, then and now, is the tight integration of DCS system components into a single package. For example, the DCS combines the controller with the operator interface, presenting them both to the customer as a single entity. This is a great advantage, especially when the DCS manufacturer integrates the two to use a single database between them. The "manufacturer envelope" contains the control system, operator workstations, operator training, technical support, maintenance, and so on.

Particularly in the early days, this total system integration was key, since there were few trained system professionals in the industrial setting. Also, working outside the single-manufacturer envelope was almost impossible because of a lack of communication standards. Each DCS manufacturer used a unique communications protocol. This advantage (from the system manufacturer's point of view) has today been largely negated by the advent of open communications protocols, such as Modbus, Ethernet TCP/IP, Ethernet H1, Foundation Fieldbus, and others.

Another key characteristic of the DCS was its optimization for specific markets. Many DCS manufacturers developed target markets that they felt would best receive their particular product. Software tools were developed that focused on that particular market, and expertise was developed in-house that made their aftermarket services particularly effective. This effect may still be felt today in many sectors in which customers have developed relationships with their service providers that are still strong enough to overcome any technology gap that may appear.

If the manufacturer envelope was the single best advantage of the DCS, it was also its single biggest disadvantage. As in any complex machine, there are always areas where the builders achieve optimum performance along with areas where perhaps another manufacturer's product would be better. But once the decision was made to align with a particular DCS brand, the customer had to accept the good with the bad and adapt to the situation.

In light of this, a trend toward hybrid control schemes has developed. This trend has been fostered by the availability of previously mentioned open-architecture communications protocols. This open architecture idea has made the integration of different manufacturers' equipment more feasible inside and outside the DCS envelope. In these hybrid arrangements, the DCS handles the more specialized or critical, analog-heavy aspects of the control scheme, and lower-cost PLCs are used for data acquisition and for the more discrete-heavy subsystems, such as material handling. This breakout is particularly apparent in the fossil power (coal, oil, and gas) industry, one of the DCS supplier's most ardent supporters, where even there, the coal yard is almost universally PLC controlled. Systems inside the plant may be 70%–90% DCS controlled, but outside the plant, the percentage is reversed.[6]

Possibly the biggest gains outside the DCS environment are in the wide variety and function of web-based reporting packages. If DCS-provided reporting packages are substandard or not specific to the customer's needs, the customer can now purchase third-party equipment and software and integrate it into the control system themselves. This ability to integrate has caused a resurgence in the DCS market. So the DCS manufacturers are adapting to the new market demands by making their systems more open and configurable.

b. The Programmable Logic Controller (PLC)

As described previously, the PLC was developed in the 1960s in response to an industry demand for a relay replacer, originally in the auto industry. Relays were used to perform most Boolean functions in those days and are still in wide use today as isolators or "interposers" between circuits of differing voltage levels. In the early days, complex Boolean logic was handled by the intricate interconnection of these relays. However, relays, being mechanical devices, tended to fail, and once the intricate wiring was in place, setting up the logic, the system was difficult to modify.

---

[6] This assertion is anecdotal, based on the author's experience in a large number of major fossil fueled power facilities in the U.S.

The PLC's purpose was to control discrete (on/off) devices, using Boolean logic. The PLC was a computer, but it was more akin to NORPAK than the DCS. The PLC was intended to be a tool for the plant electrician. It was a shorter technological leap than the DCS because of both its simpler physical design and the format of its software.

A programming language called PLC ladder logic was developed. It was optimized for the plant electrician by having the look and feel of a relay-logic-style wiring schematic, a format with which the electrician was familiar. Manufacturing management quickly realized that this would allow some of the existing plant electricians to ease into the computer world and at the same time minimize the retraining cost with respect to the DCS. As with the DCS, the PLC's biggest advantage is also its biggest limitation. The PLC offers simplicity but does not provide an inherent operator interface. That is not an issue when used strictly as a relay replacer, but in today's industrial environment, the PLC has evolved into an analog controller as well. Some PLC manufacturers (Allen-Bradley, for one) have a PLC offering (SLC-500 series) as well as an HMI/GUI offering (PanelView). There are integration advantages in such cases, but they are not as marked as that of the DCS.

The PLC program has also evolved from ladder logic, which limited the programmer, to many other languages more suited to the DCS environment. Modicon has a product called "Concept" that provides several language options that can be mixed as necessary within the same program. And products like Concept (and others) are compliant with the International Electrotechnical Commission (IEC). This is a huge leap from the typical ladder-logic program available just a few short years ago.

In summary, the PLC has come of age. When combined with good Graphic User Interface (GUI) software, it can compete even in the DCS marketplace. Its low hardware and software cost is partially offset by the added difficulty of integrating it with other control system equipment, such as the GUI. But that issue has declined in importance with the universal acceptance of such communication protocols as Modbus-Plus, Ethernet, and Devicenet, among others.

3. MAJOR CONTROL SYSTEM ELEMENTS

Refer to Figure 2-50 for the following commentary.

a. **The Physical Plant**

The physical plant is where all the input and output devices reside. These are motors, valves, instruments, motor control centers, variable-frequency drives, analyzers, and a myriad of other process control devices that must be installed, wired, and tested.

b. **The I/O Marshalling Area**

The I/O marshalling area is typically a set of cabinets where the field wiring meets the automation system wiring. A true marshalling panel contains nothing but terminals and fuse blocks, relays, timers, signal conditioners and alarm trips. It is where the field signals transition from single-pair field cables to multi-conductor cables that connect to the control system. If there is a control board that will operate the plant independently of the PLC, then the marshalling panel is where those signals are mated to the I/O wiring. In today's highly automated environment, it is relatively rare to have a fully independent control board, and in most cases the marshalling can happen inside the PLC Remote I/O cabinet. So independent marshalling cabinets are becoming rare, though marshalling itself is still a requirement.

c. **The Hardwired Control Board**

The hardwired control board is generally located in an auxiliary control room that provides an alternate method of controlling the plant in case a problem arises with the PLC-based control system or the DCS. Auxiliary hardwired control boards are gradually disappearing from the landscape, though they may still be found in power plants.

d. **The PLC**

As mentioned earlier, the PLC is a computer system that has been optimized for control applications. It has an input section that continually scans the input signals and performs de-bouncing (i.e., filtering) operations on the raw discrete signals, and scaling operations on the raw analog signals, in order to convert those raw signals into information that is useful to the operator and to higher-level software operations. All of the data that are generated are loaded into a database that is the core of the PLC program. The PLC Logic program is where the PLC programmer develops software that is specific to that particular plant and process. This program pulls information from the database, makes decisions regarding sequences, repetitive operations, continuous operations, and other operations specific to the plant and process, and loads newly generated information back into the database. The PLC Logic program monitors permissives (i.e., automatic system requests) in order to initiate start/stop or open/close commands. Those commands will not be issued unless the interlocks are met. Interlocks are personnel and/or equipment safeties that must be present for a field device to operate. If the interlocks are met, then a commanded state can be passed to the output section, where the information is converted to an electrical signal and shipped out to the end device.

e. **The HMI**

The HMI is generally located in the primary control room, an environmentally controlled space that is the nerve center of the facility. The HMI—also called a man-machine interface (MMI) or operator interface (OI)—connects the operator to the control system. The original HMI was, in fact, a control board. But the control board had an Achilles' heel: total reliance on the operator. Even under the best of circumstances, the control board is labor intensive and leaves the system open to operator error. Process repeatability is difficult to attain from one shift to the next and from one day to the next.

The HMI communicates with the PLC through its device driver, a software interpreter utility that allows the two devices to communicate. The information is loaded into a tag database that is used by all the other software utilities in the HMI software package. The GUI provides a means for the programmer to develop a set of color graphic screens that represent the process. As mentioned, these screen graphics are animated, using color changes, flashing, visibility, text boxes, and other techniques to provide information to the operator that is timely and pertinent. The Alarm Manager is a utility that 1) monitors analog values and compares them to trip settings and 2) monitors discrete values and compares them to alarm state tables. Those signals are then used to trigger audible and/or visible alerts and alarms and to initiate message displays informing the operator of process conditions that are out of specification along with providing some indication of action needed. The Trend Utility accesses real-time and historical data in order to provide information related to changes of process conditions over time. The Report Utility provides information relative to production, raw materials consumption, and other topics of interest. There can also be a Data Historian (not shown) that is either internal to the HMI, or a separate utility entirely.

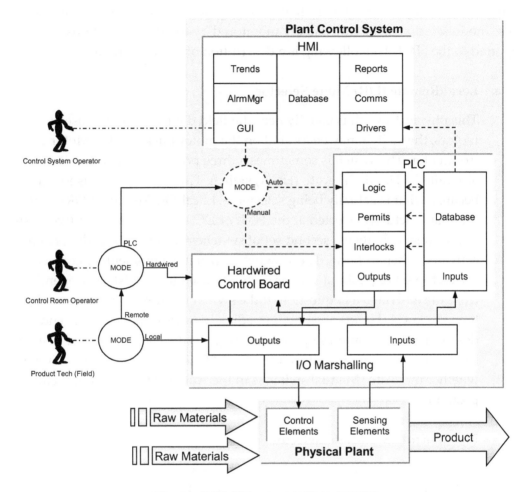

FIGURE 2-50. TYPICAL CONTROL SYSTEM

4. CONTROL MODES AND OPERABILITY

In addition to showing the major components of the control system, Figure 2-50 provides some indication of the major control modes employed by operations teams. There are generally two classes of plant operators: product technicians and control room operators. Product techs work primarily in the field. These personnel are typically in close contact with the control room operator(s) via radio or other means. If a product tech takes control of a device locally, then the control system is generally bypassed, with that individual in complete control of the local equipment. In general, the closer a control mode selection is to the field device, the more priority it has.

In the example in Figure 2-50, there are three control mode choices. Two of the node selectors are external switches (Local/Remote and PLC/Hardwired), and one is internal to the HMI (Auto/Manual). If Local is selected, then the position of the other two (upstream) switches is irrelevant. The end device in question is being controlled by the product tech at the equipment. Note that the positions of these mode-select switches are usually monitored by the PLC, and status is reported to the HMI. The following is a description of each control mode.

a. **Local/Remote (L/R) Mode Selector**

This physical switch is usually near the field device, or in the case of motors, the switch may be located on the motor bucket in the Motor Control Center. The switch is sometimes a three position switch, and it is usually called a Hand-Off-Auto (HOA) switch. This is a bit of a misnomer because what is actually being selected is Local-Off-Remote (LOR). Auto/Manual is actually selected at the HMI or DCS. If the switch is a two position L/R switch, then a second set of switches is usually included, to be activated when in Local. These switches, usually a momentary contact start/stop set of pushbuttons, are actually used to start/stop or open/close the end device in question. If the switch is a three-position switch, then the Local (Hand) position immediately starts or operates the end device. The Off position stops, or de-energizes the end device. The Remote (Auto) position passes control up to the PLC/DCS or HMI, depending on whether the Auto/Manual switch is in the Auto (PLC) or Manual (HMI) position.

b. **PLC/Benchboard Mode Selector**

This physical switch is usually in the control room, or in an auxiliary control room. The Benchboard position activates the myriad of switches and lights mounted on a benchboard. The operator then controls the facility by rotating switches, pressing buttons, and monitoring meter dials and annunciator boxes. The PLC position passes control up to the PLC or HMI, depending on whether the Auto/Manual switch is in the Auto (PLC) or Manual (HMI) position.

c. **Auto/Manual (A/M) Mode Selector**

This virtual switch is provided at the HMI. When the field switches are aligned, the control room operator can decide to operate the end device himself through the HMI (Manual Mode), or to pass operational decisions to the PLC program (Auto Mode).

If in Remote-Manual mode, the operator's selections are acted upon directly, with the PLC providing some oversight by making sure the interlocks are aligned properly. If a motor is started in Manual mode, for example, and the interlocks are subsequently lost, then the PLC will act to stop the motor without operator action. The motor should not restart until the interlocks are restored and the operator initiates a restart.

If in Remote-Auto mode, the operator passes start/stop or open/close decisions to the PLC Logic program. If the logic program detects conditions that dictate a start or stop of a device, then it is automatically started or stopped. These conditions are called permissives. In Automatic mode, the PLC logic monitors these permissives and issues commands to the end device. As in Manual mode, the interlocks must be aligned. If the interlocks are not aligned, then it should be impossible for the operator to place that device in Automatic. If in Automatic already when the interlocks are lost, then the device should immediately change to its pre-defined safe state, and go out of Auto mode. The device should not restart until the interlocks are restored and the operator puts the device back in Automatic, and the Permissive conditions align to initiate a restart.

D. The Human-Machine Interface

The modern HMI is extremely reliable, and in most cases, it eliminates the need for maintaining an expensive control board as a backup, though many fossil power plants, hydroelectric stations and nuclear facilities retain the ability to go fully manual if necessary. An HMI, in its simplest form, typically consists of a desktop PC running a commercially available Windows-based operating system. In addition to the Microsoft packages, this PC has an HMI package, with its embedded graphic user interface, historian, alarm management, and report utilities. Some leading HMI packages available are Wonderware, Intellution, RS-View, Citect, and others. These software packages have their differences, but there are many similarities.

The newer, more powerful systems boast object-oriented graphics with canned graphic elements that simplify configuration. The following are some of the main elements of this type of software package:

1. The Graphic User Interface (GUI)

The purpose of an HMI is to provide the operator with at least as much functionality as the control board. In fact, until recent years, the HMI was a control board replacer, with only enough graphics resolution to show push buttons, lights, numeric displays, and other simple graphic elements that the operator had no problem recognizing as control board elements.

Those elements are still used today. But it is also possible to provide a more graphical picture of the process and to animate that picture to reflect process conditions. Today's operator interface is likely to look like a process flow diagram,

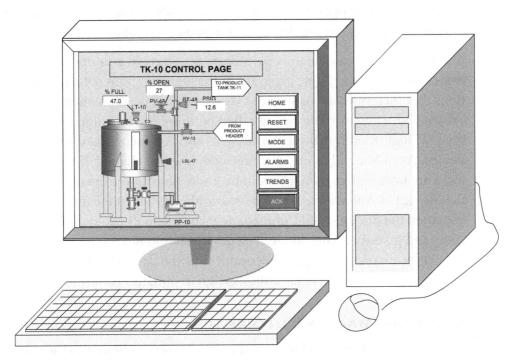

FIGURE 2-51. THE HUMAN-MACHINE INTERFACE (HMI)

with 3-D graphics, and fully animated and interactive with the operator—hence the term graphic-user interface.

But graphics are just pictures until they are linked to the outside world. To configure a GUI is to establish links between the graphic elements on the screen and calculations and analog or discrete values pulled from the PLC or from other analytical control system devices. There are two modes for the GUI software package: Development and Runtime. In Runtime mode, the graphics page is linked to a database that is in turn linked to the PLC. Data is collected, and the graphics are animated based on the content of the database. In Development mode, the graphic elements can be configured using canned graphic configuration templates. The user only needs to select the type of graphic element being configured, and a typical setup template will appear for his use in animating that graphic element.

In the screen depicted in Figure 2-52, the user has entered Development mode, and the ACK pushbutton has been opened for configuration, with its setup template in view. This is usually done by simply double-clicking on the graphic element that needs to be configured, after which a template appears that provides either pick points, where a mouse click sets or clears a status bit, or a data-entry window, where a tagname, equation, or some other string of text is entered.

In the example, the ACK pushbutton has been linked to the GUI tagname ACK_ALRM. The pushbutton is configured as a momentary pushbutton, with its initial value being "off" (false). When this button is pressed (via a mouse click or, on touch screen systems, via a touch), the numeric value stored at memory location ACK_ALRM will change from false (0) to true (1) for as long as the button is in the down position. When the button is released, the value returns to false. Two types of links, action links and animation links, are described below.

a. **Action Links**

A graphic with an action link (often displayed as a pushbutton) is one that, when actuated by the operator, results in some action on the part of the control system. If the button is pressed via mouse click while the cursor hovers over the graphic element, or via a touch, the control system interprets the button press as a request for action. There are several types of action links, some of which are included in the following list:

- *Navigation Links* let you change screens. An example of a navigation link is the HOME button. Pressing this button closes the TK-10 screen and opens the workstation's home page, which is usually a graphic overview of some sort.

- *Numeric Data Entry Links* allow the operator to enter numeric values at the keyboard and route those values to a specific local or remote location.

- *Local Command Links* change a parameter within the HMI, such as the time scale on a trend display.

- *Remote Command Links* let the operator issue a command to an external entity, such as a PLC or other device. This type of command is routed through a communication driver module that is preconfigured for the proper communication protocol.

b. **Animation Links**

A graphic with an animation link is one whose appearance can change based on some predefined condition. For example, a tank with normal level might be grey in color. That same tank may change to red if the tank level gets too high. Some types of animation links are as follows:

- *Numeric Displays* show analog values in engineering units. Frequently, these display windows contain embedded scaling calculations, eliminating the need for complex PLC programming for scaling. Animation in these displays include numeric value display, color changes for alarm indication, and sometimes a navigation box to open a popup overlay.

- *Status Displays* can be either two-state or multi-state graphics that may become visible or change color based on preconfigured conditions.
- *Alarm Displays* are usually two-state graphics that appear or flash under predefined conditions.
- *Text Displays* show different lines of text based on predefined conditions.

It is also important to remember the graphics screen is the operator's window into the process. It is not the designer's opportunity to display artistic flair. Proper screen design gives the operator the information he needs—no more, no less. It should display the equipment on the floor as accurately as possible without sacrificing clarity. For example, if the operator is able to see a tank from the control room and the tank's discharge pump is to the right of the tank, the screen should reflect that orientation whenever possible. Or the screen can reflect the P&ID orientation. Whichever approach is selected should be consistently applied.

When designing an HMI, it is important to assure operators that they cannot cause a catastrophic event by inadvertent action. A simple mouse click or pointing finger must not be allowed to initiate certain actions. To address this issue, the HMI developer need only employ a two-step approach. Use confirmation, "Are you sure?" overlays and messages on action graphics, such as pushbuttons.

A screen with a lot of vibrant colors buries the key information in "color noise" to the point that an operator may struggle to find needed information. As a guideline, the basic color scheme should be rather bland. Alarms and warnings should pop out of the background. The operator should be able to see an abnormal condition from across the room. Also, many operators have some degree of color-blindness, which calls for a high degree of contrast between a good thing and a bad thing.

ISA standard, ISA-5.5-1985, *Graphic Symbols for Process Displays*,[9] addresses this topic, but it is a bit dated. This standard recommends a black background, which has proven to provide too much contrast, and the symbols in the standard are dated. However, a very applicable recommendation in this standard is to develop a color plan ahead of time and stick to it.

To recap, here are some basic screen design guidelines:

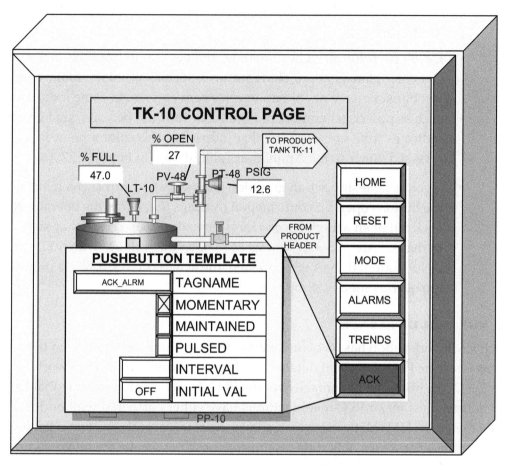

Figure 2-52. Graphical User Interface with pushbutton configuration template

- Use a lot of gray or taupe, both muted, opaque colors. The background should be slightly darker (in color value) than the equipment being depicted.

- When laying out the screen (before animation) nothing on the screen should be made markedly brighter than anything else. Shades of grey work well.

- When animating with color, normal operating conditions should be depicted with muted colors. Vibrant reds, yellows, and greens, along with flash and sound, should be reserved for alarm conditions or abnormal states.

- Avoid the temptation to make a screen too busy. The operator should be able to take in the situation at a glance. Each screen should have one central theme or focal point. If the screen is an overview screen, its focal point should be general system status and navigation. Packing a lot of action graphics (e.g., switches, sliders) on such a screen should be avoided.

- Develop a color plan ahead of time, and get buy-in from the operators. The color plan should include features, such as visibility, blink, and fast blink. Beware of animation color conventions. In some industries, such as power generation, red indicates "on" or "beware," and green indicates "off" or "safe." In other industries, this color scheme is reversed. Refer to the Animation Table shown in Figure 3-77.

- Avoid single-step action graphics. On the main control screen, have the buttons open up confirmation overlays that allow the operator to back out of an action. An *overlay* is a small pop-up screen that says, perhaps, "Are you sure you want to open this valve?" with a YES and a NO button. If the YES is pressed, the action occurs. If NO is pressed, the overlay merely disappears.

2. The HMI Database

Each HMI workstation has its own database that is usually separate from the database of the PLC. A unified database feature is common in the DCS world, giving the DCS a historical advantage over the PLC. Today, as an answer to that DCS advantage, most major PLC manufacturers have either partnered with or developed their own HMI platform, which allows them to integrate the two databases into one, but that works only if the user purchases and installs the two packages together. Most of the installed PLC base today was deployed with separate databases between the PLC and HMI. In those cases, two separate databases need to be maintained by the control systems integrator.

Some of the key words and parameters relative to the HMI database include:

a. **Tagnames**

A programmer gives each element in the database a unique tagname. Sometimes called an *alias*, this tagname is meant to be a means of quick identification of the element. Therefore, the names should be meaningful in some way. Since the HMI database and the PLC database are separate, it is beneficial to name HMI tags with the PLC tagname or PLC address to help the programmer maintain some level of organization.

b. **Scan**

As in most computer systems, HMI scan rates can be important. If the scan rate is too high, then the communication network can be choked with non-essential data transfers. If it is too low, transitions could be missed. In today's systems, the problem is more likely to be high scan rates and the resulting heavy bus traffic on the network.

To decrease bus traffic without losing information, several techniques have evolved to optimize bus activity:

- *Update by Exception.* Most HMI manufacturers have gone to an update-by-exception structure whereby information being read from the PLC, for example, is written into the database only if the data has changed more than a preset amount since the last time the data was written. For example, if the front end of the workstation software finds that the pressure in a tank has not changed since the last update, the time intensive write function that moves the data into the database is skipped.

- *Phasing.* The major HMI systems allow the programmer to define scan rates for each tag. This is called phasing and can really be useful in optimizing the HMI's network activity. For example, scanning temperatures at a high scan rate is wasteful because temperatures generally change very slowly. Often a 5- to 10-second refresh (scan) rate will be sufficient for temperatures.

- *Update by View.* The update-by-view approach assumes information not being viewed is information that does not need to be updated. Exceptions are made if an item has been configured as an alarm or as a data point in the historian. But other than those cases, only data displayed on the currently active screen is pulled from the PLC or from other workstations.

3. THE HMI ALARM MANAGER UTILITY

An alarm management utility is a standard feature of an HMI. Today's HMI does a very thorough job of alarm annunciation. Unlike its ancestor, the electronic annunciator, configuration is a snap, and it is infinitely expandable, limited only by the number of tags in the database and the scan rate. In most HMI packages, the alarm is configured as a simple database activity. For example, when setting up a new analog input, the data entry template might have scaling factors and alarm set points. If alarm set points are entered, an entry is automatically made in the alarm manager. These items are continuously monitored for the alarm conditions even when the screen is not active or in view. This increases bus activity, so alarms should be used judiciously.

4. The Historian

A historian is an archiving software utility that is frequently an add-on to the base HMI package. Sometimes a limited historian is included with the base package, but fully fledged historian utilities can be very involved products in their own right.

In short, the historian continuously scans selected HMI tags and stores the data. The refresh rates can be set individually. The method and location of the archive need to be defined. The historian is essentially a first-in, first-out buffer, so eventually the data will be overwritten. Therefore, it is important to have proper archival procedures in place to capture and manage the mass of data in a timely manner.

5. The Trend Utility

A trend chart is a graphic element that simulates a strip-chart recorder (Figure 2-53). Multiple "pens" may be configured to track data on a continuous basis. Tags that are configured as trend points are scanned continuously, overriding any other scan management systems that may be active.

There are two kinds of trends:

- A *real-time* trend tracks data continuously in the "now" time frame. As data items pass out of the frame, they are lost. The update rate of the real-time trend is much faster than that of the historical trend because the real-time trend is typically used for loop tuning, batch monitoring, continuous process monitoring, or other immediate uses. The amount of data retained is relatively small, as a high number of scans record data over a short time period, rarely more than 30 minutes, and usually more like 10.

- A *historical* trend is another window into the process. However, this is a window into the past. The historical trend accesses the data collected by the historian. The data is stored in a historical log file. Typically, the historical log files contain one day's data, closing the file at midnight and starting a new one. However, this too is configurable. The historical trend accesses these log files and displays the data in the form of a strip chart. You enter a start time, and the trend utility searches for the correct archive, pulls it up, and displays it.

Trend graphics, whether real-time or historical, work the same (Figure 2-53). A time (x-axis) scale can be set and adjusted, as can the magnitude (y-axis) scale. A vertical marker may be moved laterally across the display. Wherever the marker touches a pen track, the value at that instant of time is displayed as text.

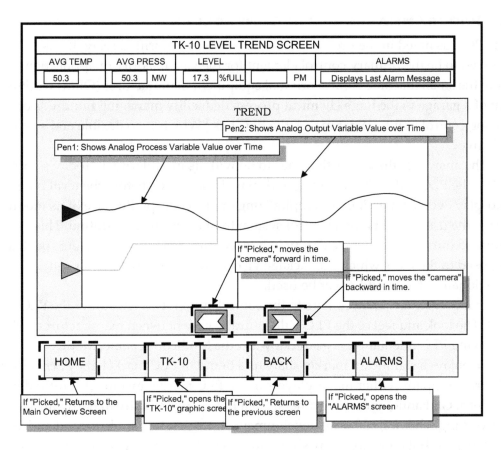

FIGURE 2-53. TREND SCREEN

6. REPORTS

Capturing data for a report is a key consideration when designing the operator interface. Most customers today require extensive material and process tracking for the products they purchase. This is particularly true in the pharmaceuticals industry, where the validation process is incredibly involved. Having the ability to generate the right report after the batch is done is critical.

Optional report writer utilities are available for most HMI systems. If a formal, polished report writer is not appropriate, most HMI systems are able to communicate with standard Windows® applications, such as Microsoft® Excel or Microsoft® Word.

Web-based reporting is becoming more prevalent, as managers want and need access to real-time information as they travel. Web portals provide secure access to the data as the control system is isolated across firewalls. The drive toward web-based systems increases the cyber-security concerns.

E. Programmable Logic Controller

The PLC, as noted in the previous section, began as a glorified relay. It has evolved to be the primary control element for the vast majority of applications, from the small machine control original equipment manufacturer (OEM) working out of a garage to the huge chemical processing facility managing hundreds of analog control points. What makes the PLC tick? Why is it so flexible and so cost-effective?

The answer to these questions has to do with its modularity. In today's vocabulary, the PLC is the ultimate plug-and-play device. Even before the term plug-and-play evolved, the PLC was capitalizing on the concept. Because of its modularity, the user could tailor a control system to fit a specific application. This afforded him tremendous flexibility in how control system dollars were spent, as opposed to the DCS, which prepackaged a control system that perhaps included functionality that would never be used.

In any case, most newcomers to the control system will encounter the PLC first. The look and feel of the PLC will naturally vary between manufacturers. The following PLC orientation is geared toward the Modicon scheme. Modicon, which enjoys a significant market share, is a brand of PLC marketed by Schneider Electric Company. Other major players in the PLC market include Allen-Bradley, Siemens, GE Fanuc, and many others. Each PLC varies slightly in hardware and networking schemes, software development strategies, terminology, and addressing schemes. But there are many similarities as well. The following discussion will be kept as generic as possible.

1. Major PLC Components

The central processing unit (CPU) is the heart of the PLC. The CPU runs software that has been written to tailor the controller to a specific intended use. The connections to external devices must be interfaced or integrated to convert information into formats that can be readily understood by all involved. New communications protocols are evolving to make this interfacing task less arduous, but it is still a major issue in the development process.

The typical PLC consists of a series of electronic modules (cards) plugged into a rack (card file) (Figure 2-54).

While there is only one CPU per PLC,[7] there can be several racks of modules. The rack that contains the CPU is called the primary rack. Subsequent racks of I/O and/or communications modules are collectively referred to as secondary or remote racks.

---

7 The exception is a hot-standby arrangement in which two CPUs are online at the same time. However, only the primary unit is in control, while the backup is merely executing logic in sync with the primary.

PART II – CHAPTER 6: THE CONTROL SYSTEM   215

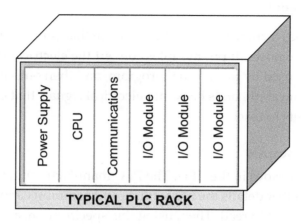

FIGURE 2-54. TYPICAL PLC RACK

a. **The Rack Power Supply**

   With the exception of analog outputs, the rack power supply typically supplies power to the PLC equipment only, not to the field devices. Most analog output modules source the signal themselves. Otherwise, I/O power must be supplied separately. This is usually a good thing because it is best to isolate the electrically noisy field devices from the power that is being distributed on the backplane to the module electronics. Most I/O modules are electrically isolated from the I/O signals either through embedded opto-isolators, which provide isolation via diodes or, in the case of discrete outputs, through relays.

b. **The Central Processing Unit (CPU)**

   The CPU is the computing element of the PLC system. The performance characteristics of the CPU define the characteristics of the system. Processing speed and user logic (programming) memory space are some of the parameters that should be considered before purchasing a PLC. The PLC vendor can provide assistance in properly sizing the CPU.

c. **The Communications Module**

   Communications modules provide special purpose data communications capability. They allow data to pass from the local PLC to other automation system equipment, such as other PLCs, directly to field devices, or perhaps to the operator interface. They provide not only a hardware interface, but also software. Frequently, there is a need for several different modules, each specific to one communications protocol.

d. **The I/O Module**

   I/O modules form the I/O device interface that connects the PLC system to the field devices. Input modules convert the electrical signals to data that can be used by the internal program, and then output modules convert program commands back into electrical signals that cause final control elements to react.

2. THE PLC PROGRAM

The PLC program resides in the CPU. The PLC manufacturer generates a program specification that details the file formats and communications parameters that a CPU is designed to read. Then, using this specification, any number of software developers can develop individual software products that can be used by programmers to create the control logic.

For example, a Modicon PLC can be programmed using Modsoft, ProWORX Nxt, Concept, or Unity, all Modicon software products. These programs differ in how they interact with the person doing the programming and in ease in printing, database management, and online editing capability. But the program that ends up on the CPU after it has been compiled must comply with the specification of the CPU.

Each of these programs has a component that sits on the PLC and a component that stays on the programmer's laptop. Programmer comments, a static copy of the program, and other information important to the human but not to the machine remain on the laptop. All else is compiled and downloaded to the machine.

a. **I/O Map**

   Each of the PLC's inputs and outputs must be mapped to a unique memory location. Each PLC manufacturer does this a little differently, but the end result is the same. All PLC software packages provide a configuration utility that allows the physical I/O point to be linked to a unique location in the PLC memory. In some cases, this mapping is inherent in the PLC system, and no work needs to be done by the programmer. But in most cases, the programmer must provide a base address for each I/O module. This base address corresponds to the first I/O point on the module. The configuration software then reserves the next contiguous set of addresses for the remaining points. More time is spent on the I/O map later in this chapter.

b. **Memory Map**

   For the most part, each element in the program is mapped to a specific memory address. There are some exceptions, such as the dynamic variable names that can be used in the Concept software package. Managing this memory map, then, can become quite a challenge.

   Spreadsheet programs, such as Microsoft® Excel, can be very helpful in this endeavor. Sometimes having a guide external to the PLC database gives programmers more flexibility in managing their memory maps and can even give them the ability to pre-configure their systems. A PLC database can be pre-configured in Excel and then imported into most PLC programs.

c. **Scaling**

   Analog inputs and outputs may need to be scaled, though it is perfectly reasonable to use unscaled integers in the calculations. Analog signals are fed to the CPU from the I/O interface, which converts electrical signals into integers. In the case of the Modicon Quantum series, the analog input modules behave differently from one module to the next. Some modules have a resolution of 16 bits, yielding an integer value between 0 and 32,767. Others have a resolution of 12 bits, yielding an integer value between 0 and 4,096. Some RTD modules produce an integer that represents temperature in tenths of a degree. So each module type must be investigated as to its particular conversion method. For more information on scaling, see Chapter 4.

d. **PLC Memory**

   As mentioned earlier, PLC user memory size is one of the parameters that must be known when the PLC is purchased. User memory allocations vary between one model of CPU and another. For example, each rung of logic consumes some memory just by virtue of its existence. Then each element on that rung consumes additional memory, the amount depending on the element. So some estimate must be made as to the amount of user memory that will be needed. User memory size is often underestimated, or ignored altogether, to someone's inevitable dismay.

e. **Documentation and the PLC Database**

   PLC information is stored in memory locations. The manufacturers give these memory locations unique alphanumeric designations (addresses).

These addresses are meaningful as they relate to the memory map, but they are meaningless in relation to the physical process and the overall control scheme. As a result, ladder logic programs can be difficult to read at best. If they are not well documented as to each rung's relationship to the overall control scheme, this relationship is very difficult, if not impossible, to manage.

Therefore, it is imperative that programmers find a way to properly document their programs. Fortunately, most software manufacturers have anticipated this need and have provided database utilities that allow the programmer to assign alias names to the memory location addresses. These alias names can be used throughout the program in the place of the alphanumeric address, thus enhancing the readability of the program. These alias names are sometimes called PLC tagnames, and the database utility is called the PLC tagname database.

In some cases, such as the GE Fanuc Logicmaster software package, the database also contains scaling information, data type information, and other configuration parameters that can be linked to calculations and special functions in the PLC program. In addition to the PLC tagname database, the programmer should make use of additional documentation capabilities that allow for rung, page, and program section comments.

f. **Programming Languages**

Several languages are available to the programmer. In the past, many of these were only available to the DCS programmer. But many PLC vendors are also offering these languages today.

Descriptions of some of the available languages include:

1. IEC Ladder Logic
   IEC Ladder Logic is an international ladder logic protocol. Most PLC manufacturers are in the process of developing a ladder logic program that is IEC-compliant.

2. IEC Function Block Logic
   A function block is a software object that can be copied and customized. A function block program consists of function block symbology as opposed to relay symbology. Ladder rungs are not used. Instead of two contact symbols in series to depict an AND function, an AND block is shown. This scheme is very powerful and is the scheme shown in most of the logic diagram examples in Part III.

Some programs that allow function block programming have utilities that allow the user to design custom blocks. Such blocks are called Derived Function Blocks, or DFBs. DFB-based programming is also very powerful, as DFBs can be developed for equipment items, pre-tested, and used multiple times in the program. If a modification is subsequently needed, a single change made to the root DFB file will change all instances of the DFB in the program. A few examples of good DFB applications are:

- Conveyors with belt slip switches, zero speed switches, pullcords, belt motor controls, and other standard instrumentation;
- Dust collectors with rotary valves, pneumatic pulsers, and blower controls;
- Tank agitator controls with level instrumentation and agitator motor controls.

3. Legacy Ladder Logic
   Legacy ladder logic is the programming language that each manufacturer has been using for years. This style of ladder logic works only with controllers for which it was designed.

4. Structured Text
   Structured Text is a programming language in which the program is text-based. Key words have specific meanings defined by the programmer. For example, the word "open" may be defined as "close output point," and "FCV10" may be linked to a specific output point, in this case, a valve. So the phrase "open FCV10" causes the contacts on the output to FCV10 to close, thus causing FCV10 to open.

5. Sequential Function Chart (SFC)
   A sequential function chart program consists of two basic elements: a *state* and a *transition*. A "state" program block executes until an exit transition is satisfied, at which time the currently active state program block is disabled and the next block is started. Figure 2-55 presents an example of an SFC program in a washing machine. When the START button is pressed, the soap feeder auger is started. The auger runs for five seconds, after which the auger stops and the water valve opens. Water fills until LSH-10 trips. Then the water valve closes and the agitator starts, and so on. SFC programs are ideal for batch processes (e.g., pharmaceuticals) and for the few inevitable sequences related to continuous processes (e.g., petroleum refining).

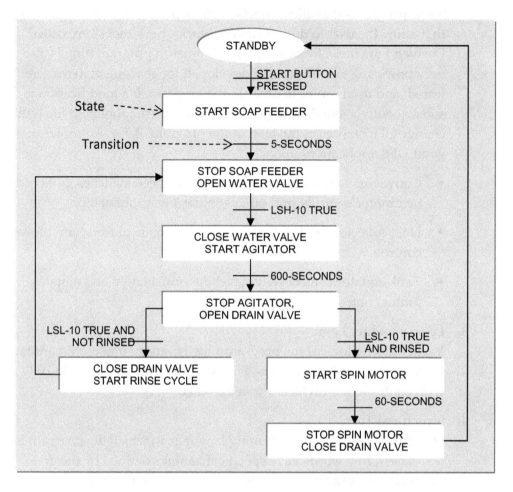

FIGURE 2-55. SEQUENTIAL FUNCTION CHART WASHING MACHINE SEQUENCE CONTROL APPLICATION

6. Continuous Function Chart (CFC)

    A CFC is a program fragment that runs continuously. Most analog logic fits well as a CFC. PID (proportional-integral-derivative) control functions are examples of CFCs. APT is a program by Siemens that combines the SFC with a continuous function chart. This is a powerful combination since the CFC is executed continuously while the SFC is sequenced. A PID control block that controls water temperature, for example, would be executed within a CFC, while the washing machine sequence is handled within an SFC.

    Figure 2-56 presents an example of a CFC. In this example, TT-14 is the water temperature signal. It is sampled by an analog input function block that converts the input signal to a digital integer value and then scales it to engineering units in degrees Fahrenheit. This value is fed to a PID function block and to two comparators. The PID block compares the actual temperature process variable (PV) to the external set point (SP) value. The PID block generates an output, which is fed to the heating element TY-14 through an analog output block. The two comparators compare the actual temperature to a high or low set point, and, if the relationship is true, set the proper analog alarm, which alerts the operator that the water temperature is out of bounds.

g. **Recommended Program Structure**

There are many ways to approach a PLC programming task. One approach that works well is to program by final control device. This method treats each final control device as a separate entity. Each pump, control valve, alarm horn, on/off valve, and so on, will have a distinct program structure that is reused as many times as there are similar devices. Interlocks, start/stop commands, and other variables are then configured to make each program block unique and specific to its associated final control device.

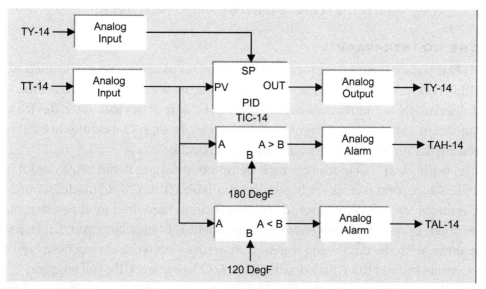

FIGURE 2-56. CONTINUOUS FUNCTION CHART WASHING MACHINE TEMPERATURE CONTROL APPLICATION

This type of programming lends itself to preliminary documentation that can be approved prior to starting work. A control detail sheet can be generated for each device, filled out to detail the control parameters of the device, and submitted for approval. Figure 2-57 presents an example of a control detail sheet.

After the device logic control detail sheet has been approved, the program "order of solve" must be considered. "Order of solve" comes into play in timing logic, where the order in which the computer executes its scan could affect the outcome of a calculation or sequence. Figure 2-58 shows a suggested format. On the first scan after power-up, the program needs to be initialized. In most cases, the software lets the programmer set initial values that will be automatically loaded at power-up. However, sometimes the initialization sequence is more involved, and programming is required.

After the system is initialized, all inputs should be scanned and processed. Data should be pulled from the operator interface or other outside "masters," such as other controllers or devices. Discrete inputs should be checked for alarm conditions. Analog inputs should be scaled (if necessary) and checked for alarm conditions. After the inputs have been processed, calculations and control algorithms should be invoked and applied to the device logic section. Then data should be packaged for transmission to the operator interface and to other users.

3. The I/O Interface

The I/O interface itself consists of one or more I/O modules. As mentioned earlier, PLC I/O modules come in a variety of configurations called form factors. I/O modules usually accommodate multiple input or output devices. Each device will be connected to a dedicated point on the module. Thus, I/O modules are characterized by the number of I/O points they can handle.

The traditional I/O module is rack mounted, meaning it slides into a slot in a card file. More form factors are becoming available. "Brick" I/O modules, such as the Genius block from GE Fanuc, are self-sufficient, hardened modules that may be mounted in remote, harsh environments. Field I/O may be wired directly to these units, with the data being transferred across a network connection. Newer I/O formats such as the Allen-Bradley Flex I/O line use a DIN-rail snap-on scheme that is compact and flexible. (A DIN-rail is a metal strip with a top-hat configuration that lends itself to a simple clamping arrangement that is designed into most electrical components, like relays, terminal blocks and other items that are mounted inside control cabinets.)

SAMPLE CONTROL DETAIL SHEET

DEVICE TAG:	M-14 AGITATOR
DESCRIPTION:	Washing Machine Agitator
LOCATION:	Inside the Washing Machine
REFERENCE:	

START CONDITIONS	RUN CONDITIONS
1. Door-Switch Interlock Made	1. Water Level Low Switch LSL-14 False
2. Water Level High Switch LSH-14 Tripped	2. "Soak" bit false.
3.	3. M14 Auxiliary Contact Made
4.	4. Emergency Shutdown False
5.	5. Agitation Timer Timing
START ACTIONS	STOP CONDITIONS
1. Start the Agitator Motor	1. Water Level Low Switch LSL-14 True
2. Start the "Agitaion Timer" - 600-seconds	2.
3.	3. M14 Auxiliary Contact Cleared
4.	4. Emergency Shutdown True
5.	5. Agitation Timer Timeout
PAUSE CONDITIONS	APPROVALS
1. "Soak" bit True.	DATE / SIGNATURE
2.	
ALARMS	
1. Command Mismatch	
2. Agitator Winding High Temp	
3	

Comments:

FIGURE 2-57. CONTROL DETAIL SHEET

a. Physical (Hardware) Address

Partitioning is done by assigning a physical address to an input or output item. An item's physical address provides information as to the point at which the signal wires attach to the control system. A good format for a physical address on a typical PLC system is as follows:

N02D01R04S05P12, where

- N02 is network node 2,
- D01 is drop 1,
- R04 is rack 4,
- S05 is slot 5, and
- P12 is point 12.

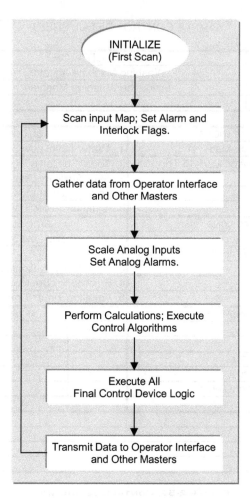

FIGURE 2-58. SUGGESTED PROGRAM FLOW OF CONTROL

The code format needs to relate to the system architecture. The physical addressing scheme should provide a unique vector to the signal's point of contact with the control system's hardware. That is why this is sometimes called a *hardware* address. Since most PLC software addresses bear little relationship to the signal's physical connection point (except in the PLC's I/O map table), the physical address is most valuable to the troubleshooter who must quickly locate a faulty module or device.

b. **Software Address**

A signal's software address is the location in the PLC memory at which the computerized version of the signal's value is found. It frequently has no relationship to the physical I/O module or rack or to any other physical device. This address is merely a memory location that has been internally mapped to a particular I/O module. As a result, the nomenclature for this address varies with the manufacturer of the control system. The following are some examples of a software address:

- Allen-Bradley PLC: N7.1 refers to an analog signal mapped to an integer storage location in Integer File 7, Word 1.

- Allen-Bradley PLC: B2.1 refers to a discrete signal mapped to a bit storage location in Boolean File 2, Word 1.

- Modicon PLC: 300001 refers to an analog input signal. In Modicon, a 3x register is always a physical analog input signal.

c. **I/O Map**

To properly size the control system and manage it throughout the project, each individual I/O point[8] must be counted and categorized by type. This is usually done in a database or spreadsheet, where each signal is mapped to its associated physical I/O address and then to its software I/O address.

There is a wide variety of signal types that might be handled by a control system. These signal types need to be classified. This is usually done within a database.

d. **Discrete (Sometimes called Digital or Binary) Signals**

Discrete signals can be provided in 24 VDC, 115 VAC, or just about any other common medium. They consume 1 bit of PLC memory. They refer to either input types or output types as follows:

- *Discrete Inputs* (I/O type DI) are generated by switches (e.g., limit switches, level switches, pressure, temperature, flow).

- *Discrete Outputs–Non-Isolated* (I/O type DO) are generated by the PLC output module. These I/O are assigned to devices, such as solenoids, alarm horns, beacons, relay coils, and annunciator windows.

- *Discrete Outputs–Isolated* (I/O type DOI) are also generated by the PLC output module. These I/O are assigned to high-current devices, such as single-phase motors, contactors for three-phase motors, and sirens.

---

[8] The term *point*, when used in reference to an I/O module, refers to a single I/O channel. A discrete input module that can accept 32 signals is referred to as a *32-point module*.

e. **Analog Signals**

Analog signals vary in value and generally consume one word, or 16 bits, in PLC memory. An example of an analog signal is a temperature signal generated by an RTD. As we have seen, an RTD signal varies across a range of a few millivolts, according to the temperature it is sensing. An RTD module converts this varying millivolt signal to a proportional integer value and stores the value at the location in PLC memory to which it is mapped.

- *Analog Inputs–Current* (I/O type code AIC) are typically 4–20 mA signals that are transmitted by, for example, pressure, temperature, or flow transmitters.

- *Analog Inputs–Voltage* (I/O type code AIV) are typically 0–10 or 1–5 VDC signals that are transmitted by various devices, such as chart recorders (when retransmitting), certain analog-to-digital (A/D) converters, and so on.

- *RTD Inputs* (I/O type code RTD) are categorized by the type of RTD. The most common RTD in use today is the 100 ohm platinum, though there are others. As has been mentioned, within those types there are other categories, such as two-wire, three-wire, or four-wire, that describe the various levels of line-loss corrections that can be employed through wiring techniques to increase signal accuracy.

- *Thermocouple Inputs* (I/O type code TCx where x is the type of thermocouple).

- *Analog Outputs–Current* (I/O type code AOC) are 4–20 mA outputs that drive transducers, positioners, and variable speed drives.

- *Analog Outputs–Voltage* (I/O type code AOV) are typically 0–10 VDC outputs that drive chart recorders, variable speed drives, and so on.

f. **Partitioning (I/O Mapping)**

The I/O interface bears directly on the amount of floor space needed for I/O termination, on cabinetry requirements, on processing power needed, and on electrical power distribution, so deriving an accurate I/O distribution early in the project is important. Each field device that generates a signal should be logged in the database and be assigned an I/O type code. This will allow an accurate I/O count to be obtained. But the device I/O count cannot be used directly to determine the I/O module count because other considerations may drive the designer to assign I/O in a way that is not the most space efficient.

	I/O TALLY WORKSHEET										
Process Area	P&ID Drawing	AI	AII	AO	RTD	T/C	DI	DO	DII	DOI	Total
Area 422	422-012	2	2	4			4				12
Area 422	422-013			3	3				2	2	10
Area 422	422-014	1				2					3
Area 423	423-001		4				16	8			28
Area 423	423-002			2					2	2	6
	SubTotal:	3	6	9	3	2	20	8	4	4	59
15%	Spares:	1	1	2	1	1	3	2	1	1	13
	I/O Grand Total:	4	7	11	4	3	23	10	5	5	72
	Points per Module	16	8	4	8	8	16	16	8	8	
	Modules:	1	1	3	1	1	2	1	1	1	12

I/O Types: **AI**-Analog Input; **AII**-Analog Input (Isolated); **AO**-Analog Output; **RTD**-RTD; **T/C**-Thermocouple; **DI**-Digital Input; **DO**-Digital Output; **DII**-Isolated Digital Input; **DOI**-Isolated Digital Output

FIGURE 2-59. I/O TALLY WORKSHEET

After the I/O type assignments are made, the I/O must be mapped to physical I/O point locations in the card set. The process of mapping the I/O is sometimes called partitioning. The partitioning process considers several factors as the I/O points are assigned to specific I/O termination point locations:

NOTE: Partitioning should be done before the control system is purchased, as the module count could be affected.

- *Redundant Equipment:* A properly partitioned system is one that takes advantage of the availability of spare field devices in order to improve fault tolerance. For example, if a set of twin pumps allows one to back up the other, then the two pumps should be loaded on different I/O modules to keep a single module fault from taking both pumps out of service. On the other hand, if the two pumps are configured in boost configuration, with both being required to run at times, then they should both be on the same module. It is also a good idea to load the modules by process subsystem. If there are several identical subsystems, then it is a good idea to load them in identical patterns on separate module sets. This practice requires more modules, but design engineering costs are reduced, and maintenance and fault tolerance are improved. It also causes natural gaps to appear in the I/O map, providing for future expansion.

- *Physical Location:* I/O points should be assigned in a logical way to facilitate checkout activities, simplify field cabling design, and reduce the effect of an I/O module (card) failure. This consideration will drive the designer to assign I/O modules to equipment items. For example, if one blower has seven discrete output (DO) points, it would be best to congregate the seven points on a single DO module, rather than spreading them across two or more. That way, if the module fails, only that blower is affected, whereas if the points are distributed across several modules, if any one of those modules fails, then the blower fails. Also, this structure will help simplify the field cabling by maximizing the multi-conductor homerun cable possibilities, and will also help in the construction checkout process, as checkout is generally done equipment item by equipment item.

- *Power Source:* It is generally a good idea, on isolated modules, to maintain a particular field voltage on all points of that module. Figure 2-45 showed an Isolated Digital Output module with a single AC I/O point loaded amid seven DC powered I/O points. This is a practice that is OK to do electrically, and is not against the NEC code, but it presents a potential hazard that should be avoided whenever possible. Electricians could get comfortable with having mostly DC circuits, which are safer to work with, and forget about that one AC circuit that is embedded in the wiring. Also, this kind of I/O practice makes routing the panel wiring difficult, as AC signal wires should not be in the same wire duct as DC signal wires.

- *Rack Power Supply:* Sometimes partitioning considerations are affected by rack backplane power budgets. Analog modules consume more power than discrete modules. The designer needs to know how much backplane power each module consumes so that a power consumption calculation can be made for the rack. If an I/O mix is driving the designer past the rack power budget, either a power supply module needs to be added, or some of the I/O modules will have to move to a different rack.

Figure 2-59 shows an example of an I/O count tally sheet. In this example, we have determined that areas 422 and 423 should be in the same I/O rack if possible. We first obtain a count from each of the five affected P&ID drawings. After the I/O tally has been obtained for each P&ID, a new total for each I/O type should be calculated to include spare capacity. In this case, 15% spare capacity is deemed sufficient. A calculation finds 15% of the subtotal for each I/O type, and the result is rounded up. Then an I/O grand total is found.

After the I/O grand total is obtained, a module count may be generated. The point density of each module should be researched and the number of modules derived. In the case in Figure 2-59, a total of 12 I/O modules is required. To expand on this example, assume a maximum rack size of 11 slots is available. It is apparent that the I/O needs to be reorganized such that it can be split into two racks.

The first rack needs to house the CPU, a power supply, and probably a network communications module. So that rack will have, at most, eight I/O slots available. The second rack will probably have nine available slots for I/O modules. Further, the designer has now made a decision to try to isolate Area 422 to one rack, and Area 423 to another. So the tally will need to be revised to reflect this new setup (Figure 2-60).

I/O TALLY WORKSHEET - RACK 1										
Process Area	P&ID Drawing	AI	AII	AO	RTD	T/C	DI	DO	DII	DOI	Total
Area 423	423-001		4				16	8			28
Area 423	423-002			2					2	2	6
	SubTotal:	0	4	2	0	0	16	8	2	2	34
15%	Spares:	0	1	1	0	0	3	2	1	1	9
	I/O Grand Total:	0	5	3	0	0	19	10	3	3	43
	Points per Module	16	8	4	8	8	16	16	8	8	
	Modules:	0	1	1	0	0	2	1	1	1	7

I/O TALLY WORKSHEET - RACK 2										
Process Area	P&ID Drawing	AI	AII	AO	RTD	T/C	DI	DO	DII	DOI	Total
Area 422	422-012	2	2	4			4				12
Area 422	422-013			3	3				2	2	10
Area 422	422-014	1				2					3
	SubTotal:	3	2	7	3	2	4	0	2	2	25
15%	Spare I/O Pts:	1	1	2	1	1	1	0	1	1	9
	I/O Grand Total:	4	3	9	4	3	5	0	3	3	34
	Points per Module	16	8	4	8	8	16	16	8	8	
	Modules:	1	1	3	1	1	1	0	1	1	**10**

9 Max!

FIGURE 2-60. REVISED I/O TALLY WORKSHEET REFLECTING NEW SETUP

The problem becomes immediately apparent as the number of I/Os required for Area 422 pushes the module count past the nine module limit. Further, reworking the I/Os to fit the rack configuration causes the module count to explode from 12 to 17. This represents a major cost impact and provides a good example of why this level of thought needs to be applied as early as the proposal stage if at all possible.

It is probably advisable to explore other alternatives, such as splitting the I/O racks by I/O type. Figure 2-61 shows Rack 1 loaded with all analog modules, while Rack 2 has all discrete modules. This split keeps the module count down, reducing it from 17 to 12, and meets slot availability criteria. This will probably be the approach of choice, provided the rack backplane has enough power supply capacity to support all the analog I/Os.

From this example, the value of using a spreadsheet for doing the I/O count and rack configuration is obvious. It is also apparent that a detailed I/O count is advisable.

F. Networking

There are basically three types of networks in use in industrial control systems: optimized/proprietary, optimized/nonproprietary, and open.

Note: A good resource for general information on networks is Omega's *Data Acquisition Systems Handbook*.[10]

1. Optimized/Proprietary Networks

Up until now, we have dealt with a single PLC as a single entity. In many applications, however, the components of a single PLC may be distributed to several locations, some of which are remote from the CPU primary. If remote racks are in use, data must flow from the remote racks to the primary. This is done through a remote I/O (RIO) local area network (LAN).

RIO LANs are considered proprietary control networks that have been optimized for deterministic data transfer between PLC components. In this context, the term deterministic implies that the communications protocol does not allow transferred data to be lost. If a data block is shipped out, the destination must send a reply to the effect that it received the message, or else an alarm is generated by the sender.

These RIO LANs are provided as native communications schemes by PLC manufacturers. In the case of a Modicon PLC, the RIO LAN is a single master network that supports up to 31 remote nodes. Each of the remote nodes must have a RIO adapter module, while the primary node with the CPU must have a RIO

I/O TALLY WORKSHEET - RACK 1										
Process Area	P&ID Drawing	AI	AII	AO	RTD	T/C	DI	DO	DII	DOI	Total
Area 422	422-012	2	2	4							8
Area 422	422-013			3	3						6
Area 422	422-014	1				2					3
Area 423	423-001		4								4
Area 423	423-002			2							2
	SubTotal:	3	6	9	3	2	0	0	0	0	23
15%	Spares:	1	1	2	1	1	0	0	0	0	6
	I/O Grand Total:	4	7	11	4	3	0	0	0	0	29
	Points per Module	16	8	4	8	8	16	16	8	8	
	Modules:	1	1	3	1	1	0	0	0	0	7

I/O TALLY WORKSHEET - RACK 2										
Process Area	P&ID Drawing	AI	AII	AO	RTD	T/C	DI	DO	DII	DOI	Total
Area 422	422-012						4				4
Area 422	422-013								2	2	4
Area 422	422-014										0
Area 423	423-001						16	8			24
Area 423	423-002								2	2	4
	SubTotal:	0	0	0	0	0	20	8	4	4	36
15%	Spares:	0	0	0	0	0	3	2	1	1	7
	I/O Grand Total:	0	0	0	0	0	23	10	5	5	43
	Points per Module	16	8	4	8	8	16	16	8	8	
	Modules:	0	0	0	0	0	2	1	1	1	5

I/O Types: **AI**-Analog Input; **AII**-Analog Input (Isolated); **AO**-Analog Output; **RTD**-RTD; **T/C**-Thermocouple; **DI**-Digital Input; **DO**-Digital Output; **DII**-Isolated Digital Input; **DOI**-Isolated Digital Output

FIGURE 2-61. I/O TALLY WORKSHEET WITH SPLIT BY I/O TYPE

head module. The primary rack is considered to be the head end, and the remote rack is considered a drop (Figure 2-62).

The various RIO modules are connected with a single RG-6 coaxial cable (or two cables if the network is redundant). A 75 ohm termination resistor (terminator) is needed at the far end of the cable to eliminate the echo effect. Also, a minimum cable length of 8.5 ft must be maintained for the Modicon RIO LAN. The frequency of the data signal is such that a minimum of 8.5 cable feet is needed for the signal to transition from 0 to 1 while traveling at light speed. A shorter cable can prevent proper operation.

Each RIO module must have a unique address, which is set by virtue of a selector switch on the module. The head communications processor controls the network, issuing calls to each of the drops in turn. If a drop does not respond after

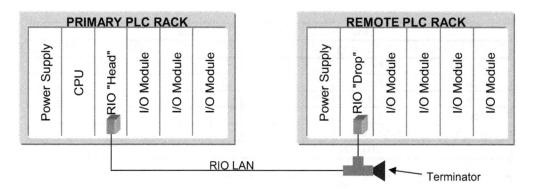

FIGURE 2-62. REMOTE I/O NETWORK

several calls, the drop is declared dead. An alarm light will activate, and polling frequency to that station will be reduced until a poll at that drop is answered. If the drop answers a call, then it is placed back online and full status is restored to it.

2. OPTIMIZED/NON-PROPRIETARY NETWORKS

Proprietary protocols connect PLC component to PLC component. However, the PLC must eventually communicate with the outside world. Before communication standards evolved, each PLC manufacturer had to develop a unique solution to this problem. Like the RIO LAN, these protocols were optimized to the specific PLC system. Unlike the RIO LAN, manufacturers made these protocols available for use with other manufacturers' equipment.

a. **Serial Communications (RS-232)**

The earliest form of data communications was *serial*, which was used with teletypes. RS-232 communications had several limitations, such as distance (fifty-feet max) and reliability (due to noise and other factors), but was, nevertheless, the primary form of data communications for many years.

The RS-232 protocol is characterized by a series of logical pulses (data bits) framed by stop bits. The pulse frequency (baud rate) must also be specified. A measure of error checking is provided by specifying parity mode: no parity, even parity, or odd parity.

Until recently, PLC manufacturers included the RS-232 protocol as the primary method for communicating with the operator interface, the printer, and many other external devices. They optimized the RS-232 protocol to their particular system by writing driver software that used the specified

RS-232 bit pattern to transmit encoded data specific to their system. The same driver software would then need to be installed on the external equipment as well in order for that equipment to be able to decode the transmitted data.

One example of such an optimized RS-232 protocol is the Modbus protocol, created by Modicon. Modbus has evolved into one of the more widely used serial protocols. Many equipment vendors such as annunciator vendors offer Modbus as standard on their equipment. The following excerpt from the Schneider Automation web site (http://www.schneider-electric.com) describes a typical Modbus network:

MODBUS
Schneider Automation uses an open data communications network called Modbus as the basis for the exchange of information among products on the factory floor. In a typical Modbus network, Modbus messages are sent over an RS-232 asynchronous serial communication link (EIA standard). This link inputs and outputs data alternately rather than concurrently. The two modes of communication within Modbus are 10-bit ASCII and 11-bit RTU.

The RS-232 protocol is still going strong today. Remote terminal units (RTUs) are standalone PLC systems that are parts of a larger PLC network connected via SCADA systems. Many times, the separate RTU nodes are connected to the central SCADA system, using modems or radios. The Modbus protocol (and other similar serial protocols) is used to connect the remote PLC to the modem or radio. The central SCADA system merely dials up the modem at the RTU and establishes communication. After the transaction is complete, the connection is broken, and the central SCADA dials up the next RTU.

b. **Optimized Local Area Networks**

Taking the RTU idea a bit further, the PLC manufacturers created LAN systems optimized to their particular systems. Modicon's Modbus-Plus, Allen-Bradley's Data Highway Plus, and Siemens's Profibus are examples of such protocols. In the case of Modbus-Plus, the scheme employed is that of peer-to-peer token passing.

With peer-to-peer token passing, as with the Modbus RTU, each node gets a turn in receiving and transmitting information. Unlike the Modbus RTU, however, the information does not need to flow to a central point. Each node is given the status of master for a period of time. This is referred to as "having the token." While a node has the token, it may broadcast global information to all the other nodes or specific information to specific nodes, or it may request information from specific nodes. When the time period is expired, the token is passed to the next node. During the time without the token, the node listens and responds to information requests from the node holding the token.

3. Non-Optimized (Open) Local Area Networks

In recent years, several events have transpired that have changed the industrial data communications landscape. Truly open protocols have emerged as viable industrial options. For example, Ethernet transmission control protocol with internet protocol (TCP/IP) has been the communications backbone of the business community for several years. This protocol has made significant inroads onto the factory floor. PLC manufacturers are now expected to be fully TCP/IP compliant.

Also, an attempt is being made to create a standardized international communication scheme for interconnecting field devices. Called FOUNDATION™ Fieldbus, this protocol allows the processing power of the DCS to be distributed to each field device, thus eliminating the need for long signal cable raceways. For more information on the Fieldbus Foundation, check online at www.fieldbus.org.

The architecture of the fieldbus scheme may vary with the vendor. In the case of Smar (www.smar.com), each field device is "smart" in that a small processor resides on the field device. A control valve, for example, may have an embedded PID controller hardwired to the transmitter. This controller controls the position of the valve and monitors its efficiency by monitoring its upstream/downstream pressure drops at specific valve positions, and communicating back to the host automation system across Foundation Fieldbus. Various asset-management software products can then analyze the information in order to predict failure modes or service needs.

4. Wireless Local Area Networks

Several wireless networking options are emerging. These wireless networks will likely become very common over the next decade. The ability to access a LAN without a physical connection is here.

a. **The "Bluetooth" Standard**

   The Bluetooth wireless standard provides wireless LAN communications for a distance of up to 30 feet. Most of us are familiar with the small Bluetooth earphones that allow us to communicate with our cell phones.

b. The "Wi-Fi" Standard (IEEE 802.11B)

The Wi-Fi wireless standard provides wireless LAN communications for a distance of up to 300 feet. Wi-Fi operates at speeds of up to 10 times that of Bluetooth. Devices that support 802.11B include laptops and other portables and handhelds. Operator control using handhelds is becoming more prevalent with small HMI screens being able to be downloaded to the handheld and the user being able to take temperature readings, stroke valves, operate pumps, and perform other activities.

5. THE ETHERNET CLIENT/SERVER ENVIRONMENT

The IT world is migrating ever so much into the controls world. In no other single area is this more apparent than in the control room, where the client/server computing model defines how information is managed and distributed across a network.

A *client* is a computer that requests data for local use. For example, a networked PC in an office is a client. It has its own software, perhaps Excel, or a CADD program that gives it a unique purpose. The PC is connected to a TCP/IP network. The software that manages information flow to and from the network is called a network client.

A *server* is a computer that monitors a network waiting for a client to request information. It serves the information up. A server may sit on more than one network, or subnet, gathering information from one and servicing information requests from the other. (A subnet is an isolated sub-network within the overall enterprise industrial network.) The information flow is bi-directional, with either subnet being able to request information and be serviced. But the server's software gives it a unique purpose. In the control room example (Figure 2-63), there are two server layers. The first is the Supervisory Control and Data Acquisition (SCADA) layer. The SCADA server has the HMI software package that contains the GUI, the alarm manager, and sometimes a data historian. The second server layer will have the Thin-Client (TC) servers that run a software package optimized for managing multiple thin-client operator stations. In that case, the TC servers are themselves clients of the SCADA servers. Physically, a server may appear to be a typical PC, or it can be in a rack-mount configuration. Servers can be optimized for high-speed data operations and high reliability and can be configured to be highly fault tolerant, with dual-core processors, dual power supplies and dual, hot-swappable mirroring Redundant Array of Independent Discs (RAID) hard drives.

a. **"Thick" Client Architecture**

A control room design with several screens, each of which is connected to a standalone PC, is called a thick architecture. While there are cases where this is still an appropriate scheme, it has some severe drawbacks. For example, if a change is made to a screen, then that change must be downloaded to each PC in the control room. Also, this is an expensive design, as each PC has its own software package that must be upgraded yearly and managed separately. Operationally, it is cumbersome, as each screen will likely have its own keyboard and mouse.

b. **"Thin" Client Architecture**

A control room design with several screens, each connected to generic boxes that all contain the same firmware and no software, is called a thin architecture. Firmware is read-only software that is loaded into a device in non-volatile memory. Firmware gives the device its character. In many cases today, a single thin client can support multiple (up to five) monitor screens with one keyboard and mouse. To support this architecture, a client-server, or pair of them, is dedicated to serving up the data to the various thin clients. Usually, the design uses a server pair, in order to provide some redundancy in case of a server crash. The thin clients can be configured to connect to Server A, as the primary, and fail over to Server B upon loss of communications.

c. **Ethernet Hardware**

An Ethernet network consists of computers, routers, hubs, bridges, gateways, and/or switches linked by optical fiber or copper (CAT5, typically, though CAT6e is recommended for the additional shielding). Usually, the backbone of the network is fiber because of the added transmission speed that is possible, and drops are CAT5 copper. The following are some simplified descriptions of the more common terms related to an Ethernet network:

- *Unmanaged Ethernet Switches* receive a packet into one port, discern the MAC (Media Access Control) address of the destination node, and distribute the packet out the proper channel to send it to the node of interest. In Ethernet communications, a *packet* is the lowest-level element on a local area network (LAN). Unmanaged switches are very cost effective, but have a minimum of onboard diagnostics.

- *Managed Ethernet Switches* receive a packet into one port, discern the MAC address of the destination node, and distribute the packet out the proper channel to send it to the node of interest, just as with unmanaged switches. But in addition, a managed switch also has an embedded router and supervisory capabilities, such as throughput limiting and a myriad of other features to give the integrator a means for managing network traffic.

- *Ethernet master* is the one managed switch among possibly several on a fault-tolerant ring topology that monitors network traffic to discern if a particular switch has failed, whereupon, the master switch begins to pass data. If all switches on the network are functioning, the master switch opens the ring to prevent data storms.

- *Routers* are used to link two subnets.

- *Repeaters* are used to extend the distance a network can cover. For example, a TCP/IP network running on CAT5 twisted pair copper cable can only cover about 100 meters. This can be extended by the use of a repeater.

- *Bridges* are used to link separate networks, either within the same communications protocol or between different protocols. For example, a device that converts Modbus-Plus token-passing network protocol to Ethernet TCP/IP is called an MBP-to-Ethernet bridge. A device that links two different TCP/IP subnets is called an Ethernet bridge.

Ethernet TCP/IP has made significant inroads into the industrial environment, though it was a slow process at first. Ethernet is a non-deterministic protocol, and industrial networks have historically been deterministic. With Ethernet, packets of information can be lost, with no reply to the sender. This can be managed, however, due to a couple of characteristics inherent in a client/server environment. A server program does not initiate anything—it merely responds to requests from client software packages, either on the same workstation or on the network. If a client asks for information, it expects a reply. If it doesn't receive the reply in a certain timeframe, the client sends the request again. This calling (polling) process continues for a specified timeout period, after which the client assumes that the server is unavailable, and an exception sequence is activated. An alarm is generated, or the client tries to find a backup server, or some other configurable action occurs. With everything operating perfectly, it may take multiple attempts for a client to have its requests serviced, but the speed of the TCP/IP network is so much higher than most other industrial networks that no appreciable delay will be noted.

It is important to use Ethernet in the appropriate way. For example, Ethernet as a Remote I/O LAN, passing control data, has historically been less acceptable than a dedicated, deterministic network for that application, as the indeterminate lag times could play havoc with control algorithms. Since Ethernet is not deterministic, the time it takes a command to propagate through the network is undefined, which can cause problems in certain control algorithms. This can be overcome, and most control system manufacturers have reacted to the super-high demand for Ethernet-ready equipment. But Ethernet still works best on the visualization layer of the network, where a classic client/server environment can be established.

Even though Ethernet is a "flat" (peer-to-peer) protocol, it is possible to emulate a fault-tolerant industrial ring configuration. This is done by virtue of one of the managed switches becoming the subnet master switch. If there is more than one managed switch on the network, the switches can be set to auto-negotiate to elect a master switch. The master makes sure the ring never closes entirely by continually monitoring the subnet, looking for stations going offline. As long as all the stations remain healthy, the master switch opens the ring at its location, thus preventing broadcast data storms.* If the master switch detects that one of the switches is offline, it closes that switch's link and allows data to pass through it, thus maintaining only one break in the network. When the faulty device is replaced, the managed switch detects it and re-opens its data bus.

*A data storm is a spike in traffic that chokes the network and causes it to crash. One way to cause a data storm is to configure an Ethernet network into a closed loop. When that occurs, packets are sent and re-sent until many copies of the same information packets are circulating on the network. This process continues until the network crashes.

6. THE INDUSTRIAL ENTERPRISE-WIDE NETWORK

Figure 2-63 shows a simplified enterprise-wide industrial network using all redundant, hot-standby technology. This network scheme uses the non-optimized, open client/server approach for the data storage/visualization layers, and the optimized, proprietary approach for the data collection layer. This narrative will describe the diagram from the bottom (field device) up.

a. **The Remote I/O (RIO) LAN**

RIO cabinets are scattered about the plant, located near their respective process areas. A process area could be a railcar unloading station, a screening plant, a scrubbing system, or a myriad of other operations that are physically adjacent. Each RIO location houses I/O modules, either rack-mounted in cabinets or distributed. The I/O modules pipe their data to an RIO communications module connected to the RIO LAN. An RIO head module in the PLC rack polls each RIO drop to collect the information. Most PLC manufacturers have their own proprietary RIO LAN for which they guarantee a certain poll rate. Those networks are usually slower than Ethernet because their communications protocols are deterministic, requiring feedback from the receiving station that a broadcast was received and readable.

b. **The Hot-Standby PLC**

Major PLC manufacturers can now compete with the DCS largely because they can now offer hot-standby processing. As mentioned earlier, this is two processors, both executing the same logic, but only one—the primary—being active, placing data on the data bus. The two processors continually monitor each other. If the primary processor fails a diagnostic test, then the backup assumes control within a specified time, and an alarm is generated. The processors are hot-swappable, allowing the technician to change the bad processor out and quickly return the system to hot-standby mode without powering down the rack.

Communications links tied to the hot-standby subsystems swap over as well, but these swapovers are sometimes not bumpless. In our example, the PLC monitors an RIO LAN and an Ethernet LAN (Subnet X). Since the RIO LAN is proprietary to the PLC manufacturer, that LAN will probably support a bumpless transfer on hot-standby swapover. But the Ethernet connection may or may not be automatic. It is up to the system administrator to understand how these networks can be affected, and to write exception-based programs to manage the transfer.

c. **Ethernet Subnet X**

Our example shows two PLC systems, each with a hot-standby processor pair. Each PLC takes data from its dedicated RIO LAN and ships information to the SCADA layer via multiple Ethernet switches, one for each PLC processor. Each half of the hot-standby processor pair has its own switch, but is cross-connected to a second switch for redundancy. The same data

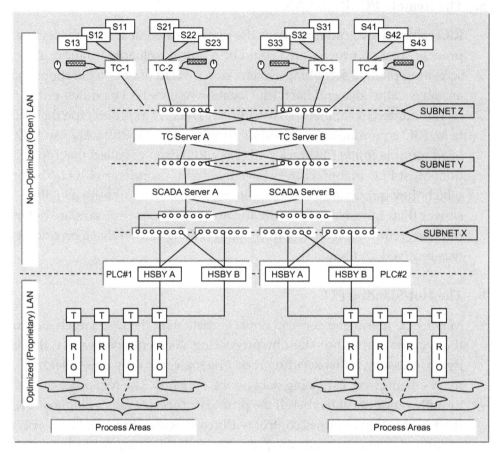

FIGURE 2-63. INDUSTRIAL NETWORK

is sent to both switches. If one switch fails, then the data from the PLC can still find its way onto the subnet through the other switch.

The Ethernet switches are themselves connected in a ring configuration that allows one switch to fail or be removed from service without affecting data flow (see Section 5c).

d. SCADA Server Pair

The SCADA server layer consists of a pair of servers configured as a primary and backup. The HMI software package sits on these servers. Their job is to poll the Subnet X PLC systems and gather information needed to populate and animate HMI screens and to stand by to serve the data up to Subnet Y upon request. The data shipped out to Subnet Y includes GUI screens and their associated data elements. The SCADA servers also take operator requests from Subnet Y and ship them to the PLCs on Subnet X.

e. **Ethernet Subnet Y**

Subnet Y is an Ethernet ring that services the four servers. It consists of two managed switches.

f. **Thin Client (TC) Server Pair**

The thin client server pair has thin client management software that manages requests for information being generated by the thin clients. These servers are also redundant, with their own associated Ethernet switches. As the operator navigates from one screen to the next, the thin client asks for the data needed to populate that screen.

G. Working with a Control Systems Integrator (CSI)

In Chapter 2, the Control System Integrator was introduced as a participant in the controls project. Many times, an industrial plant has knowledgeable control system professionals on staff. These personnel make process adjustments, conduct training, perform maintenance, provide troubleshooting services, and so on. Sometimes, however, this staff is not capable of handling large capital projects in addition to their day-to-day duties. This is where the CSI steps in. Systems integrators are, by and large, external contractors who are hired to provide programming and integration services. Sometimes they are individuals hired on an hourly basis to perform onsite changes under close supervision. Other times, a team of systems integrators works offsite to execute a turnkey project.

From the plant manager or controls engineer's point of view, how should a control systems integrator be handled? What services should a control systems integrator (hereafter, "integrator") be expected to offer, and what controls should be in place to increase the likelihood of a successful project?

1. Initial Search

Finding the right integrator is key. The first step is to determine the type of service that is needed. There are three primary categories for control systems integration services:

- *Staff Augmentation Service:* This service is used to fill a short-term labor problem. The staff augmentor is usually mated to a member of the customer's organization who provides close supervision in the performance of maintenance upgrades, production modifications, minor process revisions, and so on.

- *Turnkey Service:* The integration team is offsite and is largely unknown to the customer's organization. A project is executed off-site, and the integration team sends one or more representatives to assist in startup and training. An integration team is usually selected because of its familiarity with the equipment, its low bid, or some combination of the two. Of minimal importance to the customer is the anticipation of a long-term relationship.

- *General Integration Service:* This integrator needs to be in close proximity to the site and needs to demonstrate a desire to maintain a long-term relationship. This integrator is expected to provide rapid support on call and needs to have the capability of learning the entire plant operation.

After selecting an integrator that fits the situation, the next thing is to get a control system specification in their hands as part of the bid package.

2. Writing a Control System Specification

Writing a control system specification is one of those things a typical controls engineer does only a few times in a career—if at all. Some control system specifications are written directly to control system vendors, while others are written to CSIs that will be providing a turnkey system. The following narrative describes some of the considerations that should be made when writing a control system specification:

a. **Process Overview**

   The control system specification needs to give the prospective systems integrator an idea of what is involved in the project. A paragraph for each major control subsystem is about the right amount of detail at this stage, provided the specification is able to reference a set of Process Flow Diagrams or P&ID. The process overview should provide enough detail to reduce questions from the bidders.

b. **Operability**

   Operability issues center mainly around the HMI. A well-written control system specification will detail how the HMI should present information to the operator. The writer can either specify some or all of the items below or request that the bidders supply their recommendation for how they intend to deal with these topics, to be included as part of their bid packages.

   - *GUI Screen Graphics:*[9] If the plant has any established guidelines regarding the way the screen should be organized, or if there is a color

---

9   ISA-5.5-1985 discusses process graphics.

scheme that the integrator will be expected to comply with, it should be noted in this section.

- *A Navigation Plan* will provide a guideline for screen-to-screen maneuvers and can include a screen hierarchy showing an intuitive progression from the system overview to the device control popup overlay.

- *A Color Plan* will describe colors and textures for background, static equipment, active devices (valve color—open and closed, motor color, on and off, etc.), alarm symbols and colors, etc.

- *A Symbols Library* will provide a set of graphic symbols for each major active device in the system.

- *Alarm Manager:*[10] What is the alarm hierarchy, where alarms are classified and prioritized? What types of alarms are anticipated? For example, should the bidder anticipate two alarms per analog point or four? What about fault alarms? Mismatch alarms? What about audible alarms? For example, should an audible alarm sound if any alarm occurs or only the critical alarms? Is there a preference for how alarms are acknowledged?

- *Reports:* How many production reports are anticipated, and where will they be generated—from the production HMI, or from an engineering workstation? Are report formats available? If so, they should be included as attachments.

- *Historical Data Recorder:* Historical data collection should be detailed relative to how much trend data there is likely to be—all analog I/Os or a subset? How long should the data files remain available for recovery and display? What degree of resolution and how many samples per hour are needed? These things give the bidder an idea of how much to spend on the historian application.

- *Sequence of Events (SOE) Recorder:* A PLC-based system has a limited ability to provide SOE-type information due to the way it scans, and date/time stamping is very difficult. Some of the better PLC systems have SOE modules that can be added to their racks. These modules scan at a higher rate, and provide date/time stamps on state changes. So if a particular equipment item or subsystem is especially critical,

---

[10] ISA-RP77.60.02-2010 is a good resource for Alarm Management practice. It is focused on the power industry, but covers the topic in a broad sense for most other applications.

then that I/O can be designated for SOE recording. Those kinds of things should be considered when specifying a control system.

- *Cyber Security:* If there are any concerns regarding security issues, they should be delineated in the system specification. For example, will the equipment be in a locked room, or will the CSI supplier need to provide a lockable cabinet? Are firewalls needed? Will there be any web-based applications on the system?

- *Environmental and Ergonomics:*[11] If control consoles are being provided as a part of the specification, expectations regarding the console configuration should be included.

c. **Control**

Control schemes should be laid out for each major equipment item in the system. For example:

- *Device Control:* A *device* in this context is a control element such as a motor, valve, variable-frequency drive, actuator, solenoid, or other process equipment item that can affect plant operation if it changes state.

    - *Click Action:* As previously mentioned, it is wise to specify that all end devices react to operator commands only through a two-step process. The first action can be a mouse click on a graphic overview. That action may open a control popup overlay for the device that will have additional control buttons to select. One of the buttons on the overlay will close the overlay with no action taken. There are other schemes, with an "Are you sure?" message, for which another keystroke is needed before the selected action takes effect.

    - *Local-Off-Remote (LOR) Mode Switching:* Is the LOR switch in the field, or is it in the HMI? Sometimes a customer wants the ability to give priority to operators local to the motor (in the field), in which case the LOR switch will be in the field. Others don't want to risk trying to start up and having to send operators around to the motors to get them lined up for remote operation. If the customer wants to have a system that can be operated whenever the PLC goes offline, then the switch should be field mounted.

---

11   ISA-RP60.3-1985 is a recommended practice for control station ergonomic design.

- *Auto/Manual (A/M) Mode Switching:* Auto/Manual selection is only valid in Remote Mode. If in Local Mode, the only option is Manual, as the operator local to the motor is in control. Remote Mode passes control to the remote control system, which can either control the motor from the HMI via operator input (Manual), or from the PLC program (Auto).

- *Interlocks vs. Permissives:* Interlocks, if lost, should cause a device to switch out of Auto Mode and change to its safest state, whether on/off or open/closed. If all the device's interlocks are not made, it should be prevented from switching to Auto Mode. On the other hand, a permissive should only be active in Auto Mode. A permissive, if lost, should cause the device to move to its safe state. If the permissive is subsequently regained, the device should operate, provided the system is still in Auto. Losing a permissive should not switch the device out of Auto Mode. Also, it is important that the operator be able to see the interlock status on each device.

- *Switch Inputs:* Discrete inputs from switches generally need to be de-bounced to avoid inadvertent actions due to spurious noise or vibration-generated signals.

- *Motor Restart Inhibit Timers:* Timers should be provided to prevent motor restart within a time window. The window may be larger or smaller, depending on the size of the motor. These timers can come in with the motor controls, or they can be added to the PLC program.

- *Mismatched Timers:* If the actual state of a device differs from the commanded state, then a mismatched condition exists. If the states remain in mismatch for a timeframe, then an alarm should be generated, and the command should be removed. If the mismatch alarm is part of the interlock scheme, then the device will behave properly.

- *Continuous Control:* Analog control schemes, such as scaling, PID functions, ratio controls, and other analog algorithms, each have aspects of continuous control that can be discussed in the specification. For example, will the customer need to be able to change PID settings on the fly, or once a loop is tuned, can the settings remain in the background (i.e., hard coded)? Should the operator be able to change control and alarm setpoints? Is there a guideline for deviation alarms? Does a deviation alarm create a fault, which removes the interlock from the device, causing it to close or stop? What are the minimum controls the operator will need for each control algorithm? For more detailed information on continuous control, see Chapter 10.

- *Sequential Control:* Control systems always have sequences, even if it is only for startup and shutdown. What kind of control does the operator need, and what kind of feedback does he need to let him know sequence status? Some of the questions for sequences include: Auto/Manual, Start/Stop, Pause/Resume, Safestate. For more information, refer to Chapter 10.

- *Simulation:* What kind of simulation is needed, if any? The Functional Acceptance Test will be affected by the type of simulation that is specified. The simplest and cheapest form of simulation is simple loop-back logic that simulates feedback after a command to operate the device is issued. The loop-back logic can be added simply as an added feature of the de-bounce logic for discrete inputs.

- *Documentation:* How should the integrator document the control scheme: SAMA diagrams? Control details? Logic diagrams?[12] Narratives? For more information on documentation, refer to Chapter 10.

H. Selecting a Control System

Experienced controls engineers know that the selection of a control system is often dictated by circumstance. Choices may be limited by legacy equipment already at the site, by the availability of trained personnel, by function, etc. For example, if a plant already has a lot of old equipment, the default should be to stay with the original manufacturer's line of products to minimize the retraining cost. This default position can easily be overridden, however, if the site's historical experience with the legacy system is negative.

So make an assumption. Assume a worst-case scenario in which a legacy system exists but with a checkered past. The operations team likes the old system because they are used to it. The maintenance department hates the old system because they can't get parts, it's hard to work on, and they get very poor technical support from the manufacturer. This places the specification writer squarely in the middle. How should you, as the writer, proceed?

The PLC and the HMI should be given equal consideration. The following list, in order of precedence, should be applied to both elements individually, with the legacy and any new manufacturer being judged:

---

12 ISA-5.2-1976 (R1992) Binary Logic Diagrams for Process Operations. Also SAMA diagrams/symbols. Note: Scientific Apparatus Makers Association (SAMA) no longer exists as an organization. The functional diagram/symbols are now incorporated into ANSI/ISA-5.1-2009.

- *Legacy Benefit:* Will the retention of the legacy manufacturer be a benefit or a liability? What are the issues surrounding the existing system? Can a dollar value be placed on them, or are they solely a matter of preference?

- *Technical Support:* Is local support available? If so, how effective is it likely to be? If not, will the local sales staff be able to assist in getting the attention of the factory if necessary?

- *Local Sales Support:* Are they primarily motivated by sales, or are they motivated to help you succeed? How long have they been promoting this product? What is their customer base in the area? How does a sampling of this customer base rate the sales and technical support services of this system?

- *Investment:* How much will the system cost initially, and what is the projected long-term cost? Is a maintenance agreement a possibility? If so, how much benefit is it likely to provide?

- *Hardware*: Is the hardware accessible to a maintenance person? Is it hardy? Are replacement parts available locally, or will they need to be stored onsite? Do the power requirements fit the existing power distribution scheme, or will new power be needed?

- *Software:* Has the software been "burned in," or is it a new version? Is it user friendly and well documented? Is it scalable toward your particular future needs?

- *Training Investment*: How much retraining would be necessary? What is the cost of the training? Is it local? How often is it offered? How effective is the student likely to be after the course? In this case, again, querying the local user population can be helpful.

After the list of likely systems has been pared down, the control system should be viewed from a system-wide standpoint. The following are some issues to consider from that standpoint:

- If PLC-based, can the PLC and the HMI share a database? If not, is there a third-party software solution that would manage the two databases?

- Does the HMI have a standard interface driver that works well with the PLC, or will one need to be developed? (This was a bigger issue in previous years.)

Don't forget to consider the aftermarket for your legacy equipment. Frequently, your sales representative can point you to warehouses that store legacy equipment for resale. EBay is another option. So it is possible to recoup some of the original investment, provided your old system is serviceable.

References – Part II

1. "Type 'K' Revised Thermocouple Reference Tables." *The Temperature Handbook*, Volume MM. Stamford: Omega Engineering, Inc., 2000.

2. Earley, M. W., Sheehan, J.V., Sargent, J.S., Caloggero, J. M., and Croushore, T. M. *NEC 2002 Handbook* Quincy: National Fire Protection Association, Inc., 2002.

3. Ibid., "NEC article 344: Rigid Metal Conduit: Type RMC."

4. Ibid., "NEC Table 1: Percent of Cross Section of Conduit and Tubing for Conductors."

5. Ibid., "NEC Table 1: Percent of Cross Section of Conduit and Tubing for Conductors."

6. Ibid., "NEC Table 4: Dimensions and Percent Area of Conduit and Tubing for Conductors, Article 344: Rigid Metal Conduit (RMC), Percent of Cross Section of Conduit and Tubing for Conductors."

7. *Purged and Pressurized Enclosures for Electrical Equipment*, ANSI/NFPA 496-1998 Quincy: National Fire Protection Association [NFPA], 1998.

8. "Type 'J' Revised Thermocouple Reference Tables," *The Temperature Handbook*, Volume MM. Stamford: Omega Engineering, Inc., 2000.

9. ISA-5.5-1985, *Graphic Symbols for Process Displays*, Research Triangle Park ISA – International Society of Automation, 1985.

10. "Transactions in Measurement and Control," *The Data Acquisition Systems Handbook*, volume MM. Stamford: Omega Engineering, Inc., 2000.

PART III
TABLE OF CONTENTS

List of Acronyms – Part III 257

List of Figures – Part III 259

Introduction – Part III 267

Part III – Chapter 7: Piping and Instrumentation Diagrams (P&IDs) 271
    A. General Description 271
    B. Purpose 272
    C. Content 272
        1. Symbology (ANSI/ISA-5.1-2009) 273
        2. Symbol Identification 273
            a. Prefix 274
            b. Suffix 275
    D. Practical Application 276
        1. Tank Level: LT-10, LSH-10, LSLL-47 276
        2. Tank Fill: HV-13, ZSC-13 277
        3. Tank Discharge: PP-10 277
        4. Pump Discharge Pressure: PIC-48 277
    E. P&ID Summary 278

Part III – Chapter 8: Links to Mechanical and Civil 279
    A. General Equipment Arrangement Drawing (Civil and Mechanical) 279
        1. Purpose 279
        2. Interfaces 279
        3. Content 279
        4. Practical Application 280
        5. Equipment Arrangement Summary 280
    B. Piping Drawing (Mechanical) 281
        1. Purpose 282
        2. Interfaces 282
        3. Content (as related to I&C) 282
    C. Pump and Equipment Specifications (Mechanical) 282
    D. Links Summary 282

Part III – Chapter 9: Preliminary Engineering 283
    A. Development of a Detailed Scope of Work 284
        1. Purpose (Project Overview) 284
        2. Project Scope—I&C 284
        3. Safety Concerns 284
        4. Assumptions 284
        5. Exclusions 284
        6. Deliverables 284

      7. Milestone Schedule 285

  B. Control System Orientation 285

  C. Project Database Initialization... 288
      1. Initialize Document Control Table 289
          a. Table.. 290
          b. Query for Ordering Drawings....................................... 291
          c. Transmittal Query 292
      2. Initialize Instrument and I/O List Table 295
          a. Instrument Table 296
          b. Queries 297
          c. Reports... 300
      3. Database Summary 302

  D. Estimate and Schedule Development*... 302
      1. Cover Worksheet 303
      2. Devices Worksheet 303
      3. Count Worksheet 309
      4. Labor Worksheet 310
      5. Summary Worksheet... 314
      6. Schedule Worksheet 316
      7. Estimate and Schedule Summary 317

  E. Preliminary Engineering Summary 319

Part III – Chapter 10: Control Systems Integration (CSI) 321

  A. FEL Stage 1 – Business Planning 321
      1. Cost/Benefit Analysis... 322
      2. Control System Specification 323
      3. Functional Description 324
          a. Tank Level: LT-10, LSH-10, LSLL-47................................ 324
          b. Tank Fill: HV-13, ZSC-13 324
          c. Tank Discharge: PP-10.. 324
          d. Pump Discharge Pressure: PIC-48 324
      4. Project Estimate 325
          a. Field Device Control Elements...................................... 325
          b. Sequential Control Elements 325
          c. Continuous Control Elements 326
          d. Overview Graphic Screens.. 326
          e. Historical Data Acquisition and Reports 326
          f. Network Activities 326
      5. Project Proposal 327

  B. FEL Stage 2 – Project Definition 327
      1. Sequential Function Chart (SFC).. 328
      2. Continuous Function Chart (CFC) 328
      3. Control Narrative... 329
      4. Sequence Control Detail Sheets (SCDS) 329
      5. Device Control Detail Sheets (DCDS) 329
      6. Functional Logic Diagrams 329

  C. Control Narrative 329
      1. Sequential Function Chart (SFC).. 329
      2. Continuous Function Chart (CFC) 333
      3. SFC Control Narrative Fragment 334
          a. Powerup & Initialize 335
          b. S00 – State Zero: STANDBY 335

Table of Contents – Part III

    c. S01 – State One: AUTO-TK10 (Automatic Mode) 335
    d. S02 – State Two: FILL_TK10 (Fill the Tank). 336
    e. S05 – State Five: EMPTY_TK10 (Empty the Tank) 336
    f. S03 – State Three: TRACK_PP13 (Track the Pump) 336
   4. Sequence Step Detail Sheet (SSDS) 337
    a. Step S02 – Fill Tank 338
    b. Step S05 – Empty the Tank (Figure 3-60) 342
    c. Sequence Step Detail Sheet (SSDS) Summary 345
   5. Device Control Detail Sheet (DCDS). 345
    a. Pump PP-10 Device Logic. 346
    b. Valve HV-13 Device Logic 350
   6. Functional Logic Diagram 350
    a. Tank TK-10 Control Sequence Step 02 353
    b. Tank TK-10 Control Sequence Step 05 354
    c. Tank TK-10 Level Detect and Fill Control Logic. 354
    d. PP-10 Pump Discharge Pressure Control Logic 357
    e. PP-10 Pump Control Logic. 359
    f. Logic Diagram Summary 362
   7. Logic Diagram Standard ISA-5.1. 362
   8. FEL2 Systems Integration Summary. 363
  D. Operator Interface Specification Development – The HMI 365
   1. Animation Plan 366
    a. Colors. 366
    b. Visibility 366
    c. Flash and Beep 367
    d. Messaging. 367
   2. Screen Diagrams. 367
    a. Graphic Screen 369
    b. Control Overlays 370
    c. Component Schedule. 372
   3. Tagname Database, Device Driver, and I/O Mapping 374
   4. Finished Graphics Screen 375
   5. Alarm Manager 377
   6. Historian 378
   7. HMI Report Generation 378
  E. Network Single-Line Diagram Generation 378
  F. Other Systems Integration Tasks. 379
   1. Control System Cabinetry Design and Delivery. 379
   2. I/O Address Assignment (Partitioning). 379
    a. Hardware (HW) Address 380
    b. Software (SW) Address. 381
   3. Factory (or Functional) Acceptance Test (FAT). 381
   4. Site Acceptance Test (SAT) 384
   5. Commissioning 385
   6. Operations and Maintenance (O&M) Manual. 385
    a. Operations. 386
    b. Maintenance 386
   7. Onsite Training. 386
  G. Systems Integration Summary 387

Part III – Chapter 11: Information Management. 389

  A. Document Control 389

  B. Instrument and I/O List 390

    1. Instrument and I/O List Table 391
    2. Preliminary Design Query 391
    3. Plan Drawing Takeoff Query 392
    4. Plan Dwg Takeoff Query Report 392
    5. X-Ref Document Cross-Reference Query 393
    6. X-Ref Document Cross-Reference Report 394
  C. Database Summary 394

Part III – Chapter 12: Instrument Specifications 397

  A. Purpose 398
    1. Mechanical Designers 398
    2. Instrument Designers 399
    3. Other Users 401
  B. Interfaces 401
  C. Examples 402
    1. LT/LSH-10 402
    2. PV-48 405
  D. Summary 407

Part III – Chapter 13: Physical Drawings 409

  A. Control Room 409
    1. Environmental Issues 409
      a. Heating, Ventilation and Air Conditioning (HVAC)9 409
      b. Noise Levels 410
      c. Lighting 410
    2. Physical Arrangement 410
    3. Control Room Design Summary 411
  B. Termination Room 411
    1. Environmental Issues 411
      a. Lighting 411
      b. Heating, Ventilation, and Air Conditioning 412
      c. Piping 412
      d. Computer Floors 412
    2. Furniture and Equipment Arrangement 413
      a. Personnel Clearances 413
      b. Ingress and Egress 413
    3. Termination Room Design Summary 413
  C. Process Area (Instrument Location Plan) 413
    1. Why Produce Instrument Location Plan Drawings? 415
    2. Anatomy of an Instrument Location Plan 415
    3. Design Considerations 416
    4. Drawing Production Technique 417
      a. Step One: Initialize Drawing (Generate drawing background) 417
      b. Step Two: Spot and Classify Instruments and Instrument Groups 418
      c. Step Three: Build Cable Schedule and Size Field Junction Boxes 422
      d. Step Four: Spot Field Junction Boxes 422
      e. Step Five: Route Conduit 423
      f. Step Six: Calculate Conduit Fill 425
      g. Adjust Conduit Routings for Legal Fill 425
      h. Assign Conduit Routing Numbers 425
      i. Generate Component Schedule 426
    5. Material Takeoff 429

    D. Instrument Installation Details 433
        1. Electrical Installation Details 434
        2. Tubing Details 435
        3. Mounting Details 436
        4. Related Database Activities 436
        5. Material Takeoff 438
    E. Summary 438

Part III – Chapter 14: Instrument and Control Wiring 441

    A. Instrument Elementary (Ladder) Diagram 444
        1. Motor Elementaries 447
        2. AC Power Distribution Schematic 449
        3. DC Power Distribution Schematic 451
        4. PLC Ladder Diagram (Elementary) 452
            a. Discrete (Digital) Inputs 455
            b. Digital Outputs, Isolated 457
    B. Loop Sheet (Ref: ISA-5.4-1991)[14] 458
    C. Connection Diagrams 462
        1. Junction Box JB-TK10-1: Initial Layout 463
        2. Termination Cabinet TC-2 469
            a. DC Circuits (TS-2) 470
            b. AC Circuits (TS-1) 472
            c. Incoming Power 474
    D. Wiring Summary 479

Part III – Chapter 15: Panel Arrangements 485

    A. Procedure 486
    B. Junction Box JB-TK10-01 Arrangement Drawing ARR-002 487
        1. Set Up a Scale 488
        2. Design the Panel 488
        3. Generate a Bill of Materials 491
    C. Summary 492

Part III – Chapter 16: Procurement 493

    A. Typical Purchasing Cycle 494
    B. Material Classification 496
    C. Bulk Bill of Materials 496
    D. Detail Bill of Materials 501
    E. Procurement Summary 506

Part III – Chapter 17: Quality Control—The Integrated Design Check 509

    A. Administrative Content – Individual Checks 509
    B. Technical Content – Squad Check 510
    C. Squad-Check Roster 511
    D. Design-Check Summary 511

Part III – Chapter 18: Phase 3—Deployment 513

    A. Construction 513

       1. Kickoff Meeting 513
       2. Construction 514
  B. Pre-Commissioning.. 515
  C. Cold-Commissioning (Site Acceptance) 516
       1. Device Tests 516
       2. Subsystem Tests... 517
  D. Hot-Commissioning (Startup) 517
  E. Adjustment of Document Package to Reflect Construction Modifications 517
  F. Issue for Record 518
  G. Phase 3 Summary 518

Part III – Chapter 19: Phase 4—Support 519
  A. Warranty Support... 519
  B. Continuing Service Support... 519

References – Part III 521

LIST OF ACRONYMS – PART III

A/M	Auto-Manual Selector Switch
CFC	Continuous Function Chart
CPU	Central Processing Unit
CSI	Control Systems Integrator
CV	Control Variable
DCS	Distributed Control System
DFB	Derived Function Blocks
ENET	Ethernet
EU	Engineering Units
FEL	Front-End Loading
FJB	Field Junction Box
GA	General Arrangement Drawing
GUI	Graphic User Interface
HMB	Heat & Material Balance Diagram
HMI	Human/Machine Interface
HOA	Hand-Off-Auto Selector Switch
I/O	Input/Output
IEC	International Electrotechnical Commission
ISA	International Society of Automation
LAN	Local Area Network
LOR	Local-Off-Remote Selector Switch
mA	milliamp
MCC	Motor Control Center
mV	millivolts
NEC	National Electric Code
NFPA	National Fire Protection Association
O&M	Operations & Maintenance
OEM	Original Equipment Manufacturer
P&ID	Piping & Instrumentation Diagram
PFD	Process Flow Diagram
PV	Process Variable
RIO	Remote I/O
RTD	Resistance Temperature Detectors
RTU	Remote Terminal Unit

SAMA	Scientific Apparatus Maker's Association (now defunct, absorbed by ISA)
SCADA	Supervisory Control and Data Acquisition
SFC	Sequential Function Chart
SP	Setpoint
TC	Thin Client
VAC	Volts, alternating current
VDC	Volts, direct current
WAN	Wide Area Network
WBS	Work Breakdown Structure
XMTR	Transmitter

LIST OF FIGURES – PART III

Figure 3-1. Instrumentation and controls engineering tasks (Phases 1 – 3) 269

Figure 3-2. Typical feed tank configuration... 270

Figure 3-3. Typical P&ID symbology 274

Figure 3-4. Typical P&ID symbology showing combined automation system functions......... 275

Figure 3-5. P&ID presentation of the TK-10 subsystem..................................... 276

Figure 3-6. Basic P&ID drawing 278

Figure 3-7. TK-10 feed tank with equipment labels 280

Figure 3-8. TK-10 Feed tank area equipment arrangement................................... 281

Figure 3-9. Detailed Scope of Work... 283

Figure 3-10. Existing control system 287

Figure 3-11. Revised control system 288

Figure 3-12. List of tables 289

Figure 3-13. Document control table structure 290

Figure 3-14. Document control table, datasheet view 291

Figure 3-15. OrderDrawingsQuery (design view) 292

Figure 3-16. Document control table data 292

Figure 3-17. Transmittal query.. 293

Figure 3-18. Transmittal query design view (with criteria filter) 294

Figure 3-19. Transmittal query, datasheet view 294

Figure 3-20. Instrument and I/O list table... 297

Figure 3-21. Tagname update query, design view 298

Figure 3-22. TagnameUpdateQuery, design view, with criteria filter......................... 298

Figure 3-23. Query tagname display... 299

Figure 3-24. Reports... 300

Figure 3-25. Report wizard 301

Figure 3-26. P&ID takeoff query report 301

Figure 3-27. P&ID takeoff query report, design view....................................... 302

Figure 3-28. Finished database products 303

Figure 3-29. Cover sheet for Estimate workbook 305

Figure 3-30. Devices worksheet 306

Figure 3-31. Devices I/O assignment index table 307

Figure 3-32. Devices I/O assignment index, revised 308

Figure 3-33. Devices I/O calculator. 308

Figure 3-34. Count worksheet 309

Figure 3-35. Background data table. 310

Figure 3-36. I/O configuration worksheet 310

Figure 3-37. Labor worksheet. 311

Figure 3-38. Direct engineering labor, Phase 1 312

Figure 3-39. Direct engineering labor, Phase 2 313

Figure 3-40. Indirect engineering labor, Phase 2 313

Figure 3-41. Engineering and construction labor, Phase 3. 313

Figure 3-42. Engineering summary worksheet 314

Figure 3-43. Project cost summary table. 315

Figure 3-44. Engineering cost summary table 315

Figure 3-45. Phase 1 deliverables summary table 316

Figure 3-46. Phase 2 deliverables summary table 316

Figure 3-47. Instrument and I/O summary table 317

Figure 3-48. Schedule worksheet 318

Figure 3-49. Design schedule and staffing plan 319

Figure 3-50. Project manhour loading chart. 320

Figure 3-51. Systems Integration services checklist 322

Figure 3-52. Existing control system 327

Figure 3-53. New control system 328

Figure 3-54. TK-10 feed tank control sequence overview 330

Figure 3-55. Sequential function chart fragment. 330

Figure 3-56. Sample sequential function chart logic. 331

Figure 3-57. Sequential function chart (SFC) 332

Figure 3-58. Continuous function chart 333

Figure 3-59. Sequence step 2: "fill tank" sequence 340

List of Figures – Part III

Figure 3-60. Sequence step 5: "empty tank" sequence 343

Figure 3-61. Pump PP-10 motor controls elementary wiring diagram 347

Figure 3-62. Pump PP-10 device control detail sheet 348

Figure 3-63. HV-13 fill valve device control detail sheet 351

Figure 3-64. Sample logic diagram format 352

Figure 3-65. Logic diagram showing rat holes 353

Figure 3-66. Naming conventions for this project 353

Figure 3-67. Timing diagram for a delay timer 354

Figure 3-68. FILL_TK10 control logic 355

Figure 3-69. EMPTY_TK10 control logic 355

Figure 3-70. Device logic for TK-10 fill controls and analog alarms. 357

Figure 3-71. PP-10 device logic 358

Figure 3-72. PP-10 device on/off logic 360

Figure 3-73. One-shot, rising (OSR) edge function block 361

Figure 3-74. Pump restart inhibit signal processing 361

Figure 3-75. Sample logic diagram 362

Figure 3-76. Selected SAMA symbols now incorporated into ISA-5.1 363

Figure 3-77. Functional control diagram (ISA-5.1) 364

Figure 3-78. Animation plan 368

Figure 3-79. Preliminary screen graphics, TK-10 overview screen 369

Figure 3-80. Sample control overlays 371

Figure 3-81. Pop-up overlays 371

Figure 3-82. Animation detailing 372

Figure 3-83. Pump PP-10 control overlay 373

Figure 3-84. Animation chart 373

Figure 3-85. Final screen diagram 374

Figure 3-86. Typical data progression 375

Figure 3-87. HMI screen, pumping out in manual 375

Figure 3-88. HMI screen, filling in auto 376

Figure 3-89. HMI screen, sequence status 377

Figure 3-90. Sample alarm manager database 377

Figure 3-91. Historian sampling points 378

Figure 3-92. Simple network single-line diagram 379

Figure 3-93. Hardware addresses 381

Figure 3-94. Software addresses 382

Figure 3-95. Adding the I/O modules 382

Figure 3-96. Instrument and I/O list table, design view 391

Figure 3-97. Instrument and I/O list database, datasheet view 392

Figure 3-98. Preliminary design query 392

Figure 3-99. Plan drawing takeoff query 393

Figure 3-100. PlanDwgTakeoffQuery report 393

Figure 3-101. Plan drawing component schedule (Microsoft® Access to Microsoft® Excel). 394

Figure 3-102. Document cross-reference (X-ref) query 394

Figure 3-103. Document cross-reference report 395

Figure 3-104. Ultrasonic level transmitter 403

Figure 3-105. Instrument specification for LT/LSH/LSL-10 404

Figure 3-106. Instrument specification for Control Valve PV-48 406

Figure 3-107. Three termination room configurations 412

Figure 3-108. Sample instrument location plan drawing 414

Figure 3-109. Initialize drawing 418

Figure 3-110. Locate major equipment items 419

Figure 3-111. Locate instrument items 420

Figure 3-112. Add instrument stations 421

Figure 3-113. PlanDwgTakeoffQuery 421

Figure 3-114. PlanDwgTakeoffQuery, filtered 422

Figure 3-115. 3D to 2D and back 424

Figure 3-116. Add conduit detail 424

Figure 3-117. Recommended conduit tagging convention 426

Figure 3-118. Instrument arrangement with support data 427

Figure 3-119. Cable code cross-reference chart 427

Figure 3-120. Component schedule 428

Figure 3-121. Plan001 component schedule 429

List of Figures – Part III

Figure 3-122. Cable and conduit takeoff approach 430

Figure 3-123. Cable takeoff method 431

Figure 3-124. Conduit sizing calculator results 431

Figure 3-125. Cable takeoff by leg 432

Figure 3-126. Conduit takeoff 433

Figure 3-127. Instrument conduit installation detail 434

Figure 3-128. Instrument electrical installation detail 435

Figure 3-129. Instrument mechanical hookup detail 436

Figure 3-130. Instrument mechanical detail with throttling valve 437

Figure 3-131. Instrument mounting detail 437

Figure 3-132. Database log of details 438

Figure 3-133. Wiring design basics 441

Figure 3-134. Fabrication 442

Figure 3-135. Wiring interconnections 442

Figure 3-136. Elementary wiring diagram 443

Figure 3-137. Typical instrument elementary content 444

Figure 3-138. 4-pole relay coil with contacts 446

Figure 3-139. Four-pole relay coil with cross-references to its contacts 446

Figure 3-140. Four-pole relay contacts with cross-reference to its coil 446

Figure 3-141. Motor elementary wiring diagram 447

Figure 3-142. Motor elementary wiring diagram showing fused transformer output 448

Figure 3-143. AC power distribution elementary wiring diagram 450

Figure 3-144. AC power panel loading chart 451

Figure 3-145. DC power distribution elementary wiring diagram 451

Figure 3-146. Instrument elementary wiring diagram concept 453

Figure 3-147. Traditional ladder elementary—washing machine application 454

Figure 3-148. "Unhide Columns" window 455

Figure 3-149. Instrument and I/O list table, filter by selection 455

Figure 3-150. PLC digital input module elementary wiring diagram 456

Figure 3-151. Filtered on DOI (digital output, isolated) 458

Figure 3-152. Digital output (isolated) PLC output module elementary wiring diagram 459

Figure 3-153. Loop sheet 460

Figure 3-154. Advanced filter/sort 460

Figure 3-155. Advanced filter/sort, field selection 461

Figure 3-156. Results of advanced filter/sort 461

Figure 3-157. Creating a connection diagram 463

Figure 3-158. Terminal strip creation 464

Figure 3-159. Instrument elementary diagram, digital input module 464

Figure 3-160. Instrument elementary diagram, digital output module 466

Figure 3-161. Termination drawing setup 467

Figure 3-162. Termination chart 468

Figure 3-163. Motor elementary fragment 468

Figure 3-164. Finished termination chart 469

Figure 3-165. Junction box wiring diagram 470

Figure 3-166. Inner panel, cabinet TC2 471

Figure 3-167. DC wiring 472

Figure 3-168. Wiring diagram section of TC-1 474

Figure 3-169. AC power distribution 475

Figure 3-170. LT-10 power feed 475

Figure 3-171. Wire runs= 476

Figure 3-172. Fuse/terminal numbering sequence 477

Figure 3-173. NFPA wire color scheme 478

Figure 3-174. TC-2 wiring color scheme 478

Figure 3-175. TC-2 PLC cabinet connection diagram 480

Figure 3-176. Partial junction box diagram 480

Figure 3-177. Partial motor elementary wiring diagram 481

Figure 3-178. Power distribution information 481

Figure 3-179. Pressure control loop PIC-48 loop sheet 482

Figure 3-180. Ladder diagram for discrete modules 482

Figure 3-181. Document control table 483

Figure 3-182. Document control table and instrument and I/O list table 483

Figure 3-183. Instrument and I/O list table 484

List of Figures – Part III

Figure 3-184. Terminal block 487

Figure 3-185. Setting up a scale 489

Figure 3-186. Initial layout 489

Figure 3-187. Single-door enclosure 490

Figure 3-188. Junction box with bill of materials. 491

Figure 3-189. Finished panel arrangement. 492

Figure 3-190. Typical procurement cycle 495

Figure 3-191. Bulk materials takeoff worksheet 498

Figure 3-192. Wire and cable calculation table 500

Figure 3-193. Terminations and cabinetry 500

Figure 3-194. Conduit and conduit fittings 501

Figure 3-195. Installation detail assignment data 502

Figure 3-196. New detail sheet tally 502

Figure 3-197. Material tabulation by detail 503

Figure 3-198. Consolidated material with detail quantity. 503

Figure 3-199. Total item quantities 504

Figure 3-200. Part number and price. 505

Figure 3-201. Final bill of materials worksheet 505

Figure 3-202. Sort by description. 506

Figure 3-203. Engineering bill of materials 507

Introduction – Part III

As seen in Part I of this book, a project is an engineering deliverable in its own right. Part II focused on the state of the art of controls and automation – from basic wiring techniques to more advanced network architectures. Part III of this book focuses on the engineering and design portion of the project: Project Flow Phase 2 – System Design.

Success in Phase 2 is predicated on a well-executed Phase 1 (discussed in Part I) in which a Business Plan was developed, a Project Definition was produced, and a Project Plan was generated. Phase 1 defines the conditions that constitute a successful project. These success parameters - Schedule, Scope, and Budget – must be constituted in advance and agreed to by all participants. If those Phase 1 elements are present – even on an informal basis, then a successful Phase 2 is much more likely.

A well managed, well executed project is a wonderful thing to behold. While the object is to generate quality products and services (not to perfect the process of doing so), a well-organized, consistent approach employed project-to-project will improve the odds. If the design process itself is as repeatable as possible, then the design staff can make use of lessons learned and gradually become more proficient – and thus more quality-conscious and cost-effective. Understanding the order in which deliverables should be produced and knowing what the predecessors are to each deliverable are key to developing a repeatable engineering and design process. In any case, a "perfect" engineering design package (i.e., a set of engineering deliverables) is rarely the same from one project to the next, even between two projects for the same customer. A perfect design package—from the point of view of the engineering firm and the customer—is simply one that gets the job done according to specification and within schedule and budget. A package that perfectly satisfies either the constructor or the plant operator is likely to blow the budget, while a minimized engineering package that is cut down to fit a budget that is too tight might torpedo the project during construction or present problems for maintenance later. Content quality should never be sacrificed. But at what level of detail does increased quality become impractical?

The key is to find that specific mix of products and services provided at the proper level of detail that will meet the requirements. In order to meet the requirements, one must know what they are (following the process described in Part I). In order to determine the correct level of detail, one must know what the options are (as described in Part III).

The first step in this assessment is to understand the nuances. What is the difference between a *product* and a *service*? A product is something tangible that may be handed over to the customer or to the construction team. An example of this is a bill of materials or a drawing. The entire set of such products is called the *engineering package*. A service is work that does not necessarily result in a physical deliverable. An example of an engineering service is consulting (advising) or construction management, in which the deliverable is contributing to a smoothly functioning construction process.

The engineering project consists of several distinct elements. Figure 3-1 shows the basic set of individual products and services for a typical I&C project in a flow chart format (Note: In other parts of this book, the Electrical/Instrumentation & Controls (E/I&C) discipline is described, because typically that is how the engineering departments are organized. Part III provides information primarily regarding the "I&C" aspect of that engineering discipline's work).

The engineering package consists of two types of products:

- *Engineering products* are needed by customers, either for their records or for maintenance and upkeep of their facility. Some engineering products are necessary for procurement, and may also be needed for construction. Some engineering products are required as legal documents, to be maintained after the project. Whether issued for construction or merely issued for record, engineering products must meet or exceed customers' drafting standards and comply with their document control guidelines, as described in the project specification. Engineering packages can be further subdivided into engineering deliverables (i.e., products that must be stamped by a Professional Engineer (PE), and design deliverables (i.e., products that do not need to be stamped by a PE). The mix of engineering and design deliverables can change from project to project, and in most industrial cases, PE stamps for I&C packages are not even required.

- *Construction products* are needed for the construction of the facility, but are probably not maintained afterward. An example of such a product is an instrument location plan drawing (described herein), which is a time saver and workflow management tool for the construction team, but is generally not required for the day-to-day operation of the plant. Other examples of construction products include panel fabrication drawings, cable and conduit schedules, bills of material, etc. Sometimes, these and other construction products are generated in the project, but they are issued in the form of sketches to bypass the sometimes rigorous drawing standards of the customer, thus saving time and money.

INTRODUCTION – PART III

Engineering services are tasks that do not lead directly to products. Teaching a training class, providing procurement services, consulting, and performing a walkdown are examples of services that are part of a project's scope, but do not directly result in an engineering deliverable.

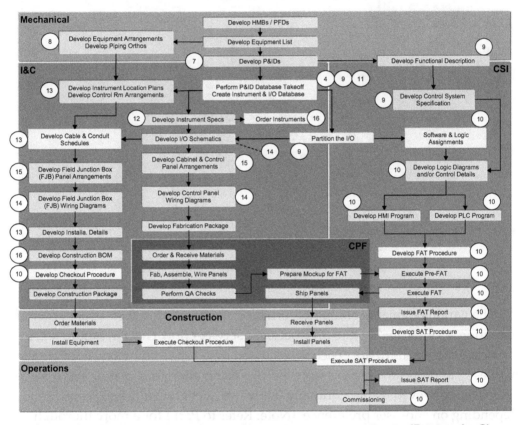

FIGURE 3-1. INSTRUMENTATION AND CONTROLS ENGINEERING TASKS (PHASES 1 – 3)

The purpose of Part III is to familiarize the reader with a range of products and services that make up a typical Instrumentation and Controls (I&C) engineering project, particularly focusing on the I&C aspect. It is organized by products and services flow-chart item number, as listed in Figure 3-1. Industry guidelines and/or standards are referenced where appropriate. In order to accomplish this, the various elements that should be considered for inclusion in the I&C engineering/project are discussed, and the function of each is illuminated with respect to the project as a whole. Methods of producing the documents are discussed, using a practical project example as the basis for these discussions. The example project is to add a feed tank, shown in Figure 3-2, and its associated equipment and instrumentation to an existing facility.

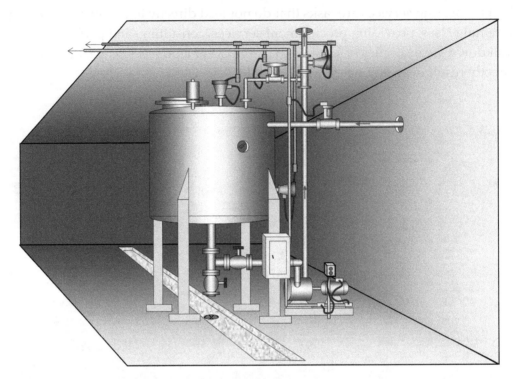

FIGURE 3-2. TYPICAL FEED TANK CONFIGURATION

The various elements of the I&C project are presented in their usual order of execution, as indicated by the products and services flow chart in Figure 3-1. In each case, the standard approach is discussed first, followed by one or more alternatives as appropriate. For example, power distribution information has historically been conveyed using a drawing. However, a chart could also serve, depending on end-user preference. (Note: Refer to Part I for a more thorough description of the four phases of a project. Part III deals with Phases 1 to 3. Phase 4, "Support," is not addressed in Part III).

Part III – Chapter 7: Piping and Instrumentation Diagrams (P&IDs)

A. General Description

Piping and instrumentation diagrams (P&IDs) are the foundation of the E/I&C (electrical/instrumentation and control) engineering package of deliverables. A P&ID is a drawing that presents selected information from multiple disciplines. It is generally "owned" by the Mechanical department and the Process subgroup. The P&ID is the daughter of the process flow diagram (PFD), a mechanical discipline product, and the heat and material balance (HMB) sheet, a process engineering product.

The PFD is simply a mechanical drawing that shows the major equipment items and their associated piping interconnections. It is essentially a block diagram showing process flow from left to right, with parallel paths and interface points clearly shown.

The HMB takes the PFD and layers on the physics (temperatures, flows, masses, etc.) of the system. The HMB provides a high level of detail on the amount of heating or cooling needed, the flow rates, the dwell times, and the throughput. From these data, equipment is selected, pumps and motors are sized, and instrumentation is selected.

The P&ID is developed out of the PFD and HMB. The P&ID is the primary means of communicating key design information between engineering disciplines. Its higher purpose is to clearly communicate the information critical to the intended plant functionality, including the functional location of instruments, pumps, motors, valves, etc. The P&ID shows a minimum of detail relative to the physical location of items, limiting such information to, perhaps, building or floor indication. Distances between items are generally not depicted at all.

The "owner" of the P&ID is usually the process engineer or whoever is serving in that capacity. This individual or group coordinates the information being placed on the drawing to make sure it is consistent with its intent. Such a coordinator is needed since information flows to this document from many sources. The following is a list of some of the different engineering disciplines that provide input to the P&ID production effort and the type of information provided:

- *Process Engineering:* drawing content, process data, vendor equipment depiction
- *Mechanical Engineering:* pipe ID (identification, not inside diameter) numbers, equipment numbers and labels, graphical depiction of equipment and piping
- *Instrumentation Engineering:* instrument numbers; graphical presentation of instruments and instrument wiring/tubing
- *Controls/Systems Engineering:* control logic depictions
- *Electrical Engineering:* motor ratings, motor wiring depictions

Using the PFD and HMB as a guide, the instrumentation P&ID designer places flow transmitters at places in the process that need flow information, level transmitters on tanks, and so on. Each item shown on the P&ID is given a unique identification (ID) number called a P&I tag, which allows the item to be tracked throughout the design.

B. Purpose

The P&ID communicates the means by which the raw materials are manipulated, as described in the HMB. It answers the question, "How will it be done?" PFDs and HMBs are purely conceptual documents. They have only a minimum of information related to the physical aspects of the piping or instrumentation of a system. The P&ID provides the link between the conceptual and the actual.

C. Content

The P&ID provides a graphical index of all the major process equipment, and all major sections of pipe instruments, and it shows their functional relationships. Pneumatic tubing is shown where the information is important to communicate controls functionality, and electrical wiring information is presented (in the form of dashed lines) to communicate functional intent to the electrical and instrumentation disciplines. Pipe specifications, pipe tag numbers, and sizes are shown. Heat tracing requirements are indicated, as are all safety devices. Control system logic is shown to a degree, as are alarms and process control elements. In short, the P&ID is the road map for the engineering design team.

To accomplish all this, a unique set of symbols and a means for identifying, or tagging, them have evolved, both of which are well documented by the International Society of Automation (ISA) in its standard ANSI/ISA-5.1-2009, *Instrumentation Symbols and Identification*.[1] Another standard, ISA-5.3-1983, *Graphic Symbols for Distributed Control/Shared Display Instrumentation, Logic, and Computer Systems*[2] applies to the depiction of items for computer display, and also provides a naming conventions for device tags.

1. Symbology (ANSI/ISA-5.1-2009)

Figure 3-3 shows a flow control loop as it might be presented on the P&ID, using typical ISA symbology. Instrument "bubbles" are placed at their functional locations in the process pipe. While the physical distance between the items could be great, they are depicted in close functional proximity. Note the (AI) and (AO) designations in the figure denote Analog Input and Analog Output, and they are shown for clarification purposes only. Those designations are not typically on a P&ID.

In this example, flow sensing element FE-10 is shown upstream of flow control valve FV-10. The flow sensing element is an orifice plate that forces the process fluid through a beveled hole that is specially sized to achieve a certain pressure drop across it for a given flow rate. A flow transmitter (FT-10) samples the pressure drop and generates an electrical output proportional to flow rate (dashed lines indicate the electrical signal). The signal is fed to a panel-mounted meter (FI-10a) and then on to an analog input point in the control system (FI-10b), where it is presented as a display to the operator. After the signal is in the control system, it is processed, and, when necessary, a control signal is generated to force a flowrate correction. This correction signal is routed to a flow control analog output point (FC-10), where the signal is converted into a 4–20 mA analog electrical signal and fed to a current-to-pneumatic (I/P) transducer that converts the electrical signal into a 3–15 psi pneumatic analog signal. This signal is then fed to the pneumatic control valve actuator to force the valve to open or close as needed to achieve the desired correction.

In practice, the two control system symbols (AI and AO) are usually combined into one that represents both control system functions, as shown in Figure 3-4.

Signal lines are differentiated to indicate the medium of the signal. As mentioned, the dashed lines indicate an electrical signal, while the solid line with two crosshatches indicates a pneumatic signal.

2. Symbol Identification

The tagging process is well documented by ISA, which presents a two-element numbering scheme of the format:

> XXXX-YYYY, where XXXX is the tag prefix that provides indication of function and YYYY is a sequential identifier to make the tag unique. Sometimes a middle element, such as a building number or a process material designator, is inserted to indicate a process area.

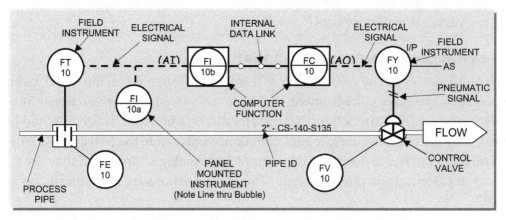

FIGURE 3-3. TYPICAL P&ID SYMBOLOGY

a. Prefix

The prefix is the important part of the identifier. In the ISA tagging method, the tag prefix letters are position dependent. The first letter indicates the physical property being measured or controlled (e.g., pressure, flow, temperature). The second or third letters are modifiers. In Figure 3-4, the F in the first position indicates a flow control item. FT in the leftmost bubble indicates the item is a flow transmitter. FI is a flow indicator, FC is a flow controller, FY is an I/P transducer, and FV is a flow control valve.

Note: An S as a second letter can be a modifier for the first letter, or it can be classified as a "succeeding" letter. This can be a bit confusing. If S is used as a succeeding letter, it applies to emergency protective primary elements. In this case, a device normally labeled PCV could also be labeled PSV if it is used as a safety device. The term xCV implies a self-actuated control valve, such as a pressure regulator. The succeeding letter combination CV should not be used in cases where the valve is not self-actuating. Thus, FCV would not be appropriate in the case of the valve shown in Figure 3-4 because FV-10 is not self-actuating. If the letter S is used as a succeeding letter, such as in LSH, it designates a switch. Here's how to tell the difference: if the device in question is generating a discrete (on/off) signal, then the S in the second position indicates the device is a switch; if the device is reacting to a variable process condition, then the S signifies a safety function.

The following are a few of the more common prefix arrangements:

- LSHH: level switch high-high
- LSH: level switch high

PART III – CHAPTER 7: PIPING AND INSTRUMENTATION DIAGRAMS (P&IDS)

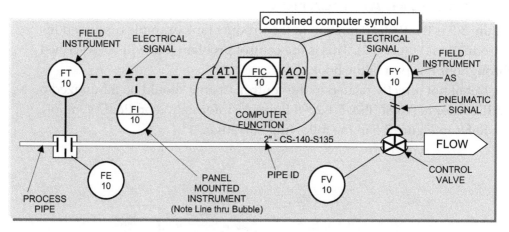

FIGURE 3-4. TYPICAL P&ID SYMBOLOGY SHOWING COMBINED AUTOMATION SYSTEM FUNCTIONS

- LSL: level switch low
- LSLL: level switch low-low
- LAL: level alarm low
- PT: pressure transmitter
- PDT: pressure differential transmitter
- AT: analyzer transmitter
- TE: temperature element
- TT: temperature transmitter
- PDSH: pressure differential switch high
- KQL: time quantity light (i.e., time is expired)
- PY: pressure transducer
- ZSO: position switch (open)
- HV: hand valve
- HS: hand switch

b. Suffix

The suffix portion of the tag is merely a sequence number unique to all the items in that control loop. In the example shown in Figure 3-4, all the items are "sequence ten" items, which indicates all of the items depicted are linked to the same control point in the process.

D. Practical Application

Figure 3-5 is the P&ID depiction of the product tank example discussed in the Introduction (Figure 3-2). This is the control problem that will be described throughout this part of the book.

Detail not directly related to the control scheme should be minimized on a P&ID. Refer to ANSI/ISA-5.1-2009 for more information on P&ID symbols.

Refer to Figure 3-5 in the following narrative.

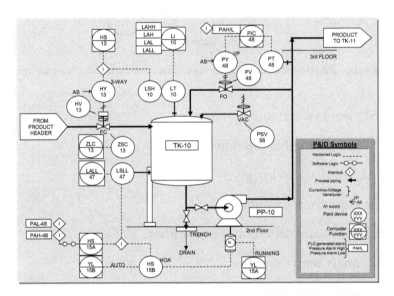

Figure 3-5. P&ID Presentation of the TK-10 Subsystem

1. Tank Level: LT-10, LSH-10, LSLL-47

Level transmitter LT-10 transmits tank level to the control system as a 4–20 mA signal. The PLC program monitors this signal, and compares it to a set of four alarm trip values. The control system then reports the tank level and the alarm trip status to the operator as represented by the LI-10 bubble. LT-10 also has a set of contacts that can be configured to provide a hardwired alarm trip based on tank level. This hardwired alarm output has been calibrated as a high-level alarm. Note the LT-10 and LSH-10 bubbles are touching to indicate they are contained in a single device. This alarm output is hardwired into the pump circuitry as an interlock such that a high level forces the fill valve closed. The term "force" in this context implies that the control system's normal operating conditions for the valve are bypassed, and the valve is made to close.

If low-low level switch LSLL-47 trips, then the pump will be stopped through a hardwired interlock.

2. Tank Fill: HV-13, ZSC-13

The tank is filled by a "wild" flow that is not controlled upstream but is allowed to run wild into the vessel as long as valve HV-13 is open. A control system command will open the valve, provided the high-level alarm generated by a switch on the level transmitter does not activate. Valve HV-13 will fail close upon loss of air supply by virtue of the pneumatics associated with the three-way solenoid, HY-13, and the position of the spring inside the valve actuator.

Closed-position limit switch ZSC-13 closes if the valve closes. A limit-closed status bit, labeled ZLC-13, informs the operator of this event. **Note:** There is a lot of discretion in how the second and third letters in a tagname may be used. The author prefers to use L in the second position (ZLC) for discrete signals, as opposed to I (ZIC), which is reserved for analog inputs. The L indicates a lamp, while I indicates an indicator. Also, an A in the second position implies an alarm signal, which is also discrete.

3. Tank Discharge: PP-10

The contents of the tank may be emptied into a trench in the floor with hand valves, or they may be directed to discharge through pump PP-10. This pump runs based on the position of hand switch HS-15B. This switch is a hand-off-auto (HOA) type of switch. In the Hand position, the pump runs provided the interlocks are set. In the Off position, the pump stops. In Auto, the pump runs based on a computer command (HS-15A), again provided the interlocks are set.

Interlocks for this pump are:

- Stop upon a low level (LSLL-47)

- Stop upon a low or high pressure (PAH-48)

If the tank fill valve closes with the pump running, there will be a risk of pulling a vacuum on the tank. Therefore, a vacuum relief valve PSV-58 has been added. This valve has an integral pilot and will open to relieve the vacuum.

4. Pump Discharge Pressure: PIC-48

Pump discharge pressure is managed by pressure control loop 48. The computerized pressure-indicating controller (PIC-48) maintains a constant pressure at the output of the pump by sampling pressure, comparing the actual pressure to a desired set point, and then modulating valve PV-48 to reduce any difference. Pressure transmitter PT-48 samples the pressure of the pump discharge as it pushes product into a downstream product tank. Pressure in the discharge line is dictated by line length, elevation, possible constrictions in the line, such as plugs or other factors. If the pressure in the line increases, indicating reduced demand downstream (perhaps because of the TK-11 fill valve closing off), then modulating valve PV-48 is opened slightly to recirculate more product back into TK-10. If downstream pressure decreases (perhaps because a downstream TK-11 fill valve opens), indicating increased demand, then the valve moves toward the closed position. This allows the constant speed motor driving the pump to run safely. When product is not needed downstream, the constant recirculation eliminates the need for an agitator.

If the control system detects a high or low pressure, and the signal remains for a certain time period, then the operator is notified, and the pump is stopped. This scheme, or portions of it, will be discussed throughout the remainder of this book.

E. P&ID SUMMARY

The P&ID is perhaps the single most important document in terms of communication of ideas. It becomes the true starting point for the design package. It is a multi-disciplinary document that provides a means for coordinating the efforts of the entire design team. Note: Before the P&ID can be released, the mechanical information has to be layered on. This is information, such as pipe number and equipment specifications. In some cases, detailed interlock notes appear. However, the minimized P&ID drawing for this project, shown in Figure 3-6, is sufficient for the controls discussions in the following sections.

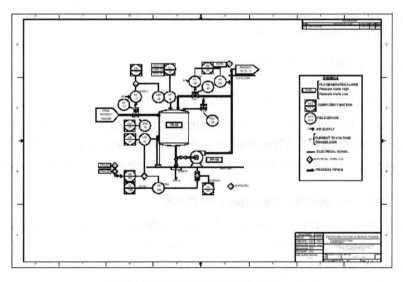

FIGURE 3-6. BASIC P&ID DRAWING

Part III – Chapter 8: Links to Mechanical and Civil

While preliminary I&C activities are ongoing, the mechanical and civil engineering groups work on several items key to the I&C design package. This chapter describes some of these items and links them to the I&C design.

A. General Equipment Arrangement Drawing (Civil and Mechanical)

General equipment arrangements (GA) are physical drawings that depict the floor arrangement of the process areas. They are derived from the structural steel drawings that show items such as column lines, elevations, and stairwells. The mechanical and civil groups team up to flesh out these drawings to include the process equipment, inline piped instruments, major pipe artifacts, safety showers, and so on.

These drawings are, for the most part, reference drawings for the I&C/CSI (I&C/Systems Integration) design team. However, it is good to coordinate the location of field junction boxes (FJBs) by getting them placed on these drawings. Doing this sometimes avoids piping conflicts since the mechanical group will see the boxes as obstacles to avoid.

1. Purpose

The purpose of the general equipment arrangement is to accurately depict major equipment items as they will appear on the process floor. More often than not, these drawings are rendered with a computerized 3-D model.

2. Interfaces

Predecessors to the GA are the P&IDs, structural drawing set, and equipment specifications. The GA is then used to generate piping orthographic drawings and instrument floor plans.

3. Content

The general equipment arrangement (often, simply "equipment arrangement") includes enough detail to depict the placement and orientation of major equipment items, such as tanks, pumps, and heat exchangers. The arrangement should also show access boundaries, personnel passages, fork truck routes, and so on.

4. Practical Application

Figure 3-7 is a representation of the product tank example discussed in the introduction to this Part, reproduced here for convenient reference.

Figure 3-8 is an example of the level of detail one might expect for an equipment arrangement depicting feed tank TK-10.

Note the North (N) arrow and the relative positions of the major pieces of equipment. The trench is shown, as are the pump and motor. The instrument junction box (depicted on the leg of the tank) is optional but recommended. This type of drawing is used as the basis for the eventual cable tray and conduit arrangements, as well as instrument locations.

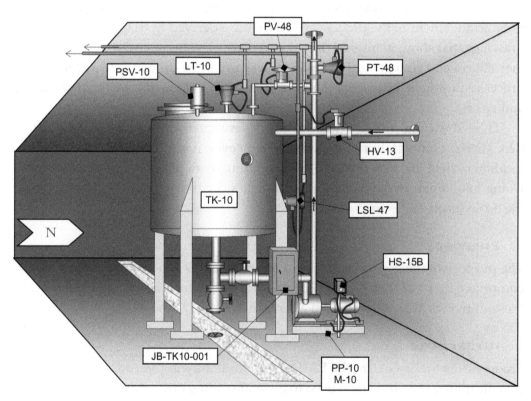

FIGURE 3-7. TK-10 FEED TANK WITH EQUIPMENT LABELS

5. Equipment Arrangement Summary

As the P&ID is to the controls design, the equipment arrangement is to the physical design. It is the foundation document for the set of physical deliverables. It is usually generated by the mechanical group in conjunction with the civil group, using their "steel" drawings as a design basis. The E/I&C/CSI team has little direct interaction with this drawing set, though as has been mentioned, it is a

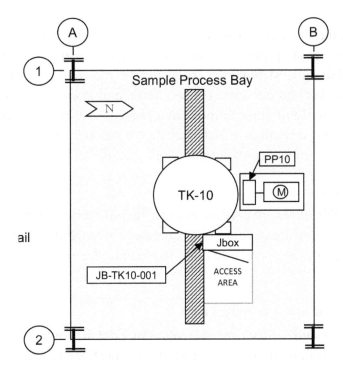

FIGURE 3-8. TK-10 FEED TANK AREA EQUIPMENT ARRANGEMENT

good idea to coordinate field junction box placement and even get junction boxes placed on the drawing if possible. Access requirements should be clearly delineated as well.

B. PIPING DRAWING (MECHANICAL)

The piping drawing set, as it affects the I&C design process, includes three primary types of drawings:

- Orthographic drawings (orthos)
- Elevation drawings
- Piping spool sheets

An *ortho* is a two-dimensional, scaled drawing that depicts piping as viewed from above. An *elevation* is a scaled drawing showing the piping as viewed from the side. A *spool sheet* is typically a sketch that shows 3-D detail of each length of pipe, including each item of material used, such as flanges, reducers, and inline instruments. This sketch should indicate slopes, elevation, and lengths but is not to scale.

1. Purpose

How do the piping drawings affect the instrument package? In order for the mechanical group to lay pipe, they must include inline instrumentation. The pipers (mechanical designers) generally request early information on the flange-to-flange dimensions of end devices, such as valves, and instrumentation, such as magnetic flowmeters and other inline items. The pipers then locate these instruments on the piping drawings, which may then be used by the I&C group to generate accurate instrument arrangements (floor plans).

2. Interfaces

Instrument specifications are predecessors to the piping drawings to the degree necessary for flange-to-flange data, the equipment arrangement, and the P&ID/HMB set. Piping drawings are used as a precursor to the instrument arrangement drawings.

3. Content (as related to I&C)

Inline instruments and sensors are shown in great detail; however, instruments mounted on the equipment are not depicted. The locations of equipment-mounted instruments may be inferred, however, since equipment orientation is indicated on the piping drawing and the equipment arrangement.

C. Pump and Equipment Specifications (Mechanical)

The mechanical (or "process") group maintains a set of files on each equipment item. For example, a sketch of each tank is made that shows the size and orientation of each flange, manhole, or other feature. Pump specifications are generated that provide anticipated flows and pressures.

D. Links Summary

Mechanical and civil drawings provide physical information that may be used to generate instrument location plans, also called instrument arrangements. These drawings are important in the generation of realistic cable and conduit schedules, among other uses. Defining interfaces to other disciplines is very important. Each of these interface points should be discussed with the discipline involved and the information flow path defined.

Part III – Chapter 9: Preliminary Engineering

Before production begins on a design package, some investigation and planning must occur. In fact, the more planning the better. For the engineering group, the practice of collecting and digesting the information prior to going into production is called Front End Loading (FEL). These FEL tasks are also called *preliminary engineering* and are critical to the success of the project. Major new discoveries made during the design production phase (Phase 3 in Part I) are usually negative events. This chapter addresses the preliminary tasks that must be accomplished before the main design task begins.

First, the project must be fully scoped, with each task delineated and quantified. This scoping process includes gaining familiarity with the existing control system and walking down the process area, noting distances, space availability, obstructions, and so on. It also requires the initialization of a database and the generation of a detailed Instrument Index and I/O list. When these tasks have been completed, a definitive estimate may be generated. Refer to Part I of this book for more information on the scoping and estimating processes.

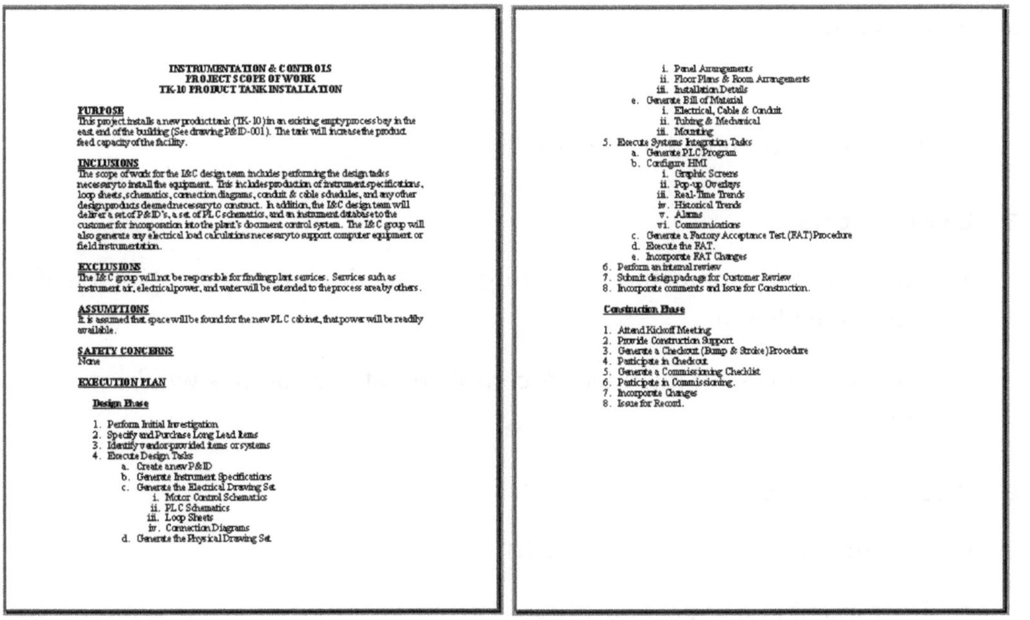

INSTRUMENTATION & CONTROLS
PROJECT SCOPE OF WORK
TK-10 PRODUCT TANK INSTALLATION

PURPOSE
This project installs a new product tank (TK-10) in an existing empty process bay in the east end of the building (See drawing P&ID-001). The tank will increase the product feed capacity of the facility.

INCLUSIONS
The scope of work for the I&C design team includes performing the design tasks necessary to install the equipment. This includes production of instrument specifications, loop sheets, schematics, connection diagrams, conduit & cable schedules, and any other design products deemed necessary to construct. In addition, the I&C design team will deliver a set of P&ID's, a set of PLC schematics, and an instrument database to the customer for incorporation into the plant's document control system. The I&C group will also generate any electrical load calculations necessary to support computer equipment or field instrumentation.

EXCLUSIONS
The I&C group will not be responsible for finding plant services. Services such as instrument air, electrical power, and water will be extended to the process area by others.

ASSUMPTIONS
It is assumed that space will be found for the new PLC cabinet, that power will be readily available.

SAFETY CONCERNS
None

EXECUTION PLAN

Design Phase

1. Perform Initial Investigation
2. Specify and Purchase Long Lead Items
3. Identify vendor-provided items or systems
4. Execute Design Tasks
   a. Create a new P&ID
   b. Generate Instrument Specifications
   c. Generate the Electrical Drawing Set
      i. Motor Control Schematics
      ii. PLC Schematics
      iii. Loop Sheets
      iv. Connection Diagrams
   d. Generate the Physical Drawing Set
      i. Panel Arrangements
      ii. Floor Plans & Room Arrangements
      iii. Installation Details
   e. Generate Bill of Material
      i. Electrical, Cable & Conduit
      ii. Tubing & Mechanical
      iii. Mounting
5. Execute Systems Integration Tasks
   a. Generate PLC Program
   b. Configure HMI
      i. Graphic Screens
      ii. Pop-up Overlays
      iii. Real-Time Trends
      iv. Historical Trends
      v. Alarms
      vi. Communications
   c. Generate a Factory Acceptance Test (FAT) Procedure
   d. Execute the FAT.
   e. Incorporate FAT Changes
6. Perform an internal review
7. Submit design package for Customer Review
8. Incorporate comments and Issue for Construction.

Construction Phase

1. Attend Kickoff Meeting
2. Provide Construction Support
3. Generate a Checkout (Bump & Stroke) Procedure
4. Participate in Checkout
5. Generate a Commissioning Checklist
6. Participate in Commissioning.
7. Incorporate Changes
8. Issue for Record.

Figure 3-9. Detailed Scope of Work

A. Development of a Detailed Scope of Work

A scope of work is simply a list of tasks (Part I, Chapter 3). If a definitive estimate is required, with an accuracy of, say, ±10%, then the scope of work should be rather detailed. A detailed scope of work describes performance expectations for each line item and provides reference to any standards or practices that might apply. The following is a very basic scope of work for our imaginary project:

1. Purpose (Project Overview)

This project is to install a new product tank (TK-10) in an existing empty process bay in the east end of the building (see drawing P&ID-001). This tank will increase the product feed capacity of the facility in order to support new production requirements that will take effect in March.

2. Project Scope—I&C

The I&C Design Team will produce design documents necessary to construct and maintain the TK-10 subsystem. This task includes the deliverables listed in Section 6. The existing plant P&ID set will be modified as necessary to incorporate the new P&ID.

3. Safety Concerns

The product is caustic. See MSDS sheets provided.

4. Assumptions

- Existing drawings are accurate.
- Tank orientation will be like TK-09.

5. Exclusions

- This task does not include updating existing drawings, except as specifically noted herein.
- Procurement is not included. A complete Bill of Materials will be provided.

6. Deliverables

- Instrument Specifications (2 each)
- Instrument Loop Sheets
- PLC Schematics
- Connection Diagrams

- Conduit & Cable Schedules
- New P&ID
- Revised P&IDs

7. MILESTONE SCHEDULE

(Note: A milestone schedule is simply a list of major events with either a date or a duration. If the start date is unknown, then durations relative to the Notice to Proceed (NTP) date is usually sufficient.)

The project will follow this milestone schedule:

- NTP   Notice to Proceed.
- NTP + 1 week   Publish drawing list and instrument index.
- NTP + 3 weeks   Issue preliminary engineering documents for customer review.
- NTP + 4 weeks   Receive customer review comments.
- NTP + 5 weeks   Issue shop drawings to panel fabricator.
- NTP + 6 weeks   Issue detail drawing set for customer comment.
- NTP + 7 weeks   Receive customer review comments.
- NTP + 10 weeks   Issue all drawings for construction.

Note: In this case, the schedule does not include Control Systems Integration (CSI) or construction activities. These can be layered on in a similar fashion.

The scope description shown above can be coupled with economic sections, and together, they could be submitted as the proposal, and become the contract under which the project will be executed. Laying out assumptions and exclusions is very important both to the engineering company and to the end user. It should state clearly those things that may be contentious, forcing debate on those items early rather than late.

B. CONTROL SYSTEM ORIENTATION

One of the first tasks of the design team is to get familiar with the control system used at the plant or, if a new control system is needed, to scope out and specify the new system. As already stated, our test-case scenario involves expanding an existing programmable logic controller (PLC) control system.

So how would one go about obtaining the necessary information? The place to start is a meeting with the technical group at the plant who maintains the existing

PLC system. These folks should be fully apprised of the upcoming project and will probably already have a strategy in mind. In fact, in all likelihood, this group would normally do the systems integration for a small addition such as this, but for our purposes, we will assume they are too busy.

The following is a checklist of items to be determined during the initial control systems meeting:

- *Network Diagram:* Does one exist? How can we obtain a copy?

- *PLC:* What kind? What software? Has it been customized? Are there any specific loadables we will need? Where is the central processing unit (CPU) located? Where will our new rack be located? May we obtain a copy of the software? Who will be our contact?

- *Human-Machine Interface (HMI):* What kind? What software? Do written specifications exist? May we obtain a copy of the software? Where is the HMI? Who will be our technical contact?

- *Operations:* Do operating procedures exist, and may we borrow a copy? Who will be our operations contact? What about alarms and reports? Are there any specific requests regarding this project?

- *Maintenance:* Do maintenance manuals or procedures exist, and may we obtain copies?

- *Logic Diagrams:* Are they required? If so, may we see some examples? Shall we generate and submit logic diagrams before doing further work (recommended)?

- *Purchasing:* Shall we do the purchasing, or will the plant purchase per our recommendations? Who are the plant's local control system vendors? Do they provide good service?

As a result of our control systems meeting, we have made the following discoveries:

- A network diagram does exist and will be provided.

- The PLC is a generic model 444, running generic ladder logic software. The CPU is in the equipment room adjacent to the control room, and our new rack will be mounted on the wall outside the control room in the hazardous process area in a small PLC cabinet. The cabinet will need to be purged.

- The HMI is running GenericHMI brand software, version 3.5. Written guidelines do not exist, so the existing screens will be used as a guide. The

HMI is on a workstation in the control room. This project will add one graphic screen and some alarms to that workstation.

- Operations procedure(s) will need to be written for this project.

- Maintenance manuals will need to be generated to cover the new work.

- Logic diagrams are required. They will be submitted prior to beginning the programming task.

- Purchasing will be done by the plant using material lists provided by us. The discovery process reveals that the existing control system consists of a single PLC and HMI (Figure 3-10).

PLC drop 1 resides in the equipment room adjacent to the control room. Most of the I/O signals are brought to this location. Drop 2 resides in the motor control center (MCC) room. The CPU and Ethernet data communications module are contained in drop 1, as is a remote I/O master module.

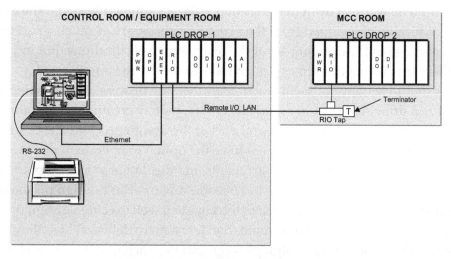

FIGURE 3-10. EXISTING CONTROL SYSTEM

The network will need to be modified to accommodate the new rack (Figure 3-11). The remote I/O local area network (LAN) will need to be extended to the new site. A tap will be installed in the MCC room to extend the network to the site. The terminating resistor will move to the new site as well. I/O modules will be added as they are identified.

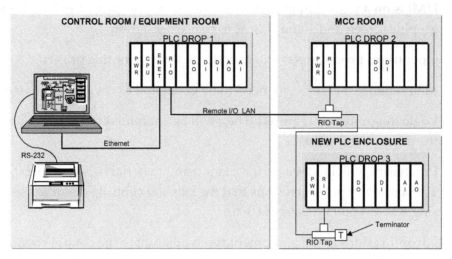

FIGURE 3-11. REVISED CONTROL SYSTEM

C. Project Database Initialization

Information management is a major consideration for every I&C project. To the experienced instrument designer (see chapter 12.A.2), the I/O list and the drawing list, which are developed from the project database, are as much a part of the design package as the drawing set or the instrument specifications. But in years past, these lists were not used as design tools to the degree possible today. Today's multi-user environment has transformed the process of developing the I/O list and drawing list from a clerical activity, merely creating lists for ultimate delivery to the customer, to an integral part of the design process. The designers and engineers in today's environment have the option to operate in real time. For example, designers can now log instruments into the database as they place them on a loop sheet. Thus, the database has become an important design productivity and communications tool that is one of the keys to a well executed design process. The database, if properly used, is more than just a mere deliverable to the customer: it is the "glue" of the design package that ties each of the various elements together and goes far toward guaranteeing a quality product. (See Chapter 4 in Part II for a broader discussion of information management.)

This chapter provides an example of a Microsoft® Access database that has all the elements necessary for a typical instrumentation project.[1] Access is called a relational database because relationships can be configured among data tables. It is a multi-user database that allows data access by more than one person at a time. Data changes are stored every time the user leaves a record, so collisions between

---

1. For an in-depth look at how to generate an accurate bid estimate and create a project schedule using Microsoft® Excel, refer to the CD-ROM entitled, "Software Tools for Instrumentation and Control Systems Design" included with this book. For more information, see page 523.

users are avoided. This also eliminates the risk of losing data upon a power failure or other calamity. Please review Chapter 4 before continuing.

For the typical I&C project, two primary types of information must be managed by the project database: what we will call "document" (information relative to the documents, but not the content), and "instrument" (information relative to the engineering elements themselves). With Access, each of these types of information can be maintained in the same database. But since the types of information related to each task are quite different, for our purposes they will be kept in separate tables (refer again to Figure 3-12 reproduced below).

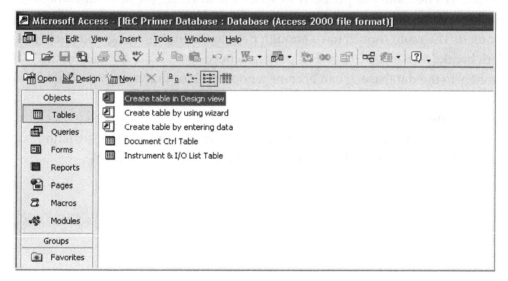

FIGURE 3-12. LIST OF TABLES

As the name suggests, the document control table manages the drawings, specifications, calculations, transmittals, memoranda and other project documents. The instrument and I/O list table manages the instruments. Each table has associated queries and reports. The document control table is described in the following sections; the instrument and I/O list table is then described in Section 2.

1. INITIALIZE DOCUMENT CONTROL TABLE

Drawings and specifications are the primary products of the design team. As such, their production should be properly managed. Document control is one of the more high-profile tasks that a design team performs, and their performance hinges, many times, on the document control database. More than a mere list, this table is an important project management tool.

Each document is tracked by revision. The table should have one record per document revision. When a drawing is acquired from the customer for revision, for example, it will have a document number and a revision number assigned by the customer. Each time a revision is completed, the change must be approved and recorded. The next revision will then be done under the next higher revision number. Depending upon the document control policy of the customer, it is possible to have multiple revisions of the same drawing in the same project.

So, the number of records in the table correlates with the number of document revisions. Note that the number of revisions does not necessarily reflect the number of documents, as multiple revisions are sometimes needed for a single document.

How many fields should there be per record? There should be several descriptive fields as necessary to properly describe the drawing (or other documents). There should also be several date fields, to allow milestone attainment to be recorded. The table should have enough data to facilitate sorting and packaging. Work breakdown structure (WBS) packages need to be built. By adding the WBS numbers to the database, it can become a good tool for managing scheduled tasks.

a. Table

Figure 3-13 shows a recommended document control table structure, while Figure 3-14 shows project data contained by data table.

Field Name	Data Type	Description
ID	AutoNumber	Record Number (Automatically Assigned)
Document#	Text	Document ID Number
Rev#	Text	Revision Number
Title Line1	Text	Line 1 Title (Ex: Dave's Plastic Factory)
Title Line2	Text	Line 2 Title (Ex: Product Line 2 - Extrusion)
Title Line3	Text	Line 3 Title (Ex: Junction Box 3A)
Title Line4	Text	Line 4 Title (Ex: Wiring Diagram)
Short Title	Text	Abbreviated Title
Type Code	Text	Document Type Shortcode (Ex: "Elem"="Elementary", "Conn"= "Connection Diagram", etc.)
Package	Text	Design Package ID
Comments	Text	
PrelimReq?	Date/Time	When was a print requested for this document?
PrelimRcd?	Date/Time	When was a print rec'd from client?
Ordered	Date/Time	When was drawing file ordered from client?
Received	Date/Time	When was drawing file receieved?
IFIR	Date/Time	When was the drawing file Issued For Internal Review?
IFA	Date/Time	When was the drawing Issued For Approval?
IFC	Date/Time	When was the drawing Issued For Construction?
IFR	Date/Time	When was the drawing Issued For Record?
Approved	Date/Time	When was the drawing approved by the customer?

FIGURE 3-13. DOCUMENT CONTROL TABLE STRUCTURE

Each of the fields listed represents a column of information in the data table. This table supports several activities, some of which are:

- Ordering drawings

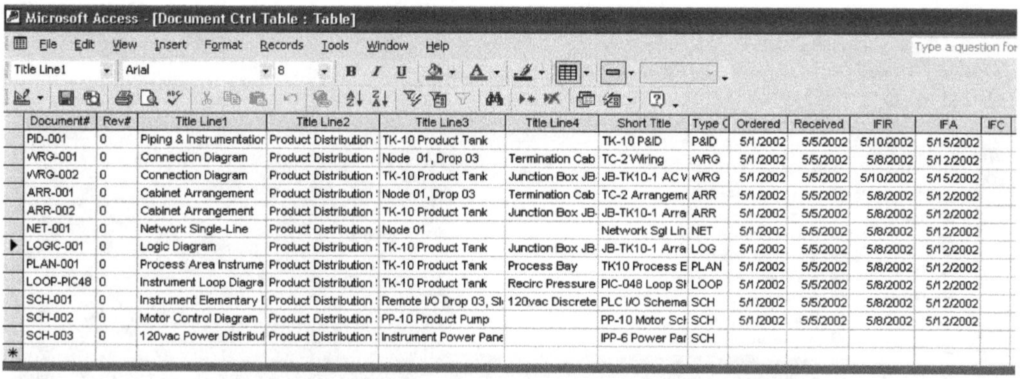

PART III – CHAPTER 9: PRELIMINARY ENGINEERING

FIGURE 3-14. DOCUMENT CONTROL TABLE, DATASHEET VIEW

- Referencing drawings
- Issuing drawings (generating transmittals)

Each of these activities may demand access to specific information that is different among them. That is where a query or a form comes into play. In this example, a set of queries for ordering drawings will be created.

The information in the "Description" field appears at the bottom of the screen in Datasheet View, so it is a good idea to provide examples of the types of data envisioned for each field. The numerous date/time fields are for document tracking. Switching to datasheet view, as shown in Figure 3-14, gives us access to the data.

b. Query for Ordering Drawings

In Access, the Order Drawings Query fetches information from the document control table that relates specifically to ordering drawings. Other information is ignored.

In Figure 3-15, the first drawing record has been entered using Order Drawings Query. Note that only seven of the available 18 fields of data are included in this query. Imagine that there are several drawings in this database. If a new set of drawings needs to be ordered, the drawings could be entered using this query. Then, by filtering for the date ordered, the specific set of drawings could be listed.

Let's put a few records in the data table (Figure 3-16).

Notice the short title is designed to group like drawings together. For example, all P&ID drawings will group up, as will arrangements and plans, logic and loops. This can be helpful, particularly if it is difficult to

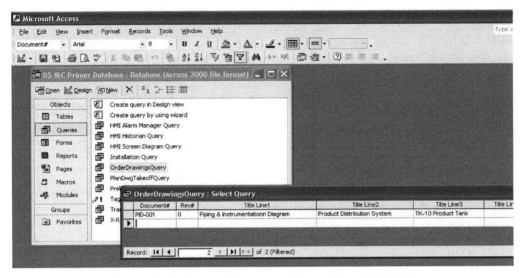

FIGURE 3-15. ORDER DRAWINGS QUERY (DESIGN VIEW)

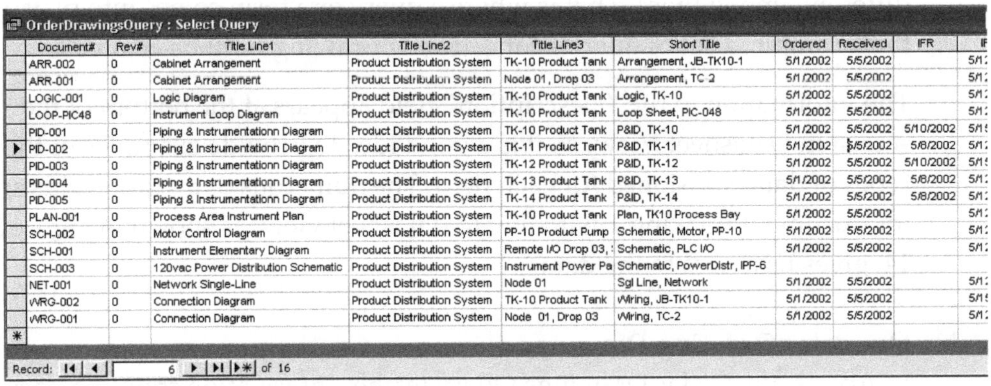

FIGURE 3-16. DOCUMENT CONTROL TABLE DATA

determine drawing type from the drawing numbering scheme, and frequently the titles are inconsistent in terms of formatting. Also, notice the date fields. These fields are useful in generating transmittals, as we will see next.

c. **Transmittal Query**
Another key activity in the document control process is document tracking. The transference of documents should be a formal process that is tracked very carefully. Each time a drawing or drawing package is issued, it should be accompanied by a coversheet that describes the contents of the document package, its purpose, who originated it, and its destination. This coversheet is called a *transmittal* and is given a unique tracking num-

ber that is maintained in a separate database. Transmittal logs are usually maintained by the project manager or by the discipline lead.

When drawings are to be transmitted (shipped), a different subset of available fields is needed. The transmittal query lets the designer flag the drawings to be shipped by logging the date in the proper issue field. The shipping clerk may then filter on that date and generate the drawing list to be attached to the transmittal.

Attached to the transmittal is a list of documents included in the package. The document control database should be used to generate this list (Figure 3-17).

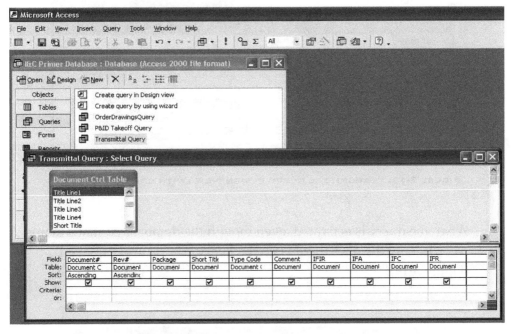

FIGURE 3-17. TRANSMITTAL QUERY

The various date fields are included to allow sorting and filtering. In Figure 3-18, notice "#5/15/2002#" listed in the criteria section of the Issue for Approval (IFA) field. The "#" sign tells the software that the information is a date. Any drawing with a date of 5/15/2002 is selected by this query for display, as shown in Figures 3-16. For example, if the purpose of the transmittal is to issue the package for approval, then the user enters the date of issue in the IFA (Issue for Approval) column for each document that will be a part of the package. Then, a filter can be injected into the criteria parameter that will filter out any records that do not have the proper date in the proper field.

In Figure 3-18, only documents to be issued for approval on 15 May 2002 will be displayed when the query is invoked. To run the query, the user presses the key on the Access main menu.

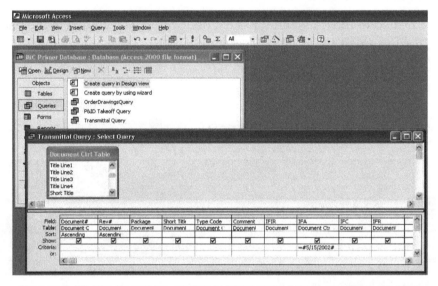

FIGURE 3-18. TRANSMITTAL QUERY DESIGN VIEW (WITH CRITERIA FILTER)

When the query is activated, even though the database contains many records, only the ones that match the query criteria are displayed (Figure 3-19).

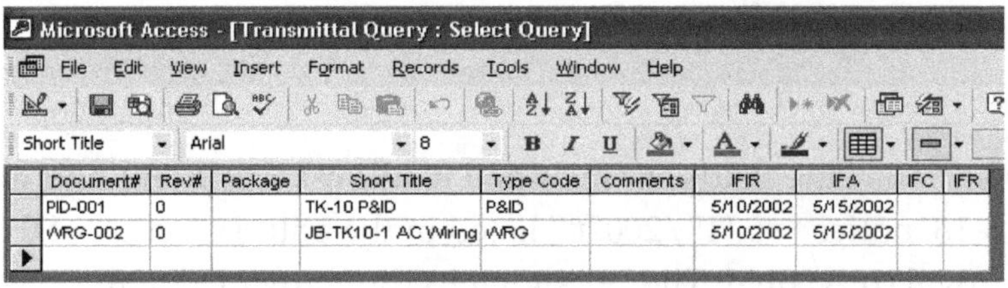

FIGURE 3-19. TRANSMITTAL QUERY, DATASHEET VIEW

This allows the user to view only the information he needs. Note that the Package field is empty. In an actual database, this field would contain a WBS activity number, used to link the package to the project schedule.

Thus, a spreadsheet-like list of drawings to be issued is generated. This list may be printed as displayed, or a report may be linked to this query and generated to improve the appearance of the printed information.

2. Initialize Instrument and I/O List Table

The I&C discipline handles a large number of items, and designers must be very capable in the area of data management. There are a number of activities that any instrument database should be able to accommodate:

- P&ID data takeoff
- I/O list management
- Document cross-reference indexing
- Inventory management
- Checklists

See Part II, Chapter 4, for more information on the data collection process.

In the database, each activity may have one or more associated queries, forms, and/or reports. But each of these must find its way back to the instrument and I/O list table, where the data actually resides. The instrument and I/O list table serves the following purposes:

- *Maintains* an inventory of all instruments and devices associated with the project, regardless of type. For example, a level transmitter (wired) and a relief valve (unwired) should both be listed. Information is provided, such as device tag, description, manufacturer, and specification number.

- *Describes* each instrument or device and ties it to its process function.

- *Acts as a cross-reference* by listing all drawings and related documents associated with each instrument or device. Installation details, instrument loop sheets, P&IDs, and other documentation related to the instrument or device in question should be presented. Even computer tagnames may be listed if they are known.

- *Acts as an I/O list* by incorporating I/O type, software address, and hardware address.

There are several advantages of combining all these functions into a single table, but the primary advantage is the shared data. Data items may be shared among users, allowing the information to be entered once in the table, but used for several different purposes through the use of queries and reports.

Before much of the data can be entered, a set of P&IDs should be in an advanced design stage. These P&IDs are used as the basis for the table. Other information may be culled from the logic narrative or logic diagrams when they become available. The instrument and I/O list table should be considered a living element that is continuously modified throughout the life of the project. Each instrument is tracked throughout the design process. For example, any time the instrument appears on a drawing, that drawing number is logged against the instrument. At the end of the project, the user can generate a list of all instruments that are associated with a particular process area (instrument location plan drawing), on a particular I/O module (I/O list), or in a particular junction box.

a. **Instrument Table**

The instrument and I/O list table is where the data "lives." Data may be entered, using a query or form, but the table is where the information is actually stored. So proper configuration of the data table is key to the utility of the database.

Database fields are cheap. The number and type of fields should reflect the moves being made in the design (Figure 3-20). Empty fields are OK, but if the data has been confirmed to be not applicable to a particular item, then it is wise to put NA in that location. An empty field implies unfinished work. The following are some issues addressed by the following database structure:

- If ISA-5.1-1984 (R1992), *Instrument Symbols and Identification* is followed, instrument tagnames will be multi-segmented. The first segment relates to the type of measurement or the process connection. The final segment is a sequential number that makes the item unique. Many times a modifier is injected between the first and final segments to tie the item to a production line or other grouping. This database logs each segment separately to allow for the greatest flexibility in sorting. It is possible to automatically combine the segmented tagnames into a single field by using an update query. The field set aside for this purpose is the Tagname field.

- Control system data are logged in I/O Type, HWAddress (hardware address), and SWAddress (software address) fields. If the item is not computerized, "NA" should be logged in each of these fields.

- Process description is addressed with three fields: Equipment, Service, and Description. From left to right, they get more specific. Using one of the examples shown in the field list, "TK02 Product Tank Product Level Transmitter" would be the end result of the descriptions shown.

Splitting this into three segments allows greater flexibility in sorting and filtering.

- Related documents are listed, as is other related information. For example, all associated drawings are listed, as are the junction box, the terminal strip, and the group of terminals being used.

Field Name	Data Type	Description
ID	AutoNumber	Record Number
TagPre	Text	Instrument Tag Prefix (Ex: FT, PT, LSH, HV, etc.)
TagMod	Text	Instrument Tag Modifier (Ex: Building No, Process ID, Plant, etc.)
TagSuf	Text	Instrument Tag Suffix (Sequential element; Ex: 0001, 001B, etc.)
Tagname	Text	Instrument Tagname (Created from three fields by "Update" Query...)
I/O Type	Text	Computer I/O Interface Type (Ex: DI, DO, AI, AO, RTD, NA, etc.)
HW Address	Text	Computer Hardware Address (Ex: N02D01R02S11P15, Node 02, Drop 01, Rack 02, Slot 11, Point 15)
SW Address	Text	Computer Software Address (Ex: 100021, N14:2, etc.)
Equipment	Text	Related Equipment Designation (Ex: TK02 Product Tank, HE02 Heat Exchanger, etc.)
Service	Text	Related Service (Ex: Product Level, Steam Supply Pressure, etc.)
Description	Text	Instrument Type (Ex: Level Transmitter, Pressure Switch, etc.)
Spec#	Text	Instrument Specification Number
P&ID#	Text	P&ID Drawing Number
Schematic#	Text	Schematic (Elementary) Wiring Diagram
LoopSheet#	Text	Instrument Loop Sheet
MarshCab#	Text	Marshalling Cabinet ID Number
MarshCabArg#	Text	Marshalling Cabinet Arrangement Drawing Number
MarshCabWrg#	Text	Marshalling Cabinet Connection Diagram Drawing Number
MarshTerm#	Text	Marshalling Cabinet Terminal Strip & Terminal Number (Ex: TS2-15,16)
FieldJBox#	Text	Field Junction Box ID Number
FieldJBoxArg#	Text	Field Junction Box Arrangement Drawing Number
FieldJBoxWrg#	Text	Field Junction Box Connection Diagram Drawing Number
FieldTerm#	Text	Field Junction Box Terminal Strip & Terminal Number (Ex: TS4-01,02)
MechDetail	Text	Mechanical Installation Detail (Tubing, etc)
ElecDetail	Text	Electrical Installation Detail (Conduit, etc.)
MountDetail	Text	Mounting Installation Detail (Instrument Stand, brackets, etc.)

FIGURE 3-20. INSTRUMENT AND I/O LIST TABLE

b. **Queries**

As has been mentioned, a query is a database utility that allows the information stored in the data table to be formatted for a specific purpose. In effect, the query takes a picture of a predefined slice of the database and presents it to the user. The user may then use this picture to develop a report, to view data sets, or even to enter new information.

There are several types of queries. Two will be discussed here: *Select* query and *Update* query.

1. Update Query
   An update query may be used to make global database changes. The database presented here maintains the segmented instrument tagname in separate fields for example. These separate data items need to be combined into a single tagname. The Tagname Update Query does this automatically. An update query may be created by going to the query menu and selecting Create Query in design view. Once in

design view, the particular data table to be used as the source of data should be selected. Then, the fields needed for the query should be selected from the list of fields in the selected table.

The default query style is Select Query, so the type of query will need to be configured. While in design view, the user may select main menu item Query, and pick the Update Query option (Figure 3-21).

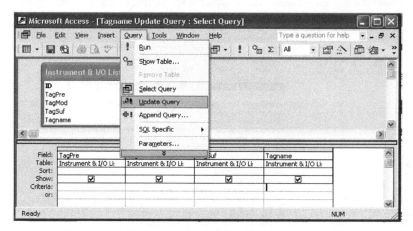

FIGURE 3-21. TAGNAME UPDATE QUERY, DESIGN VIEW

The query structure will change. The Sort parameter will change to the Update To parameter. To automatically generate the tagname from its various elements, the command [TagPre] & "-" & [TagSuf] should be entered in the Update To parameter of the Tagname field (Figure 3-22).

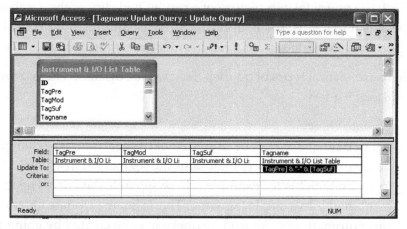

FIGURE 3-22. TAGNAME UPDATE QUERY, DESIGN VIEW, WITH CRITERIA FILTER

In this database, only two of the three tagname fields will be used, so only those two should be included in the command phrase.

To run the query from design view, the user may simply press the key at the top of the page. To run the query from the query menu page, the user may double click on the query name. This query may be activated at any time by any user and should be activated frequently to capture data entry changes to the tag element fields. The query, when activated, will fetch data from the fields in brackets and insert a dash between the elements. Data in this field will appear as displayed in Figure 3-23.

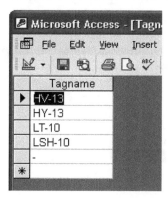

FIGURE 3-23. QUERY TAGNAME DISPLAY

2. Select Query
   As mentioned, the select query is the default Access query format. This type of query allows the user to preselect classes of data for display. Once displayed, the data may be modified just as if the data was being edited in the table. In fact, any modifications made in the query are actually made in the table. When a query is saved, only the query structure is saved. The query itself contains no data.

   A selected query may be created by going to the query menu and selecting Create Query in design view. Once in design view, the particular data table to be used as the source of data should be selected, and then the fields that are needed for the query should be selected from the list of fields in the selected table.

   To run the query from design view, the user must simply press the key at the top of the page. To run the query from the query menu page, the user double clicks on the query name.

c. **Reports**

Reports may be generated based on a query, a form, or a table. For example, an I/O list report may be tied to the I/O list query. The query fetches just the data needed for the report, while the report formats the data for presentation. In some cases, the query itself may be used as a report. But field names are usually truncated or encoded. The report utility lets the user display more meaningful field names, or even add text strings between data elements to make the report more readable. The report also gives the user the ability to segregate records into groups, count them, create group headings, and so on.

To create a report (Figure 3-24), the user selects Reports from the Access menu, then selects "Create report by using wizard."

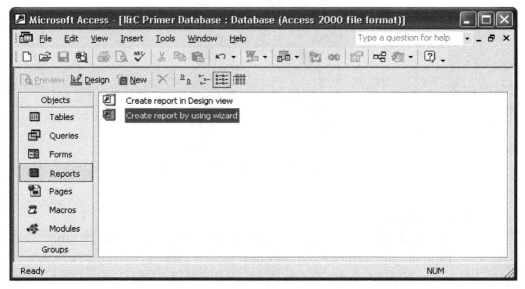

FIGURE 3-24. REPORTS

The Report Wizard window assists in the creation of the report (Figure 3-25). First, under the Tables/Queries pick box, the user selects the table or query that will feed data to this report. In this case, the report is tied to the P&ID takeoff query. Using the arrow keys to the right of the Available Fields pick box, the user moves the desired fields into the Selected Fields window. Then, when all the fields are in place, the user selects Next, then continues through the wizard screens until the report is generated. Based on the data previously entered, the report will appear as shown in Figure 3-26.

PART III – CHAPTER 9: PRELIMINARY ENGINEERING   301

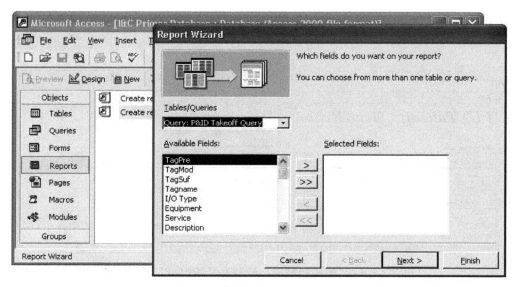

FIGURE 3-25. REPORT WIZARD

PID Takeoff Query Report

Tagname	I/O T	Equipment	Service	Description	PID#	Spec#
HV-13	NA	TK-10 Product Feed	Product Supply	Feed Valve	PID-001	
LT-10	AI	TK-10 Product Feed	Product Level	Level Transmitter	PID-001	
LSH-10	HW	TK-10 Product Feed	Product Level	Level Switch	PID-001	NA
-					PID-001	

FIGURE 3-26. P&ID TAKEOFF QUERY REPORT

If changes to this structure are desired, pressing the key will open the report in design view (Figure 3-27).

Design view of the report utility breaks the report into the various sections of the report page. The report header shows up on the first page only, while the page header shows up on each page. The text labels in this area default to the field names, but they may be modified by the user if necessary.

The detail section is where the data is displayed. The labels in this section need to match the field names exactly. Of course, the page footer can be made to display whatever the user wants it to display. The default is to display time, date, and page number.

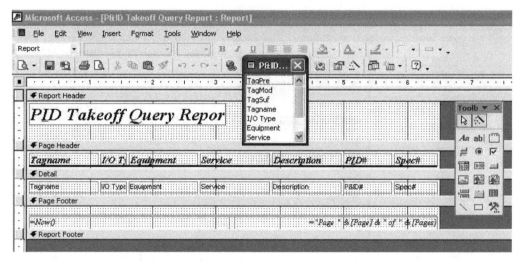

FIGURE 3-27. P&ID TAKEOFF QUERY REPORT, DESIGN VIEW

3. Database Summary

The database is one of the most important tools available to the designer. As an information management tool, it is unparalleled. It can be a wonderful productivity tool as well. To gain the greatest positive benefit for the least investment, it is important each design team member be fully database capable. The database should be in the background at all times, and each member of the design team should continually update it as new information is learned or as work is accomplished. If the database is only updated at the end of the project, then all the time spent entering the data is lost as overhead time. None of it will be recouped as a productivity or quality improvement.

Figure 3-28 is a view of several products of a finished database.

D. Estimate and Schedule Development*

Before the team goes too far in the design, a detailed estimate and project schedule should be generated in a spreadsheet program, such as Excel. An in-depth discussion of estimating and scheduling is provided in Part I.

The numbers presented below, particularly labor rates and costs, are fictional and should not be used to develop an actual estimate. This estimate does not include systems integration, which, for this hypothetical project, will be done on a time and material (T&M) basis.[2]

---

2. For an in-depth look at how to generate an accurate bid estimate and create a project schedule using Microsoft® Excel, refer to the CD-ROM entitled, "Software Tools for Instrumentation and Control Systems Design" included with this book. For more information, see page 523.

FIGURE 3-28. FINISHED DATABASE PRODUCTS

1. COVER WORKSHEET

The cover sheet (Figure 3-29) should be filled out with as much detail as possible. The heading information is particularly important, as it will automatically appear at the top of each of the succeeding worksheets.

2. DEVICES WORKSHEET

The devices worksheet provides a place to tally instruments that are identified during the P&ID takeoff process. If the instrument database is to be used, the I/O

Document#	Rev#	Title Line1	Title Line2	Title Line3	Short Title	Ordered	Received	IFR	IF
ARR-002	0	Cabinet Arrangement	Product Distribution System	TK-10 Product Tank	Arrangement, JB-TK10-1	5/1/2002	5/5/2002		5/1
ARR-001	0	Cabinet Arrangement	Product Distribution System	Node 01, Drop 03	Arrangement, TC-2	5/1/2002	5/5/2002		5/1
LOGIC-001	0	Logic Diagram	Product Distribution System	TK-10 Product Tank	Logic, TK-10	5/1/2002	5/5/2002		5/1
LOOP-PIC48	0	Instrument Loop Diagram	Product Distribution System	TK-10 Product Tank	Loop Sheet, PIC-048	5/1/2002	5/5/2002		5/1
PID-001	0	Piping & Instrumentationn Diagram	Product Distribution System	TK-10 Product Tank	P&ID, TK-10	5/1/2002	5/5/2002	5/10/2002	5/1
PID-002	0	Piping & Instrumentationn Diagram	Product Distribution System	TK-11 Product Tank	P&ID, TK-11	5/1/2002	5/5/2002	5/8/2002	5/1
PID-003	0	Piping & Instrumentationn Diagram	Product Distribution System	TK-12 Product Tank	P&ID, TK-12	5/1/2002	5/5/2002	5/10/2002	5/1
PID-004	0								
PID-005	0								
PLAN-001	0								
SCH-002	0								
SCH-001	0								
SCH-003	0								
NET-001	0								
WRG-002	0								
WRG-001	0								
*									

Tagname	Equipment	Service	Description	P&ID#	PlanDwg	PlanItem	Elevation	Cables	Routing
YS-TK10-15B	PP-10 Product Pump	Motor Controls	Run Status	PID001	MCC001				
HS-TK10-15B	PP-10 Product Pump	Motor Controls	Start Cmmd	PID001	MCC001				
LSH-TK10-10	TK-10 Product Feed Tank	Product Level	Level Switch	PID001	PLAN001	01	2H	2A	A1-PLAN001
LT-TK10-10	TK-10 Product Feed Tank	Product Level	Level Transmitter	PID001	PLAN001	01	2H	1B,3A	D1-PLAN001,A1-
PV-TK10-48	TK-10 Product Feed Tank	Recirc Pressure	Throttling Valve	PID001	PLAN001	02	2H	NA	NA
PY-TK10-48	TK-10 Product Feed Tank	Recirc Pressure	I/P Transducer	PID001	PLAN001	02	2H	1B	D1-PLAN001
PT-TK10-48	PP-10 Product Pump	Discharge Press	Pressure Transmitter	PID001	PLAN001	03	2H	1B	D1-PLAN001
ZSC-TK10-13	TK-10 Product Feed Tank	Product Supply	Valve Closed Status	PID001	PLAN001	04	2H	3A	A1-PLAN001

P&ID#	TagPre	TagSuf	Tagname	I/O Type	HW Address	TagMod	SW Address	Equipment	Service	Description	Spec#
PID001	PT	48	PT-TK10-48	AI	N01D03R01S08P01	TK10	AI0100	PP-10 Product Pump	Discharge Press	Pressure Transmitter	
PID001	LT	10	LT-TK10-10	AI	N01D03R01S08P02	TK10	AI0101	TK-10 Product Feed Tank	Product Level	Level Transmitter	
PID001	PY	48	PY-TK10-48	AO	N01D03R01S10P01	TK10	AO0100	TK-10 Product Feed Tank	Recirc Pressure	I/P Transducer	NA
PID001	ZSC	13	ZSC-TK10-13	DI	N01D03R01S06P01	TK10	DI0100	TK-10 Product Feed Tank	Product Supply	Valve Closed Status	NA
PID001	YS	15B	YS-TK10-15A	DI	N01D03R01S06P02	TK10	DI0101	PP-10 Product Pump	Motor Controls	Switch in Auto	NA
PID001	YS	15A	YS-TK10-15B	DI	N01D03R01S06P03	TK10	DI0102	PP-10 Product Pump	Motor Controls	Run Status	NA
PID001	LSLL	47	LSLL-TK10-47	DI	N01D03R01S06P05	TK10	DI0104	TK-10 Product Feed Tank	Low Level	Level Switch	
PID001	HS	15B	HS-TK10-15B	DOI	N01D03R01S04P02	TK10	DO0102	PP-10 Product Pump	Motor Controls	Start Cmmd	NA
PID001	HY	13	HY-TK10-13	DOI	N01D03R01S04P01	TK10	DO0101	TK-10 Product Feed Tank	Product Supply	Solenoid, 3-Way	NA
PID001	LSH	10	LSH-TK10-10	HW	NA	TK10	NA	TK-10 Product Feed Tank	Product Level	Level Switch	NA
PID001	PSV	58	PSV-TK10-58	NA	NA	TK10	NA	TK-10 Product Feed Tank	Pressure	Relief Valve	
PID001	PV	48	PV-TK10-48	NA	NA	TK10	NA	TK-10 Product Feed Tank	Recirc Pressure	Throttling Valve	

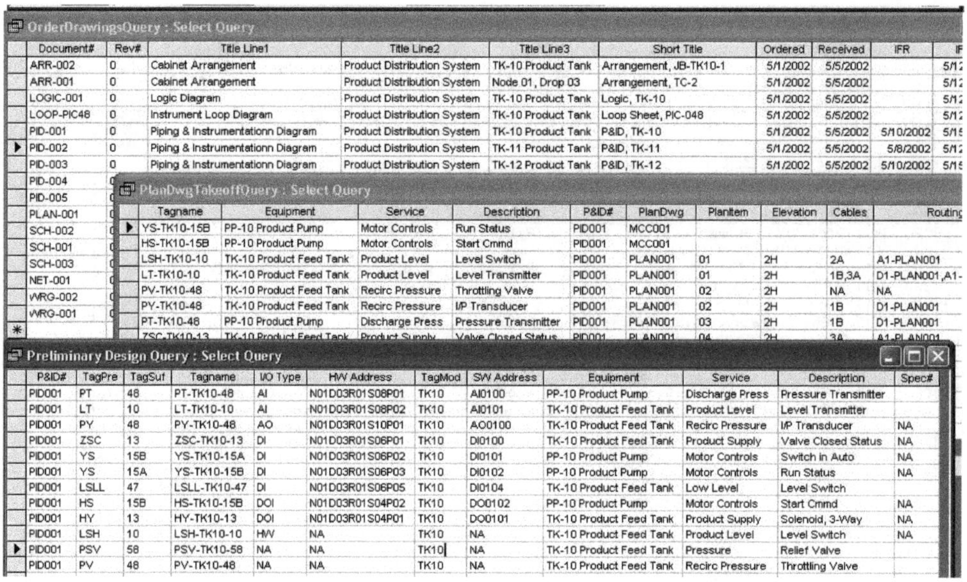

FIGURE 3-28 (CONTINUED). FINISHED DATABASE PRODUCTS

list should have already been created. If that is the case, filling out this worksheet will be an easy matter of extracting information from the database.

If the team is using the devices worksheet, the P&ID takeoff is accomplished by simply logging the instruments into the worksheet. The I/O count is automatically calculated, as is the estimated cost of the instruments.

Using the P&ID provided in Figure 3-5 (reproduced on page 169), the project estimate may be developed by first filling in the device data table (devices worksheet) as shown in Figure 3-30. There are seven items, which cost a total of $13,900. There will be incidental installation expenses of about $310.

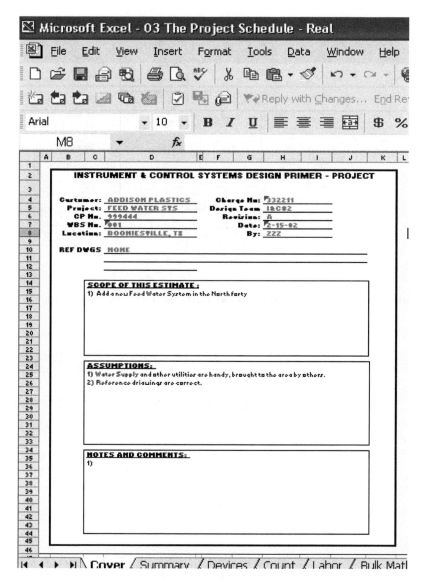

FIGURE 3-29. COVER SHEET FOR ESTIMATE WORKBOOK

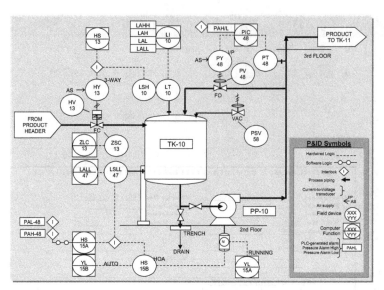

FIGURE 3-5. (REPRODUCED). P&ID PRESENTATION OF THE TK-10 SUBSYSTEM

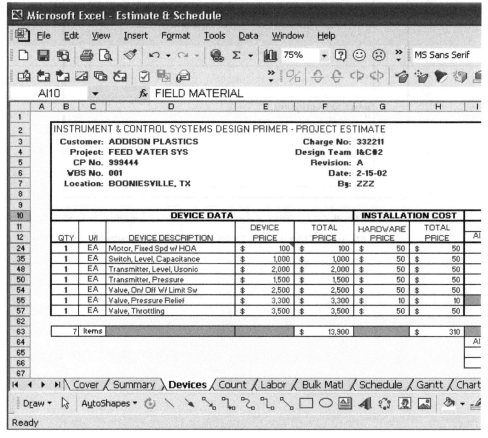

FIGURE 3-30. DEVICES WORKSHEET

After the instruments have been entered, the user should hide, not delete,* the intervening instruments that had a quantity of zero. Prices should be checked, and then the I/O assignment index should be evaluated for accuracy (Figure 3-31).

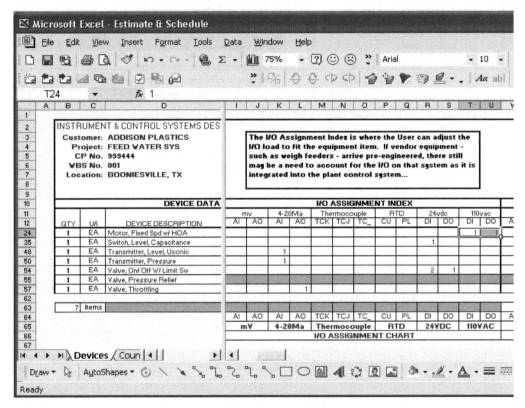

FIGURE 3-31. DEVICES I/O ASSIGNMENT INDEX TABLE

In our example, the default configuration for a motor is one digital input and one digital output as shown in the cells highlighted. Investigation reveals this customer wants to operate motors in either remote mode, with the PLC, or local mode, bypassing the PLC. The customer wants interlocks hardwired so they are in the circuit regardless of mode. That requires a change to the motor logic.

The new scheme will require three digital inputs and two digital outputs. So the I/O assignment index should be revised (Figure 3-32).

The resulting I/O count is displayed in the I/O calculator table (Figure 3-33). So the seven new instruments require a total of 12 new I/O points. Six of the seven new instruments require conduit connections, and two require tubing connections for instrument air.

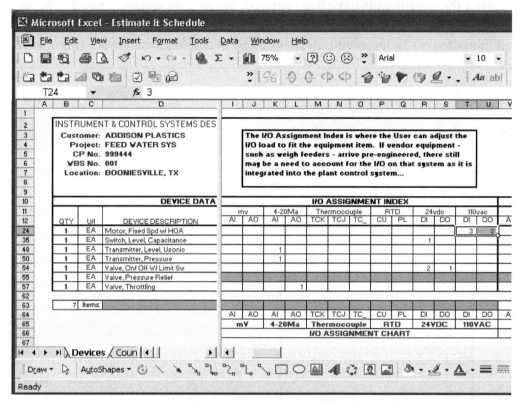

FIGURE 3-32. DEVICES I/O ASSIGNMENT INDEX, REVISED

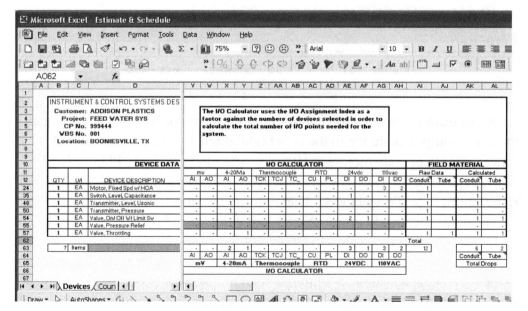

FIGURE 3-33. DEVICES I/O CALCULATOR

3. COUNT WORKSHEET

The next step is to configure the I/O map using the count worksheet (Figure 3-34). This worksheet fetches the I/O configuration from the I/O calculator and uses that information to calculate number of modules, including percentage spare, module cost, number of elementary (PLC schematic) drawings, and number of loop sheets. The calculator uses information input into the background data table and two parameters (%Spare and the number of proportional-integral-derivative [PID] loops) above the heading in the I/O configuration worksheet.[3]

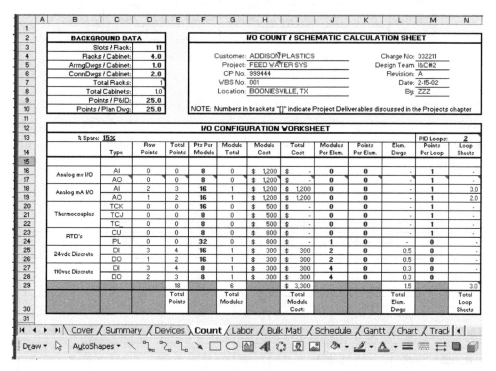

FIGURE 3-34. COUNT WORKSHEET

The first step is to fill in the background data table (Figure 3-35). The default number of slots per rack is 10. The rack being used in this project is a 10-slot rack. Only one rack will be in this cabinet. Total racks and total cabinets are automatically calculated, though in this case we know those numbers. The Points per P&ID value is needed only if a number of P&IDs need to be estimated, as might be the case in a budgetary estimate. Likewise, the point density on the instrument location plans helps estimate the number of location plan drawings that will be needed. In this case, neither calculation is applicable.

---

3. Caution: Hiding information on a spreadsheet does not prevent data from being rolled up and totaled. The user should use care to hide only records with a quantity of zero.

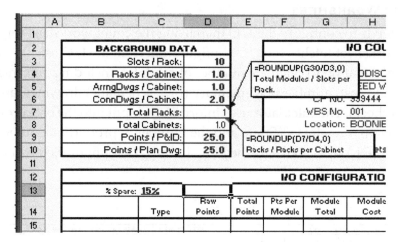

FIGURE 3-35. BACKGROUND DATA TABLE

Next, points per module, module cost, modules per elementary, points per elementary, and points per loop data should be verified in the I/O configuration worksheet (Figure 3-36). We will assume for our purposes that the default values are accurate.

	Type	Raw Points	Total Points	Pts Per Module	Module Total	Module Cost	Total Cost	Modules Per Elem.	Points Per Elem.	Elem. Dwgs	Points Per Loop	Loop Sheets
Analog mv I/O	AI	0	0	8	0	$ 1,200	$ -	0	0	-	1	-
	AO	0	0	8	0	$ 1,200	$ -	0	0	-	1	-
Analog mA I/O	AI	2	3	16	1	$ 1,200	$ 1,200	0	0	-	1	3.0
	AO	1	2	16	1	$ 1,200	$ 1,200	0	0	-	1	2.0
Thermocouples	TCK	0	0	16	0	$ 500	$ -	0	0	-	1	-
	TCJ	0	0	8	0	$ 500	$ -	0	0	-	1	-
	TC_	0	0	8	0	$ 500	$ -	0	0	-	1	-
RTD's	CU	0	0	8	0	$ 800	$ -	0	0	-	1	-
	PL	0	0	32	0	$ 800	$ -	1	0	-	0	-
24vdc Discrete	DI	3	4	16	1	$ 300	$ 300	2	0	0.5	0	-
	DO	1	2	16	1	$ 300	$ 300	2	0	0.5	0	-
110vac Discrete	DI	3	4	8	1	$ 300	$ 300	4	0	0.3	0	-
	DO	2	3	8	1	$ 300	$ 300	4	0	0.3	0	-
			18		6		$ 3,300			1.5		3.0
			Total Points		Total Modules		Total Module Cost:			Total Elem. Dwgs		Total Loop Sheets

% Spare: 15%   PID Loops: 2

I/O CONFIGURATION WORKSHEET

FIGURE 3-36. I/O CONFIGURATION WORKSHEET

So an estimate of $3,300 is made for I/O modules, an elementary PLC schematic plus part of another, and three loop sheets. This leads to the labor worksheet.

4. LABOR WORKSHEET

The labor worksheet calculates engineering labor based on a list of deliverables. This worksheet consists of five tables (Figure 3-37):

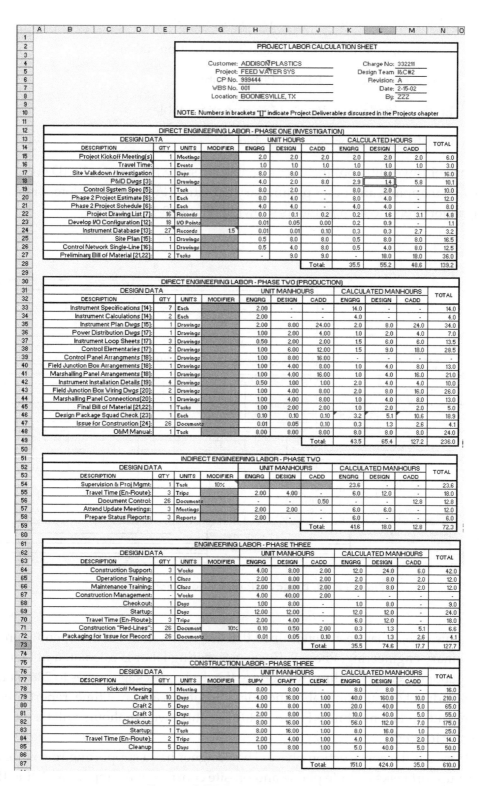

FIGURE 3-37. LABOR WORKSHEET

- Direct engineering labor—Phase 1 (investigation) table
- Direct engineering labor—Phase 2 (production) table
- Indirect engineering labor—Phase 2 table
- Engineering labor—Phase 3 (construction) table
- Construction labor—Phase 3 table

Phase 1 (investigation) includes orientation, site walk downs, P&ID generation, I/O mapping, and other preliminary activities (Figure 3-38). Note: Bracket designations refer to item numbers shown in Figure 3-1, page 269.

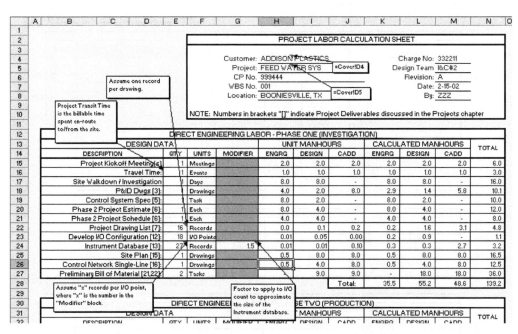

FIGURE 3-38. DIRECT ENGINEERING LABOR, PHASE 1

In this case, it has been determined that 139 manhours will be expended in preliminary engineering. Phase 2 direct engineering labor is calculated in Figure 3-39. Labor expended to directly produce the design package comes to 236 manhours. Indirect engineering labor includes such items as travel time, report preparation, and update meetings (Figure 3-40). Seventy-two manhours will be set aside for indirect tasks.

Phase 3 is the construction phase. The designer has a supporting role at this time. Support tasks include phone and on-site construction support, assisting the CSI group with operator training, maintenance training, and checkout and startup support (Figure 3-41).

Part III – Chapter 9: Preliminary Engineering

	DIRECT ENGINEERING LABOR - PHASE TWO (PRODUCTION)									
DESIGN DATA					UNIT HOURS			CALCULATED HOURS		TOTAL
DESCRIPTION	QTY	UNITS	MODIFIER	ENGRG	DESIGN	CADD	ENGRG	DESIGN	CADD	
Instrument Specifications [14]:	7	Each		2.00	-	-	14.0	-	-	14.0
Instrument Calculations [14]:	2	Each		2.00	-	-	4.0	-	-	4.0
Instrument Plan Dwgs [15]:	1	Drawings		2.00	8.00	24.00	2.0	8.0	24.0	34.0
Power Distribution Dwgs [17]:	1	Drawings		1.00	2.00	4.00	1.0	2.0	4.0	7.0
Instrument Loop Sheets [17]:	3	Drawings		0.50	2.00	2.00	1.5	6.0	6.0	13.5
Control Elementaries [17]:	2	Drawings		1.00	6.00	12.00	1.5	9.0	18.0	28.5
Control Panel Arrangements [18]:	-	Drawings		1.00	8.00	16.00	-	-	-	-
Field Junction Box Arrangements [18]:	1	Drawings		1.00	4.00	8.00	1.0	4.0	8.0	13.0
Marshalling Panel Arrangements [18]:	1	Drawings		1.00	4.00	16.00	1.0	4.0	16.0	21.0
Instrument Installation Details [19]:	4	Drawings		0.50	1.00	1.00	2.0	4.0	4.0	10.0
Field Junction Box Wiring Dwgs [20]:	2	Drawings		1.00	4.00	8.00	2.0	8.0	16.0	26.0
Marshalling Panel Connections[20]:	1	Drawings		1.00	4.00	8.00	1.0	4.0	8.0	13.0
Final Bill of Material [21,22]:	1	Tasks		1.00	2.00	2.00	1.0	2.0	2.0	5.0
Design Package Squad Check [23]:	1	Each		0.10	0.10	0.10	3.2	5.1	10.6	18.9
Issue for Construction [24]:	26	Documents		0.01	0.05	0.10	0.3	1.3	2.6	4.1
O&M Manual:	1	Task		8.00	8.00	8.00	8.0	8.0	8.0	24.0
						Total:	43.5	65.4	127.2	236.0

Figure 3-39. Direct engineering labor, Phase 2

	INDIRECT ENGINEERING LABOR - PHASE TWO									
DESIGN DATA					UNIT HOURS			CALCULATED HOURS		TOTAL
DESCRIPTION	QTY	UNITS	MODIFIER	ENGRG	DESIGN	CADD	ENGRG	DESIGN	CADD	
Supervision & Proj Mgmt:	1	Task	10%				23.6		-	23.6
Travel Time (En-Route):	3	Trips		2.00	4.00	-	6.0	12.0	-	18.0
Document Control:	26	Documents		-	-	0.50	-	-	12.8	12.8
Attend Update Meetings:	3	Meetings		2.00	2.00	-	6.0	6.0	-	12.0
Prepare Status Reports:	3	Reports		2.00	-	-	6.0	-	-	6.0
						Total:	41.6	18.0	12.8	72.3

Figure 3-40. Indirect engineering labor, Phase 2

	ENGINEERING LABOR - PHASE THREE									
DESIGN DATA					UNIT HOURS			CALCULATED HOURS		TOTAL
DESCRIPTION	QTY	UNITS	MODIFIER	ENGRG	DESIGN	CADD	ENGRG	DESIGN	CADD	
Construction Support:	3	Weeks		4.00	8.00	2.00	12.0	24.0	6.0	42.0
Operations Training:	1	Class		2.00	8.00	2.00	2.0	8.0	2.0	12.0
Maintenance Training:	1	Class		2.00	8.00	2.00	2.0	8.0	2.0	12.0
Construction Management:	-	Weeks		4.00	40.00	2.00	-	-	-	-
Checkout:	1	Days		1.00	8.00	-	1.0	8.0	-	9.0
Startup:	1	Days		12.00	12.00	-	12.0	12.0	-	24.0
Travel Time (En-Route):	3	Trips		2.00	4.00	-	6.0	12.0	-	18.0
Construction "Redlines":	26	Document	10%	0.10	0.50	2.00	0.3	1.3	5.1	6.6
Packaging for 'Issue for Record':	26	Documents		0.01	0.05	0.10	0.3	1.3	2.6	4.1
						Total:	35.5	74.6	17.7	127.7

	CONSTRUCTION LABOR - PHASE THREE									
DESIGN DATA					UNIT HOURS			CALCULATED HOURS		TOTAL
DESCRIPTION	QTY	UNITS	MODIFIER	SUPV	CRAFT	CLERK	ENGRG	DESIGN	CADD	
Kickoff Meeting	1	Meeting		8.00	8.00	-	8.0	8.0	-	16.0
Craft 1	10	Days		4.00	16.00	1.00	40.0	160.0	10.0	210.0
Craft 2	5	Days		4.00	8.00	1.00	20.0	40.0	5.0	65.0
Craft 3	5	Days		2.00	8.00	1.00	10.0	40.0	5.0	55.0
Checkout:	7	Days		8.00	16.00	1.00	56.0	112.0	7.0	175.0
Startup:	1	Task		8.00	16.00	1.00	8.0	16.0	1.0	25.0
Travel Time (En-Route):	2	Trips		2.00	4.00	1.00	4.0	8.0	2.0	14.0
Cleanup	5	Days		1.00	8.00	1.00	5.0	40.0	5.0	50.0
							-	-	-	-
						Total:	151.0	424.0	35.0	610.0

Figure 3-41. Engineering and construction labor, Phase 3

An attempt is also made to derive construction manhours. The construction labor table produces a very rough idea of the E & I (electrical and instrumentation) construction task by craft.

5. Summary Worksheet

Derived estimate values are automatically collected and displayed in summary format on the project cost summary sheet (Figure 3-42). This worksheet contains six tables.

PROJECT COST SUMMARY SHEET

Customer: ADDISON PLASTICS	Charge No: 332211
Project: FEED WATER SYS	Design Team: I&C#2
CP No. 999444	Revision: A
WBS No. 001	Date: 2-15-02
Location: BOONIESVILLE, TX	By: ZZZ

ENGINEERING COST SUMMARY

ENGINEERING DETAIL		Mhrs	AVG RATE	Cost	TOTAL
Phase One Projection	Engineering	35	$90	$ 3,194	
	Design	55	$70	$ 3,861	$ 9,483
	CADD & Clerical	49	$50	$ 2,428	
Phase Two Projection	Engineering	85	$90	$ 7,655	
	Design	83	$70	$ 5,836	$ 20,486
	CADD & Clerical	140	$50	$ 6,995	
Phase Three Projection	Engineering	36	$90	$ 3,196	
	Design	75	$70	$ 5,219	$ 9,297
	CADD & Clerical	18	$50	$ 883	
		575			$ 39,266

PROJECT COST SUMMARY

CATEGORY	AVG RATE	COST
ENGINEERING	NA	$ 39,266
CONSTRUCTION	$38	$ 23,180
INSTRUMENTS	NA	$ 13,900
MATERIAL	NA	$1,557
PROJECTED COST:	$	77,903

INSTRUMENT SUMMARY

Item Total:	7
Instrument Cost:	$ 13,900
Installation Hardware Cost:	$ 310

I/O SUMMARY

I/O TYPE		COUNT
Analog mv I/O	AI	0
	AO	0
Analog mA I/O	AI	3
	AO	2
Thermocouples	TCK	0
	TCJ	0
	TC_	0
RTD's	CU	0
	PL	0
24vdc Discrete	DI	4
	DO	2
110vac Discrete	DI	4
	DO	3
Total:		18

PHASE 1 DELIVERABLES SUMMARY

Project Kickoff Meeting(s)	1	Meetings
Travel Time:	1	Events
Site Walkdown / Investigation	1	Days
P&ID Dwgs [3]:	1	Drawings
Control System Spec [5]:	1	Task
Phase 2 Project Estimate [6]:	1	Each
Phase 2 Project Schedule [6]:	1	Each
Project Drawing List [7]:	16	Records
Develop I/O Configuration [12]:	18	I/O Points
Instrument Database [13]:	27	Records
Site Plan [15]:	1	Drawings
Control Network Single-Line [16]:	1	Drawings
Preliminary Bill of Material [21,22]:	2	Tasks

PHASE 2 DELIVERABLES SUMMARY

Instrument Specifications [14]:	7	Each
Instrument Calculations [14]:	2	Each
Instrument Plan Dwgs [15]:	1	Drawings
Power Distribution Dwgs [17]:	1	Drawings
Instrument Loop Sheets [17]:	3	Drawings
Control Elementaries [17]:	2	Drawings
Control Panel Arrangements [18]:	-	Drawings
Field Junction Box Arrangements [18]:	1	Drawings
Marshalling Panel Arrangements [18]:	1	Drawings
Instrument Installation Details [19]:	4	Drawings
Field Junction Box Wiring Dwgs [20]:	2	Drawings
Marshalling Panel Connections[20]:	1	Drawings
Final Bill of Material [21,22]:	1	Tasks
Design Package Squad Check [23]:	1	Each
Issue for Construction [24]:	26	Documents

FIGURE 3-42. ENGINEERING SUMMARY WORKSHEET

For the purposes of this estimate, total project cost is $77,903 for the I&C department. This is broken down in the project cost summary table (Figure 3-43).

PROJECT COST SUMMARY		
Category	Avg Rate	Cost
Engineering	NA	$ 39,266
Construction	$38	$ 23,180
Instruments	NA	$ 13,900
Material	NA	$1,557
Projected Cost:	$	**77,903**

FIGURE 3-43. PROJECT COST SUMMARY TABLE

The engineering cost table (Figure 3-44) applies average dollar-per-hour rates to the manhours to derive engineering cost. These manhour values are extracted from the Labor Worksheet (Figure 3-38 for example).

ENGINEERING COST SUMMARY					
Engineering Detail		Mhrs	Avg Rate	Cost	Total
Phase One Projection	Engineering	35	$90	$ 3,194	
	Design	55	$70	$ 3,861	$ 9,483
	CADD and Clerical	49	$50	$ 2,428	
Phase Two Projection	Engineering	85	$90	$ 7,655	
	Design	83	$70	$ 5,836	$ 20,486
	CADD and Clerical	140	$50	$ 6,995	
Phase Three Projection	Engineering	36	$90	$ 3,196	
	Design	75	$70	$ 5,219	$ 9,297
	CADD and Clerical	18	$50	$ 883	
		575			$ 39,266

FIGURE 3-44. ENGINEERING COST SUMMARY TABLE

Deliverables are also derived from the worksheets (Figures 3-45 and 3-46). Finally, the instrument and I/O summaries (Figure 3-47).

Following this estimation process does not guarantee a valid estimate. The numbers presented here must be tailored to the specific design organization and customer. However, this structured estimating process is valuable in that it may be tweaked from project to project until the estimator becomes extremely proficient and accurate. A side benefit is the generation of a very detailed design basis that may be used to evaluate the execution of the project.

PHASE 1 DELIVERABLES SUMMARY		
Project Kickoff Meeting(s)	1	Meetings
Travel Time:	1	Events
Site Walkdown / Investigation	1	Days
P&ID Dwgs [3]:	1	Drawings
Control System Spec [5]:	1	Task
Phase 2 Project Estimate [6]:	1	Each
Phase 2 Project Schedule [6]:	1	Each
Project Drawing List [7]:	16	Records
Develop I/O Configuration [12]:	18	I/O Points
Instrument Database [13]:	27	Records
Site Plan [15]:	1	Drawings
Control Network Single-Line [16]:	1	Drawings
Preliminary Bill of Material [21,22]:	2	Tasks

FIGURE 3-45. PHASE 1 DELIVERABLES SUMMARY TABLE

PHASE 2 DELIVERABLES SUMMARY		
Instrument Specifications [14]:	7	Each
Instrument Calculations [14]:	2	Each
Instrument Plan Dwgs [15]:	1	Drawings
Power Distribution Dwgs [17]:	1	Drawings
Instrument Loop Sheets [17]:	3	Drawings
Control Elementaries [17]:	2	Drawings
Control Panel Arrangments [18]:	-	Drawings
Field Junction Box Arrangements [18]:	1	Drawings
Marshalling Panel Arrangements [18]:	1	Drawings
Instrument Installation Details [19]:	4	Drawings
Field Junction Box Wiring Dwgs [20]:	2	Drawings
Marshalling Panel Connections [20]:	1	Drawings
Final Bill of Material [21,22]:	1	Tasks
Design Package Squad Check [23]:	1	Each
Issue for Construction [24]:	26	Documents

FIGURE 3-46. PHASE 2 DELIVERABLES SUMMARY TABLE

6. SCHEDULE WORKSHEET

Most projects are driven either by time or by cost. The ideal schedule, in terms of executing the engineering design, is often different from the schedule required by the customer. However, the designer must have some idea of staffing capabilities when the estimate is presented. The schedule worksheet attempts to automate the scheduling process to a degree by linking the schedule categories to labor values derived while doing the estimate.

The default schedule worksheet based on the manhour estimate is shown in Figure 3-48.

INSTRUMENT SUMMARY

Item Total: 7
Instrument Cost: $ 13,900
Installation Hardware Cost: $ 310

I/O SUMMARY

I/O Type		Count
Analog mV I/O	AI	0
	AO	0
Analog mA I/O	AI	3
	AO	2
Thermocouples	TCK	0
	TCJ	0
	TC_	0
RTDs	CU	0
	PL	0
24 VDC Discrete	DI	4
	DO	2
110 VAC Discrete	DI	4
	DO	3
	Total:	18

FIGURE 3-47. INSTRUMENT AND I/O SUMMARY TABLE

The schedule activities mimic the estimate. In fact, task and budget data are pulled directly from their associated estimate sheets. The spread field is then calculated. The spread calculation sums the values in the Week X fields and subtracts that from the budgeted amount. So the two fields are equal.

The hours in the spread column must be spread across the schedule from week to week. The estimator should have an optimum team size in mind and should spread the hours based on the number of hours per week the team will be working this project. For our purposes, assume a two-person design staff. Figure 3-48 shows the design schedule for this project, as the estimator has spread the available hours he expects to expend each week. Note that the budget hours and the spread hours are now different values.

As shown in this schedule, the project life cycle is 10 weeks in length, given the staffing level provided. The manhour loading chart on the chart worksheet is shown in Figure 3-50.

7. ESTIMATE AND SCHEDULE SUMMARY

Developing a good estimate and a realistic schedule is as important for the design team as it is for the project management team. To be useful as a design aid, the estimate should be detailed enough to use as a scope of work while being rela-

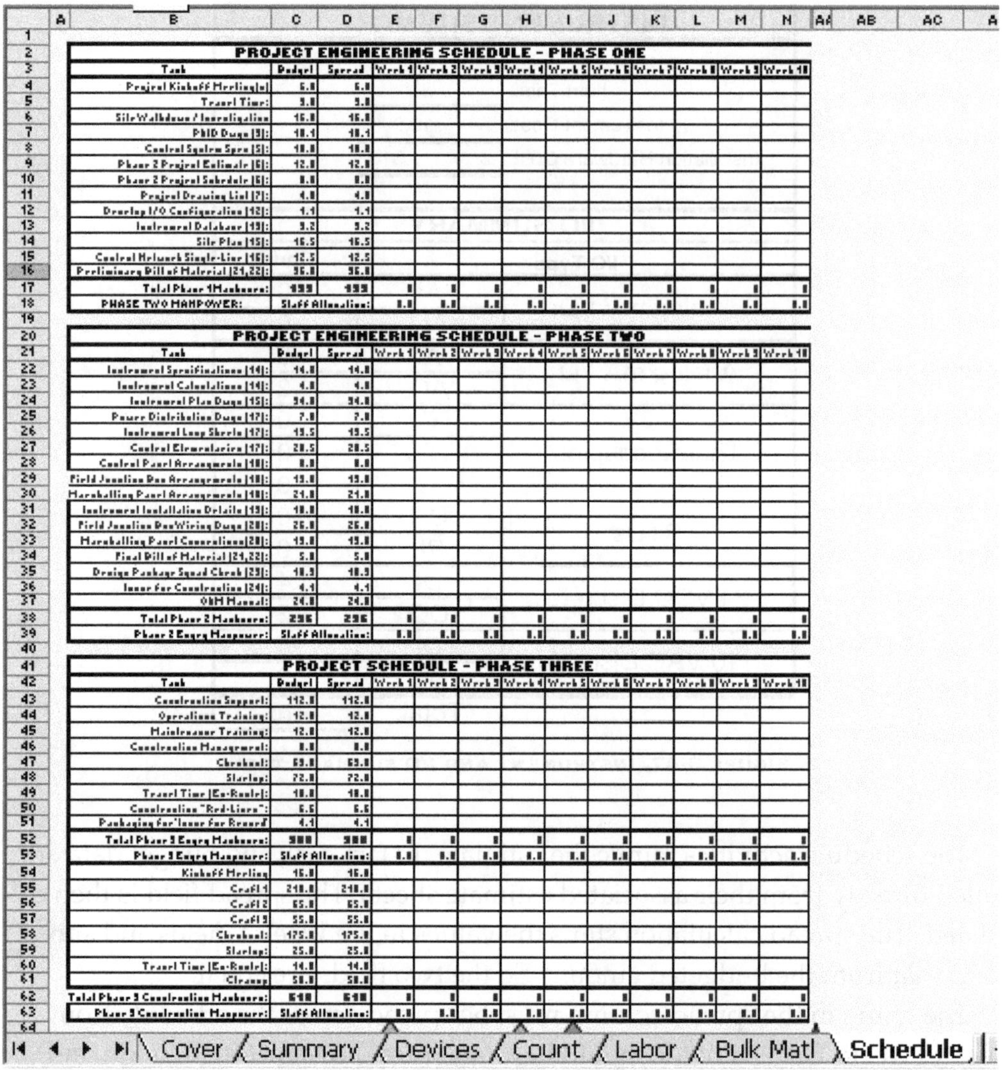

FIGURE 3-48. SCHEDULE WORKSHEET

tively easy to produce. The estimating tool presented here uses a relatively small amount of user input to generate a large amount of information.

Also, it is important that the schedule reflect the estimate. This schedule is absolutely tied to the estimate through data links. The user simply spreads the hours as necessary to fit the time frame allotted and the resources available.

Thus, a team-friendly estimate and schedule may easily be generated. Progress reports to the customer may be in an entirely different, less detailed format that is probably not very useful to the design team. The project manager can always reorganize the numbers generated here into a less detailed report for the customer, while it is difficult, or even impossible, to do the reverse. **Note:** This workbook is available online at www.ISA.org.

PROJECT ENGINEERING SCHEDULE - PHASE ONE

Task	Budget	Spread	Week 1	Week 2	Week 3	Week 4	Week 5	Week 6	Week 7	Week 8	Week 9	Week 10	Week 11
Project Kickoff Meeting(s):	6.0	0.0	6										
Travel Time:	3.0	0.0	3										
Site Walkdown / Investigation	16.0	0.0	16										
P&ID Dwgs [3]:	10.1	2.1	8										
Control System Spec [5]:	10.0	0.0	10										
Phase 2 Project Estimate [6]:	12.0	0.0	12										
Phase 2 Project Schedule [6]:	8.0	0.0	8										
Project Drawing List [7]:	4.8	0.8	4										
Develop I/O Configuration [12]:	1.1	0.1	1										
Instrument Database [13]:	3.2	0.2	3										
Site Plan [15]:	16.5	0.5		16									
Control Network Single-Line [16]:	12.5	0.5		12									
Preliminary Bill of Material [21,22]:	36.0	0.0			36								
Total Phase 1 Manhours:	139	4	71	64	0	0	0	0	0	0	0	0	0
PHASE TWO MANPOWER:		Staff Allocation:	1.8	1.6	0.0	0.0	0.0	0.0	0.0	0.0	0.0	0.0	0.0

PROJECT ENGINEERING SCHEDULE - PHASE TWO

Task	Budget	Spread	Week 1	Week 2	Week 3	Week 4	Week 5	Week 6	Week 7	Week 8	Week 9	Week 10	Week 11
Instrument Specifications [14]:	14.0	0.0		12			2						
Instrument Calculations [14]:	4.0	0.0				4							
Instrument Plan Dwgs [15]:	34.0	2.0			24		8						
Power Distribution Dwgs [17]:	7.0	-1.0				8							
Instrument Loop Sheets [17]:	13.5	0.5			12		1						
Control Elementaries [17]:	28.5	0.5			24		4						
Control Panel Arrangements [18]:	0.0	0.0											
Field Junction Box Arrangements [18]:	13.0	0.0				12	1						
Marshalling Panel Arrangements [18]:	21.0	1.0				16	4						
Instrument Installation Details [19]:	10.0	0.0				8	2						
Field Junction Box Wiring Dwgs [20]:	26.0	0.0				24	2						
Marshalling Panel Connections [20]:	13.0	0.0				12	1						
Final Bill of Material [21,22]:	5.0	1.0					4						
Design Package Squad Check [23]:	18.3	2.3					16						
Issue for Construction [24]:	4.1	0.1					4						
O&M Manual:	24.0	0.0					16	8					
Total Phase 2 Manhours:	236	7	0	12	72	72	65	8	0	0	0	0	0
Phase 2 Engrg Manpower:		Staff Allocation:	0.0	0.3	1.8	1.8	1.6	0.2	0.0	0.0	0.0	0.0	0.0

PROJECT SCHEDULE - PHASE THREE

Task	Budget	Spread	Week 1	Week 2	Week 3	Week 4	Week 5	Week 6	Week 7	Week 8	Week 9	Week 10	Week 11
Construction Support:	42.0	10.0							8	8	8	8	
Operations Training:	12.0	0.0								12			
Maintenance Training:	12.0	0.0								12			
Construction Management:	0.0	0.0											
Checkout:	3.0	0.0									3		
Startup:	24.0	0.0									24		
Travel Time (En-Route):	18.0	2.0							8		8		
Construction "Red-Lines":	6.6	0.6										6	
Packaging for 'Issue for Record'	4.1	0.1										4	
Total Phase 3 Engrg Manhours:	128	13	0	0	0	0	0	8	40	8	43	10	0
Phase 3 Engrg Manpower:		Staff Allocation:	0.0	0.0	0.0	0.0	0.0	0.2	1.0	0.2	1.2	0.3	0.0
Kickoff Meeting	16.0	0.0						16					
Craft 1	210.0	0.0						40	80	60	30		
Craft 2	65.0	0.0						5	20	20	20		
Craft 3	55.0	0.0						3		40	12		
Checkout:	175.0	7.0						8	40	40	80		
Startup:	25.0	9.0									16		
Travel Time (En-Route):	14.0	-2.0						8			8		
Cleanup	50.0	0.0										50	
Total Phase 3 Construction Manhours:	610	14	0	0	0	0	0	80	140	160	166	50	0
Phase 3 Construction Manpower:		Staff Allocation:	0.0	0.0	0.0	0.0	0.0	2.0	3.5	4.0	4.2	1.3	0.0

FIGURE 3-49. DESIGN SCHEDULE AND STAFFING PLAN

E. PRELIMINARY ENGINEERING SUMMARY

To execute a task, it is necessary to first have a plan. To develop a *good* plan, there must be some investigation. The set of preliminary engineering tasks is designed to yield a project plan that is needed for a properly managed and scoped project. The tasks presented here are generally universal. Most of the time, a scope of work is generated.

Whether it is done formally, as a set of documents, or not, the control system must be evaluated, documents must be collected, an I/O list must be generated, and a schedule must be prepared. The tools presented here merely offer a way of

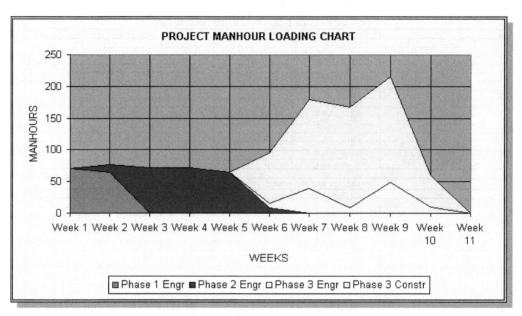

FIGURE 3-50. PROJECT MANHOUR LOADING CHART

approaching these tasks. The user, however, may adapt these techniques without the risk associated with new and untested methods. The key is to develop a system that lets the user conduct a post-mortem after the project is completed. To do that, the original estimate should be organized to line up with the execution strategy in the scope of work.

Part III – Chapter 10: Control Systems Integration (CSI)

Core CSI tasks include software development, HMI configuration, and data communications. In addition, systems integrators will naturally develop a keen understanding of plant operations during the software development process. That makes them the best candidates for the authorship of an operations and maintenance (O&M) manual (to describe the new controls from the operations and the maintenance perspectives) and to lead the training evolution. Figure 1-9 (Part I) provides an overview of a CSI project plan.

Many control systems integrators are also control cabinet fabricators. This is a natural fit, as the control equipment is needed for testing, and that is best done after fabrication. Customers should insist upon a Factory Acceptance Test (FAT) prior to shipment to demonstrate software capability and scope compliance. The FAT may consist of some degree of process simulation, from "loop-back" simulation to process simulation software coupled with physical I/O simulation. A good rule of thumb is that an hour of FAT will save at least two hours of trouble onsite, and that ratio can be much higher.

While this is not a chapter on programming per se, some programming concepts are discussed. We will leave the programming details to other books, of which there are many. Rather, we take a specific real-world situation and apply good Front End Loaded (FEL) design techniques to develop an engineered solution. This is done within the context of the overall project as outlined in the engineering package classification flow chart in the Part III Introduction (Figure 3-1). The CSI activity (item 10 on the flow chart) represents a broad category that includes software development, data communications, and operability issues.

This chapter presents some tools and techniques for getting buy-in from all parties *before* the programming or installation begins. Figure 3-51 is a simple checklist that provides a starting point.

Note: The scope of supply presented in the figure does not include the PLC program or the HMI program because those items are not included in this book. Only the items discussed in Part III are checked.

A. FEL Stage 1 – Business Planning

FEL Stage 1 (Part I, Chapter 1) is where the project estimate is built and the proposal submitted. Prerequisites are the P&ID (Figure 3-5), an I/O count

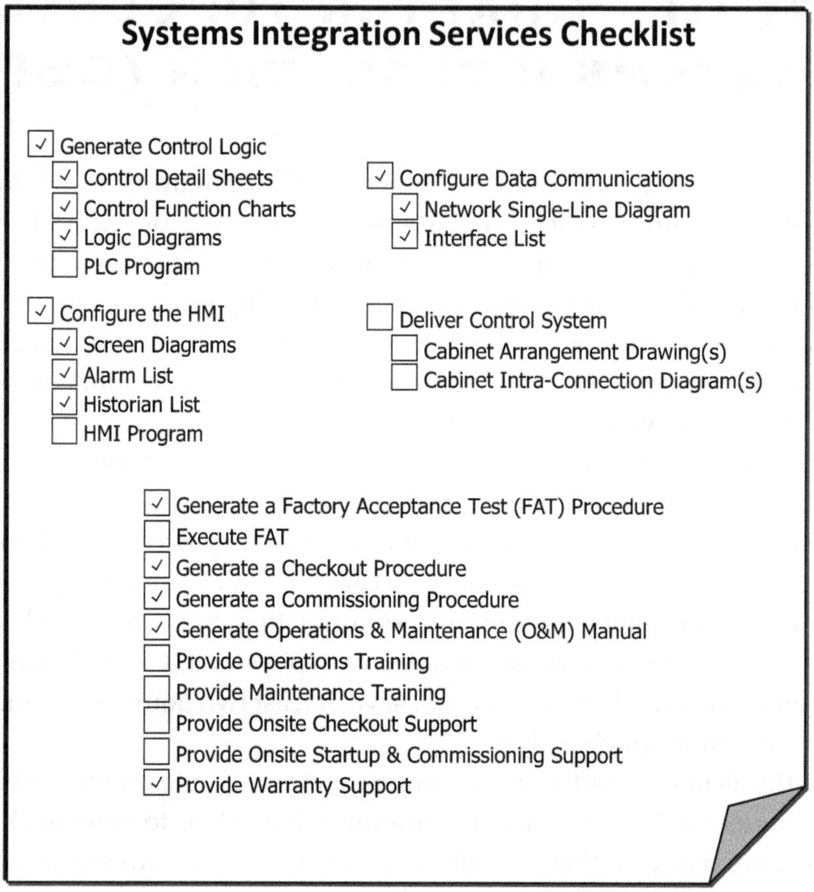

Systems Integration Services Checklist

- [x] Generate Control Logic
  - [x] Control Detail Sheets
  - [x] Control Function Charts
  - [x] Logic Diagrams
  - [ ] PLC Program
- [x] Configure the HMI
  - [x] Screen Diagrams
  - [x] Alarm List
  - [x] Historian List
  - [ ] HMI Program
- [x] Configure Data Communications
  - [x] Network Single-Line Diagram
  - [x] Interface List
- [ ] Deliver Control System
  - [ ] Cabinet Arrangement Drawing(s)
  - [ ] Cabinet Intra-Connection Diagram(s)
- [x] Generate a Factory Acceptance Test (FAT) Procedure
- [ ] Execute FAT
- [x] Generate a Checkout Procedure
- [x] Generate a Commissioning Procedure
- [x] Generate Operations & Maintenance (O&M) Manual
- [ ] Provide Operations Training
- [ ] Provide Maintenance Training
- [ ] Provide Onsite Checkout Support
- [ ] Provide Onsite Startup & Commissioning Support
- [x] Provide Warranty Support

FIGURE 3-51. SYSTEMS INTEGRATION SERVICES CHECKLIST

(Figure 3-47), a Project Schedule (can be inferred from Figure 3-48), and the Control System Specification (Section 2, this chapter). A budgetary project estimate and proposal can sometimes be built from just the P&ID, but a definitive estimate is only made possible with the addition of the other items.

1. COST/BENEFIT ANALYSIS

For a systems integrator, as for any engineering service provider, a cost/benefit analysis of the project should be done prior to even generating a proposal for the customer. The process of generating a proposal is time consuming and expensive. And once the proposal is submitted, the commitment is implied, and it is much more difficult to back out of the project. The decision to pursue a contract should be based on a thorough evaluation prior to bid submittal, as backing out after the customer awards the project is unprofessional and could affect the bidding company's reputation. But even at that, it is better to back out early than to fail.

For the systems integrator, this analysis should cover, at minimum, the following topics:

- *Resource Suitability:* Do we have the right staff mix to be competitive in the bid and to be successful if the project is awarded to us?

- *Resource Availability:* If this project is awarded, and our staff is suitable to the task, are those resources available in the time window being discussed?

- *Strategic Need:* Are we in dire need of a project? In such a case, we may be willing to take on a project that would be otherwise unsuitable.

- *Customer Reliability:* The customer's ability to pay and his payment history. A customer who requires a big investment to get him to pay his bills is a customer to avoid.

- *Customer Relationship:* Is this customer a strategic client, one for which a long-term relationship is either ongoing or desirable? The answer to this question will dictate the tone of the proposal, and may induce a bid even if the project is inconvenient in terms of workflow or other considerations.

In the end, a decision to submit a proposal is a decision to participate, and that decision should not be considered lightly.

2. Control System Specification

A control system specification is usually provided as a part of the customer's Request for Proposal (RFP). There should be enough information provided in this document to allow the integrator to gain an understanding of the scope of the task. Information should begin with a functional description and include at least the following:

- General overview of the task
- Applicable industry standards and protocols
- Hardware and software preferences
- Preferences, if any, such as software tagnames, device names, etc.
- Preferred documentation formats
- FAT requirements
- Checkout and commissioning expectations

3. Functional Description

As mentioned, a functional description is usually provided as a part of the control system specification. The following is a fragment of the functional description provided for our feed tank example project (refer to Figures 3-2 and 3-5):

a. **Tank Level: LT-10, LSH-10, LSLL-47**
   The tank level (LT/LSH-10) is displayed on the HMI. Its alarm output is hardwired such that a high level forces the fill valve closed, regardless of PLC command. If level switch low-low LSLL-47 trips, then the pump will be stopped through a hardwired interlock.

b. **Tank Fill: HV-13, ZSC-13**
   The tank is filled by a "wild" flow rate that is not controlled upstream but is allowed to run wild into the tank as long as valve HV-13 is open. An control system command will open the valve provided the high level alarm generated by a switch on the level transmitter does not activate. Valve HV-13 is a fail-closed valve that will stay open as long as tank level is below the LSH-10 set point. Closed-position limit switch ZSC-13 informs the operator of this event.

c. **Tank Discharge: PP-10**
   The contents of the tank may be emptied into a trench in the floor with hand valves, or they may be directed to discharge through pump PP-10. This pump runs based on the position of HOA hand switch HS-15B. In the Hand position, the pump runs, provided the interlocks are set. In the Off position, the pump stops. In Auto, the pump runs based on a control system command (HS-15A), provided the interlocks are set.

   The interlocks for this pump are

   - Stop upon a low level (LSLL-47)
   - Stop upon a low or high pressure (PAH/L-48)

   If the tank fill valve closes with the pump running, there will be a risk of pulling a vacuum on the tank. Therefore, a vacuum relief valve PSV-58 has been added. This valve has an integral pilot and will open to relieve the vacuum.

d. **Pump Discharge Pressure: PIC-48**
   Pump discharge pressure is managed by pressure control loop PIC-48, which maintains a constant pressure at the output of the pump by sampling pressure, comparing the actual pressure to a desired set point, and

then modulating valve PV-48 to reduce any difference. Pressure transmitter PT-48 samples the pressure of the pump discharge as it pushes product into a downstream product tank. If the pressure in the line increases, indicating reduced demand downstream, then modulating valve PV-48 is opened slightly to recirculate more product back into TK-10. If downstream pressure decreases, indicating increased demand, then the modulating valve moves toward the closed position. This allows the constant speed motor driving the pump to run safely. When product isn't needed downstream, this constant recirculation eliminates the need for an agitator.

If the control system detects a high or low pressure and the signal remains for a certain time period, then the operator is notified, and the pump is stopped.

4. Project Estimate

Estimating a design project is mostly about defining the tasks that must be accomplished and making assessments about how the team will execute them. An estimating tool is included in the CD that accompanies this book. It is a tool to help estimate the I&C design task, but it also can be adapted to CSI requirements. Whether that tool is used, or another one, it is advisable to approach the estimate in such a way that a post-mortem is facilitated. The only way to get good at estimating is to do several, and then compare the actual results to the expectations.

The following is a good approach to estimating CSI project tasks:

a. **Field Device Control Elements**
Sometimes an I/O count is used, but some find it more beneficial to base the CSI estimate on the number and types of instruments and automated end devices. Each item type will have its own requirements in terms of control requirements, graphic display, alarm features, etc. Deriving a level of effort needed for each device type, and then multiplying that level of effort by the total number of like devices, will yield a value that should be included as a portion of the CSI project estimate.

b. **Sequential Control Elements**
All control projects include sequences, even if they are only the startup or shutdown sequences. Sequences present their own set of issues for the programmer and should be considered as separate elements of the estimate. Usually, a sequence will have its own graphic display to provide a

control point for the operator to manage the sequence, or at least monitor it.

c. **Continuous Control Elements**
Most control projects include some aspect of continuous monitoring or control. If these continuous control points are part of the development of a control scheme, then some time should be included for them. For example, analog inputs will all need to be conditioned. They also will need to be scaled, and the likelihood is that alarms will need to be generated using comparators. Ratio control, PID control, and PID Feed-Forward (a form of predictive control) are some examples of continuous control elements that will require programming time.

d. **Overview Graphic Screens**
Some of the HMI estimate is bound up in the Field Device, Sequential, and Continuous Control schemes already mentioned. For example, most end devices (valves, motors, pumps, etc.) will have dedicated control overlays (popups) that appear when the device is selected by the operator. These pop-up displays have detailed animation and action graphics that are focused on that end device. Overview graphic screens, however, provide a P&ID-type view of the process, with multiple devices and displays that provide the operator with immediate view of the subsystem as a whole. Overview graphic screens should be counted and an estimate generated for each.

e. **Historical Data Acquisition and Reports**
Configuration of the data historian should be covered in the estimate. Considerations include the number of points, frequency of sample, etc. Also in this category are the production and material consumption reports.

f. **Network Activities**
Some of the aspects of the network that affect project estimates are:

- The number of network LANs being used
- The number of nodes on each network
- The types of data transfers
- The number of switches, routers, hubs, and other network-related equipment to configure and test

5. Project Proposal

The proposal should reflect much of the information presented in the RFP. An effective way to do that is to base it on the RFP and make a statement to the effect that "this project conforms to the RFP with the following exceptions," and then list the exceptions. For more information on what should be in a proposal, refer to Part I, Chapter 3.

B. FEL Stage 2 – Project Definition

FEL Stage 2 tasks are sometimes done after the contract is awarded. Occasionally, a customer will request that an engineering study be performed in order to obtain a more definitive project budget, in which case FEL2 and FEL3 become a small project that precedes the primary one. In either case, FEL2 is where the project gets fully defined, taking all the preliminary information provided in the RFP and fleshing it out to remove any unknowns and to make final decisions on control schemes and approaches.

For our example project, the preliminary engineering process (described in Chapter 9) revealed the existing control system is a PLC, and our scope of work requires the addition of a new I/O rack, as the existing system is full (Figure 3-52). The rack requires a new enclosure to be mounted outside the existing equipment room. Because the existing PLC drop 2 has room for new motor controls, our new pump motor will be terminated there.

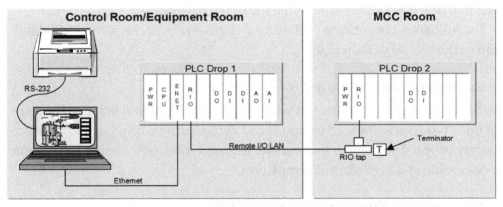

Figure 3-52. Existing control system

The plant systems group has designated the new rack as drop 03 on the remote I/O network. The new scheme is shown in Figure 3-53.

This chapter explores several sets of documents generated that describe the actions that will be taken by the CSI programming staff. These documents

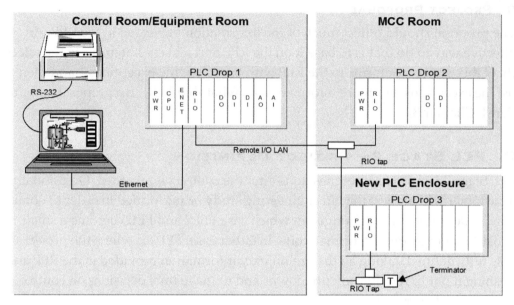

FIGURE 3-53. NEW CONTROL SYSTEM

enhance the project team's ability to manage the programming task, while spreading the risk to the entire CSI management team. It is in everyone's interest to have a clearly defined set of tasks that have been fully vetted. In addition, these documents will form the basis of the deliverables (training manuals and project documentation manuals) that will be prepared later. So, a time investment in these documents in the early stages of the project, while time consuming initially, pays off many times over through the life of the project.

The following constitute a preliminary engineering package from the standpoint of the CSI set of deliverables:

1. SEQUENTIAL FUNCTION CHART (SFC)

One SFC drawing per sequence. It will provide information relative to the steps that must be taken in the startup and shutdown sequences, plus any other process or safety oriented sequences that may be required. Each SFC should be presented for discussion at a high state of completion.

2. CONTINUOUS FUNCTION CHART (CFC)

One CFC drawing per continuous control function. A continuous function is usually an analog control loop or perhaps a set of related loops. Only a representative sampling need be presented in this early stage, though any loops of particular interest or complexity should be developed and presented. The CFCs that are developed should be presented for discussion at a high state of completion.

3. Control Narrative

The Control Narrative is an important document that presents the control scheme in a clear language that can be understood by those not in the controls profession.

4. Sequence Control Detail Sheets (SCDS)

Each sequence step in the SFC should be detailed using an SCDS per step. The preliminary engineering package presented in FEL2 needs only a representative sampling of SCDSs to provide insight as to how these sequence steps will be documented and programmed. The full suite of SCDSs will be generated in Phase 2.

5. Device Control Detail Sheets (DCDS)

One sheet per device type showing how a typical device will operate in each device category. A DCDS provides interlock information as well as any minor device related sequences that need to occur.

6. Functional Logic Diagrams

Logic diagrams can sometimes be omitted if the integrator does a good job with the detail sheet set described above. Logic diagrams are notoriously poor conveyors of information when it comes to sequences, leading the programmer to a "spaghetti code" result in the software. Sequences are much better handled in the SFC/SCDS format. However, device logic and CFC logic work well in the logic diagram format.

C. Control Narrative

A Control Narrative is a document that describes all the major controls aspects of the project in clear text. The narrative will pull from and provide reference to the document set, in particular the P&ID's. The Control Narrative is a key component of the control scheme documentation and is used as the basis for the programming tasks.

1. Sequential Function Chart (SFC)

After the initial concept has been approved in Project Phase 1 (see Part I), it is time to flesh out the controls a bit. The sequential function chart, as the name suggests, describes control sequences. An associated name for this type of control is *state logic*. It is logic that changes state based on transitions. It is characterized by a program fragment that executes until an exit condition is satisfied. This fragment is frequently referred to as a program block. A program block in an SFC is called a *state* (S), and the exit condition that causes the block to stop functioning is called a *transition* (T).

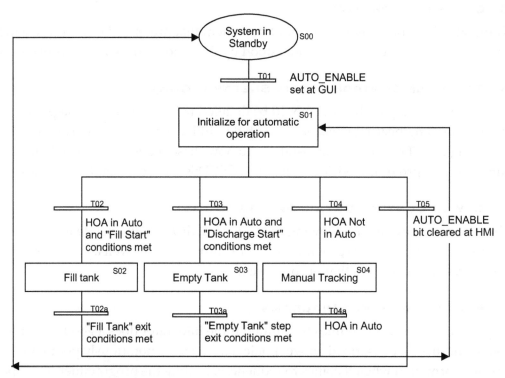

FIGURE 3-54. TK-10 FEED TANK CONTROL SEQUENCE OVERVIEW

In the SFC fragment in Figure 3-55, the program is based on a sequence of events, in this case starting a pump and activating a ride-through timer to give the pressure time to build before activating a low-pressure monitoring function. After a program block's exit conditions are met, as defined by the transition definition, then that section of logic stops executing. Until that time, however, the logic executes with every machine cycle.

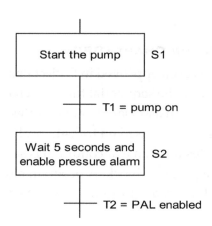

FIGURE 3-55. SEQUENTIAL FUNCTION CHART FRAGMENT

State S1 starts the pump. The start command is held true until the pump is confirmed to be running at transition T1. When T1 goes true, state S2 activates. After the timer has been confirmed off, then the low-pressure alarm is enabled.

A more involved sequence is depicted in Figure 3-56, in which transition T1 activates state S1. The system remains in state S1 until either transition T2, T3, or T4 become true. If T2 is true, then the system enters state S2. From that point, either or both state S2A and state S2B may be activated, depending on the condition of transitions T2A and T2B. The double line indicates an AND function, whereby S2A AND S2B may be activated at the same time. Exit conditions for state S2A are defined by transition T2A1, which, if true, will return the system to state S1 or transition T2A2, which will return the system to state S2. State S3 is actually a "macro" state, which has its own subsequence that must be completed before its T3A exit condition is met. A separate SFC must be developed for this subsequence.

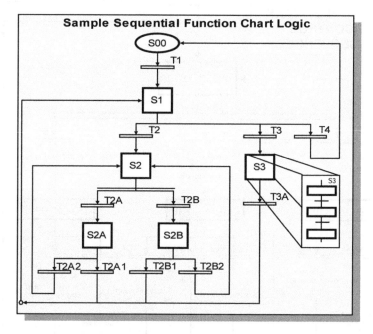

FIGURE 3-56. SAMPLE SEQUENTIAL FUNCTION CHART LOGIC

This sample chart (Figure 3-56) is similar to the overview diagram discussed previously (Figure 3-54). But the overview diagram is merely a first cut. It should now be fleshed out to be more representative of the ultimate control scheme.

The SFC shown in Figure 3-57 provides more detail. It depicts the various control modes envisioned by the integrator for control of the TK-10 product tank and its associated equipment. While the SFC is a useful tool in communicating intent, it is rarely a standalone device. The SFC is usually accompanied by a control narrative, a sequence step control detail sheet, a CFC, or some combination of these. To provide a basis for comparison, each method is provided in the sections that follow.

Figure 3-57 shows the same four modes of operation as in the overview diagram. An additional path is provided for the fill and empty companion sequences to reset immediately if the Auto Enable permissive signal is lost. The system also resets if the sequence step's exit conditions are met while the other sequence is inactive. If the other step is active, then the system does not reset but instead allows the other step to continue until done. The tracking step is activated when the operator disables automatic mode and will exit only when the operator reverses that action.

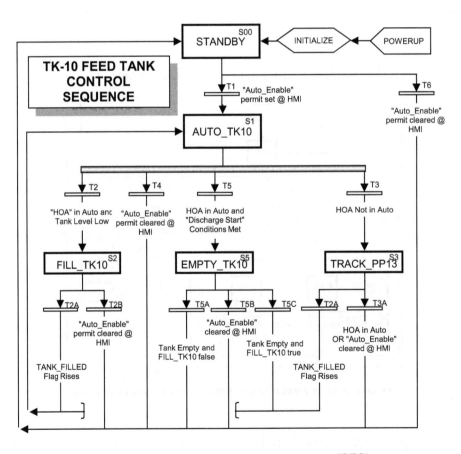

FIGURE 3-57. SEQUENTIAL FUNCTION CHART (SFC)

2. CONTINUOUS FUNCTION CHART (CFC)

The CFC describes control activities that are continuous. The SFC just discussed does not describe continuous activities very well. Data acquisition and control activities need to occur regardless of the step in a sequence. The SFC may disable analog alarms generated by the CFC when certain steps are active, but the data acquisition and control functions continue to execute. Figure 3-58 is a simplified CFC for this application.

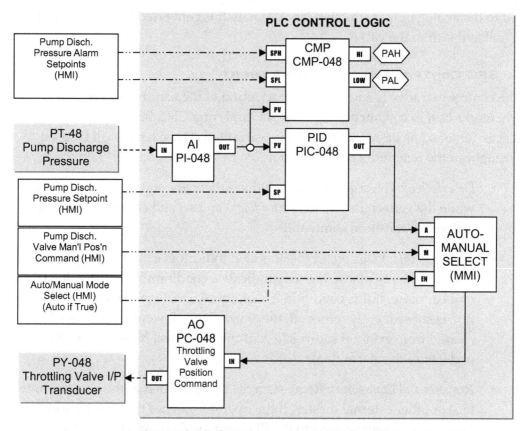

FIGURE 3-58. CONTINUOUS FUNCTION CHART

Pressure transmitter PT-48 sends an analog (4–20 mA) signal to the PLC. An analog input (AI) software function block accepts the signal and converts it to an integer value between 0 and 4096 (12-bit resolution) or perhaps to some floating-point number that represents engineering units.

This numeric representation of the signal is fed to an analog alarm comparator (CMP) block and to a PID control block, where a calculation is performed to compare the actual value (PV; process variable) to the set point (SP). The PID block then varies its output as necessary to reduce any deviation. The operator loads the set point (i.e., desired operating pressure) at the HMI. The set point information is communicated to the PLC via data communications and is linked through programming to the PID block.

The output of the PID block is fed to a software switch function that passes either the PID block output or a manual valve position command loaded by the operator at the workstation. The auto/manual command signal originates at the HMI. The operator presses a button on the screen, which sets or clears a bit. If the bit is true, then the switch transmits the automatic signal. If the bit is false, then the switch transmits the manual signal. The output of the auto/manual switch is fed to the analog output function block, where it is converted back to a 4–20 mA signal and sent to the valve.

3. SFC Control Narrative Fragment

The control narrative is a text-based description of the control system. The narrative works best in conjunction with an SFC and/or a CFC. Before proceeding, a discussion of some basic terminology is warranted. These terms will be used throughout the remainder of the chapter:

- *Descriptives:* The adjectives *true* and *false* describe states of a bit or flag, while the verbs *raise* and *lower (for flags)* and *set* and *clear (for bits)* describe actions or transition commands.

- *Flags and Bits:* Flags are PLC-generated indicators called "semaphores." For example, an alarm flag may indicate a condition has existed for 10 seconds or more. If this condition is met, the alarm flag is raised. If the condition is subsequently removed, the alarm flag is lowered. Bits are set or cleared from external sources (e.g., if the high-level bit is set for 10 seconds, raise the alarm flag).

- *Requests and Commands:* Requests (req) are generated at the operator interface or other external master. If the operator pushes the start button, the HMI issues a start request to the PLC system. Commands (cmd) are generated by the PLC in response to requests. Between the request and the command lies the PLC logic that monitors safeties and system conditions and, based on rules set forth in the program, allows or disallows the action. If the request is deemed to be valid by the rules, then a command is issued to honor the request.

- *Permissives (and Permits):* Permissives (and permits) are discrete elements (flags or bits) that must be set in order to operate in automatic mode. A permissive is similar to an interlock in that loss of signal causes the device to return to its safe condition (off, closed, etc.). However, unlike the interlock, the permissive, when lost, does not cause the device to change mode from automatic to manual. An example of the use of a permissive is a conveyor belt that starts and stops based on the level of a bin downstream of it.

- *Interlocks:* Interlocks are discrete elements (flags or bits) that must be set in order to operate in ANY mode. Loss of signal causes the device to return to its safe condition (off, closed, etc.), plus if that device is in automatic mode, it will switch to manual mode, requiring operator intervention before restarting. An example of the use of an interlock is a conveyor motor that overheats, opening its thermal overload contacts. Once the motor cools, the overload contacts will reclose. If the belt remained in automatic mode, it would then restart, provided its permissives were still met. However, if the overloads are wired as interlocks, the belt will remain off, and an alarm will be generated to alert the operator.

The following is a fragment of the control narrative for this project relative to the SFC (refer to Figure 3-57):

a. **Powerup & Initialize**
On power-up, the system ensures all automatically controlled devices are initialized to their predefined safe state. All status flags are cleared, and calculations are initialized. When the initialization routine has been completed successfully, the program raises the INIT status flag.

b. **S00 – State Zero: STANDBY**
When the INIT status flag rises, or if the AUTO_ENABLE flag falls, the system enters state zero. In this state, the STANDBY flag is raised, and any other active sequence flags are lowered. Alarm conditions are monitored, and the AUTO_ENABLE permissive is checked. If the operator presses the Auto Mode Enable button on the HMI, the AUTO_ENABLE permissive rises, providing the exit conditions for this state. As the program block is exited, the STANDBY flag is lowered.

c. **S01 – State One: AUTO-TK10 (Automatic Mode)**
When the STANDBY flag falls with the AUTO_ENABLE permit active, state 01 is activated. This program step raises the AUTO-TK10 status flag and monitors each of the control mode flags to determine what action to take: Should the tank be filled? Should the tank be emptied? Should the pump status be tracked? The following (d, e, and f) are descriptions of the three automatic operating modes, any of which may be active at any given time.

d. **S02 – State Two: FILL_TK10 (Fill the Tank)**
   If the tank level is low as indicated by the low-level trip LAL-10, and if it is in automatic mode, then raise the FILL_TK10 flag and fill the tank by opening valve HV-13. Keep the valve open until the tank is filled as indicated by high-level trip LAH-10 or until the AUTO_ENABLE permissive is lost. Lower the FILL_TK10 flag and exit the step.

e. **S05 – State Five: EMPTY_TK10 (Empty the Tank)**
   If the tank level is high as indicated by the high-level trip LAH-10, and if it is in automatic mode, and if the OK_TO_DISCH permissive is active from downstream, and if the HOA switch at the pump is in the AUTO position, then raise the EMPTY_TK10 flag and empty the tank by starting pump PP-10. Maintain the pump until the low-level trip LAL-10 indicates the tank is empty or until one of the permissives is lost. During the exit process, lower the EMPTY_TK10 flag.

   As mentioned earlier, pump PP-10 is a fixed-speed pump that uses a recirculation loop to manage its discharge pressure. Since this control scheme does not control the downstream demand, the discharge pressure can vary considerably, depending on how many users are taking product. Control valve PV-48 throttles as necessary to maintain a reasonable discharge pressure. If the discharge pressure falls below a certain level for a time period, then it is assumed there is a leak, and the pump is stopped. If the pressure rises beyond a set point, indicating a closed block valve or a malfunctioning recirculation valve, then the pump is stopped.

f. **S03 – State Three: TRACK_PP13 (Track the Pump)**
   If in automatic mode, and if the pump's HOA switch is *not* in the AUTO position, then set or clear the PLC output to the pump to mimic the pump's RUN status. The input to the PLC is true only when the switch is in the AUTO position, so the switch position in the other two states must be inferred by the RUN status input from the motor. By mimicking the pump output, the user is able to switch to automatic mode while the pump is running, and provided the interlocks are set and the AUTO_ENABLE permissive is active, the pump is maintained in run mode. **Note:** Switching through the off position from Hand to Auto may cause the pump to shut down anyway. This functionality depends on the speed with which the switching is done and the sensitivity of the motor controls.

   During all three valid modes of automatic operation, the AUTO_TK10 flag remains raised. This flag is lowered only if the operator removes the

AUTO_ENABLE permissive, after which the system returns to standby mode.

4. Sequence Step Detail Sheet (SSDS)

The SSDS (Figure 3-59) is text-based and augments the SFC by describing activities and conditions being executed and monitored in each sequence step in more detail than is possible in an SFC. Each sequence step is a small standalone program that could be as simple as opening a valve, or as complex as executing its own embedded sequence. It also provides a means for describing how unusual circumstances are handled while the step is active, and for describing indirect activities, such as enabling/disabling analog alarms.

The SSDS is particularly useful in batch operations that have many steps. Not all of the steps in our example system have been included, but enough detail is presented here to get the basic concepts across. Note that these detail sheets relate to the sequencing of the process only. They do not directly drive output devices. Device logic is described on device logic detail sheets, which are described in the following section. Device logic detail sheets describe all the controls related to the functionality of a device. Some of the functionality is likely related to status flags that are raised or lowered by the sequence steps.

The SSDS consists of several sections that describe most of the conditions likely to be encountered. The state of the step is determined by conditional logic. For each condition, there is an associated action. For example, the "start condition" section is paired with a "start action" section. If the start conditions are met, how should the system respond? It is a cause-and-effect relationship.

Condition sections are enumerated using *case* designations. Each new line item within a case is logically ANDed with the other line items listed against that case. For example, condition X may be met by either case A or case B. Case A has two conditions, A1 and A2. Both must be true for case A to be satisfied. Case B may only have one condition or several. Each condition is logically ANDed in order to satisfy case B. The overall condition X is satisfied if either the case A OR the case B conditions are met.

Action sections are enumerated using *time* designations. If a condition is met, then there may be four actions to take. The first two actions may happen simultaneously, and so each would be designated T1. The third action may depend on an external event triggered by the previous actions. This would be designated T2. The fourth action may depend on a subsequent event, and so would be designated T3.

The following are descriptions of the various sections in Figure 3-59:

- *Start Conditions*: What are the conditions needed to activate this sequence step? Note: Conditions are designated Ax, Bx, and so on, where each A item is logically ANDed with each subsequent A item. The A group is logically OR'd with the B group, each element of which is likewise ANDed.

- *Start Actions*: Once the step is activated, what does it do?

- *Run Condition*: Sometimes the conditions needed to start a function are different from those needed to maintain it in its active state. Starting a pump and keeping it running, for example, can require two different streams of logic. This section defines the conditions needed to maintain the step in its active state.

- *Run Actions*: When the run conditions are satisfied, what happens?

- *Alarms*: What sequence alarms are activated during this sequence step?

- *Sequence Safety Interrupts*: Are there any conditions that may require immediate system response and subsequent operator interaction?

- *Safety Interrupt Actions*: What actions should the system automatically take to achieve a safe state? Once the safe state is achieved, the system waits for resume conditions to be met.

- *Pause Conditions*: What conditions are necessary to have the system enter a pause state? A pause is usually initiated by the operator.

- *Pause Actions*: What actions should the operator take to achieve the pause state?

- *Resume Conditions*: What operator actions are necessary to exit the pause or safe state?

- *Resume Actions*: What actions are necessary to return from a pause condition?

- *Exit Conditions*: What conditions must be met to exit the step?

The following are some of the sequence control data sheets required for this task:

a. **Step S02 – Fill Tank**
   The sequence step detail sheet in Figure 3-59 reflects the system activity when Sequence Step 02 (S02) is activated (refer to Figure 3-59 if necessary). This step fills the tank when the start and run conditions are met and stops filling at other times, as defined by the safety interrupt, pause, and exit conditions.

For the programmer, each condition described on the detail sheet has its own status flag that is raised if the conditions are met. This gives a structure to the program that ties it closely to the detail sheet. Each sequential step has its own status flag that is raised when the step is active and lowered when the step is exited. A logical statement that controls this flag is:

X == (Start or Run) and not (Exit), where X is the step active status flag.

Note: The "==" symbol means "is true if." So, "X" (the step active status flag) is raised if either the Start or Run flag rises, and if the Exit flag is not raised.

So for this step,

FILL_TK10 == (S02_Start or S02_Run) and not(S02_Exit). The FILL_TK10 flag signifies the step is active.

1. Start (or Open) Conditions
   Conditions are given alphanumeric case designations. Case A has four elements that must all be true. There is no Case B. If any of the case conditions are met, then the start condition is satisfied. Logically, the statement reads, "If A1 and A2 and A3 and A4 occur, then raise the Step Active status flag and execute the start actions."

   In our example, Case A conditions are met if the AUTO_TK10 step is active, if the SRC_RDY permissive is active, if the FILL_TK10 flag is false (indicating the step is not already in run mode), and if the tank level has not been low (LAL10T) for the delay time. This is a useful format for communicating with a general audience once the basic nomenclature has been described. For the programmer, these statements may be reduced to a single logical statement that results in the raising of the FILL_TK10 status flag in the form of:

   X == (A1 and A2 and A3 and A4) or (B1 and B2), where X is the status flag and Ax and Bx are individual elements described in the start conditions section.

   So, for this particular step,

- S02_START == AUTO_TK10 and SRC_RDY and not (FILL_TK10 or LAH10T or S02_EXIT or S02_RUN).

2. Start Actions
   If the start conditions are met, then the FILL_TK10 step active flag is raised as described previously. Also, the S02_START flag is raised. This flag is raised only long enough to satisfy the run conditions

described below. Run actions lower this flag, so it is a good signal to use as an open command for "fill valve HV-13." The device logic for "fill valve HV-13" is described later.

SEQUENCE STEP DETAIL SHEET

SEQUENCE	S02: FILL TANK	DATE:	5/2/2002
DESCRIPTION:	TK-10 FILL SEQUENCE	DOCUMENT#:	SCD-002
LOCATION:	Building 14, Bay 13	REVISION#:	A
REFERENCE:		CHECKED BY:	

START CONDITIONS (S02_SRT)		PAUSE CONDITIONS (S02_PSD)	
A1	Flag AUTO_TK10 is True	A1	Permit "SRC_RDY" lost (Not Ready to Send)
A2	Permit SRC_RDY Active (Ready to Send)		
A3	Flag FILL_TK10 is False		
A4	Flag LAL10T is True (Tank Level Lo)		

START ACTIONS		PAUSE ACTIONS	
T1	Raise the FILL_TK10 Step Active Flag	T1	Raise the S02_PSD flag
T1	Raise the S02_SRT flag		

RUN CONDITIONS (S02_RUN)		RESUME CONDITIONS (S02_RSM)	
A1	S02_START is True	A1	Permit SRC_RDY Active (Ready to Send)
B1	S02_RUN is True		

RUN ACTIONS		RESUME ACTIONS	
T1	Raise the "S02_RUN" flag	T1	Lower the S02_PSD flag
T2	Lower the "S02_SRT" flag	T1	Raise the S02_RSM flag for 5 seconds

ACTIVE PROCESS ALARMS		EXIT CONDITIONS (S02_EXIT)	
1	HV13_MM Command Mismatch	A1	Flag LAH10T rises
2		B1	Permit AUTO_ENABLE falls
3			
4			

SEQUENCE SAFETY INTERRUPTS		EXIT ACTIONS	
1	LAHH10 True	T1	Raise the "S02_EXIT" flag
		T2	Wait for ZLC10 true…
		T3	Lower all "S02_XXXX" Status Flags.
SAFETY INTERRUPT ACTIONS		T4	Lower the "FILL_TK10" Status Flag
1	Raise the "S02_SAFE" flag		
2			
3		REVISIONS	

Comments	REV	DATE	DESCRIPTION
This sequence step controls the open/close postion of the TK-10 Feed Tank Fill Valve. In automatic mode, this is done as necessary to maintain tank level between the LAH and LAL level setpoints.			
		APPROVALS	
		DATE	SIGNATURE

FIGURE 3-59. SEQUENCE STEP 2: "FILL TANK" SEQUENCE

3. Run Conditions
   Starting and running are two different things. The step remains in run mode based, perhaps, on different criteria than those needed to start the step. In this case, run conditions are met if either of two cases is satisfied: In case A, the S02_START flag is raised, or in case B, the S02_RUN flag is raised.

   To begin, the S02_RUN flag is not raised. S02_START rises, causing the RUN actions to occur. One of the run actions causes the S02_RUN flag to rise, at which point the S02_START flag is lowered. So the run conditions are maintained through the transition.

   The logical equation (logical statement) for the valve command bit is modified as follows:

   S02_RUN == not(S02_EXIT) and (S02_START OR S02_RUN)

4. Process Alarms
   Alarms are processed continuously. However, some may only be active at certain times. For this step in the process, only one alarm is closely monitored. HV13_MM is the command mismatch alarm, which is embedded in the device logic of the valve. If the valve's commanded state differs from its actual state for a prescribed time period, then this alarm activates.

5. Sequence Safety Interrupts
   If a level Hi-Hi alarm is triggered (TK10LAHH), the system enters a safe state that requires operator intervention before proceeding. For the LAHH alarm to trip, the system must have ignored the LAH trip, and the valve must not have closed.

6. Safety Interrupt Actions
   The safe-state flag S02_SAFE rises. As a result of this, the valve's device logic closes the valve and holds it closed until the operator cycles the AUTO_ENABLE flag to false. This, in effect, resets the system, meeting the step's exit conditions. Then fill step S02 is exited and needs to be reentered via the start conditions.

7. Pause Conditions
   The system pauses if the system upstream is no longer ready to send product, as signified by the loss of the SRC_RDY permissive.

8. Pause Actions
   Step-paused flag S02_PSD is raised. The device logic of the valve monitors this flag and closes the valve if it rises.

9. Resume Conditions

   The system resumes if the system upstream is again ready to send product, as signified by the activation of the SRC_RDY permissive.

10. Resume Actions

    Step-paused flag S02_PSD is lowered. This causes valve HV-13 to reopen.

11. Exit Conditions

    Exit conditions are met if the operator disables auto mode.

12. Exit Actions

    At T1, the S02_EXIT flag is raised. The system waits for the valve to close as signified by the ZLC10 bit rising. Then all S02_XXXX sequence status flags are lowered, and the FILL_TK10 step active flag is lowered.

b. **Step S05 – Empty the Tank (Figure 3-60)**

   Refer to the SFC (Figure 3-57) if necessary. This sequence step detail sheet deals with step 05, Empty Tank. In this step, if the tank level is high, as indicated by the high-level trip LAH-10, and if it is in automatic mode, and if the OK_TO_DISCH permissive is active from downstream, and if the HOA switch at the pump is in the AUTO position, then the EMPTY_TK10 flag is raised, and the tank is emptied by running pump PP-10. The pump runs until the low-level trip LAL10 indicates the tank is empty or until one of the permissives is lost. Again, to see the full effect of the activity described on this sheet, it is necessary to refer to the device logic detail sheet, which is shown later.

   1. Start Conditions

      There is only one start case with several conditions: The AUTO_TK10 step must already be active. If the DEST_RDY permissive is active (indicating the folks downstream are ready for product), if the EMPTY_TK10 flag is false (indicating the step is not already in run mode), if the tank level has not been high (flag LAH10T raised) for the delay time, and if the HOA is in auto (YL15B), then the Start Conditions are met.

   2. Start Actions

      If the start conditions are met, then the EMPTY_TK10 step active flag is raised as described previously. Also, the S05_START flag is raised. This flag is raised only long enough to satisfy the run conditions described below. Run actions lower this flag, so it is a good signal to use as a start command for discharge pump PP-10. The device logic for this pump is described later.

SEQUENCE STEP DETAIL SHEET				
SEQUENCE	S05: EMPTY TANK		DATE:	5/2/2002
DESCRIPTION:	TK-10 DISCH. SEQUENCE		DOCUMENT#:	SCD-005
LOCATION:	Building 14, Bay 13		REVISION#:	A
REFERENCE:			CHECKED BY:	
START CONDITIONS (S05_SRT) || PAUSE CONDITIONS (S05_PSD)			

START CONDITIONS (S05_SRT)		PAUSE CONDITIONS (S05_PSD)	
A1	Flag AUTO_TK10 is True	A1	Permit "DEST_RDY" lost (Dnstrm Not Ready)
A2	Flag EMPTY_TK10 is false		
A3	Permit "DEST_RDY" (Dnstrm Ready) is Set		
A4	Flag LAH10T is raised (Level High)		
A5	YL15B Bit is True (HOA in Auto)		
START ACTIONS		PAUSE ACTIONS	
T1	Raise the EMPTY_TK10 Step Active Flag	T1	Raise the S05_PSD flag
T1	Raise the S05_SRT flag		
RUN CONDITIONS (S05_RUN)		RESUME CONDITIONS (S05_RSM)	
A1	S05_SRT is Up	A1	Permit SRC_RDY Active (Ready to Send)
B1	S05_RUN is Up		
RUN ACTIONS		RESUME ACTIONS	
T1	Raise the "S05_RUN" flag	T1	Lower the S05_PSD flag
T2	Lower the "S05_SRT" flag		Raise the S05_RSM flag for 5 seconds
ACTIVE PROCESS ALARMS		EXIT CONDITIONS (S05_EXIT)	
1	PP10_MM Command Mismatch	A1	Flag LAL10T rises
2		B1	Permit AUTO_ENABLE falls
3			
SEQUENCE SAFETY INTERRUPTS		EXIT ACTIONS	
1	LALL10 True	T1	Raise the "S05_EXIT" flag
		T2	Wait for YL15B to fall...
SAFETY INTERRUPT ACTIONS		T3	Lower all "S05_XXXX" Status Flags.
1		T4	Lower the "EMPTY_TK10" Status Flag
2		REVISIONS	

Comments	REV	DATE	DESCRIPTION
1) If HOA is not in Auto, then it is in either Hand or Manual. If in Manual, then pump is running. If pump is running, then PLC sets its output, so if HOA switched back to Auto, the pump will continue to run, provided interlocks are set. 2) The "S05_SAFE" flag is a Safety Interrupt. It inhibits pump operations. Once set, it will stay set until operator cycles the Auto Enable switch at the HMI.			
	APPROVALS		
		DATE	SIGNATURE

FIGURE 3-60. SEQUENCE STEP 5: "EMPTY TANK" SEQUENCE

3. Run Conditions

Run conditions are met when either of two cases is satisfied. In case A, the S05_START flag is up or, in case B, the S05_RUN flag is up. To begin, S05_RUN is down. S05_START is raised by Start Actions, causing the run actions to occur. One of the run actions causes the S05_RUN flag to rise, at which point the S05_START flag is lowered. So the run conditions are maintained through the transition.

4. Process Alarms
   Alarms are processed continuously. However, some may only be active at certain times. For this step in the process, only one alarm is closely monitored: the PP10_MM is the command mismatch alarm, which is embedded in the device logic of the valve. If the valve's commanded state differs from its actual state for a prescribed time period, then this alarm activates.

5. Sequence Safety Interrupts
   If a level Low-Low alarm is triggered (TK10LALL true), the system enters a safe state that requires operator intervention before proceeding. For the LALL alarm to trip, the system must have ignored the LAL trip and the pump must not have stopped, or there is a leak somewhere.

6. Safety Interrupt Actions
   The safe-state flag S05_SAFE rises. As a result, the valve's device logic closes the valve and holds it closed until the operator cycles the AUTO_ENABLE flag false. This, in effect, resets the system, meeting the step's exit conditions. Then, step S05 is exited and needs to be reentered via the start conditions.

7. Pause Conditions
   The system pauses when the system upstream is no longer ready to send product, as signified by the loss of the DEST_RDY permissive.

8. Pause Actions
   Raise step-paused flag S05_PSD. The device logic of the valve monitors this flag and closes the valve if it rises.

9. Resume Conditions
   The system resumes when the system upstream is again ready to send product, as signified by the activation of the DEST_RDY permissive.

10. Resume Actions
    Step-paused flag S05_PSD is lowered. This causes pump PP-10 to restart.

11. Exit Conditions
    Exit conditions are met if the user disables auto mode (permit AUTO_ENABLE falls).

12. Exit Actions
   At T1, the S05_EXIT flag is raised. The system waits for the pump to stop, as signified by the PP10_RNG flag falling. Then all S05_XXXX status flags are lowered, and the EMPTY_TK10 step active flag is lowered.

c. **Sequence Step Detail Sheet (SSDS) Summary**
   The SSDS provides the controls team with a tremendous amount of flexibility in describing the program flow of control. The programmer must remember to consider each category of activity for every new step in the process (i.e., to program "by the numbers." The use of status flags (semaphores) makes the status of each sequence step very accessible. This is helpful when troubleshooting. The flags may also be used to trigger process actions, such as opening valves and starting pumps. But as already stated, the actual device logic is dealt with elsewhere.

   The use of so many flags can slow down the programming process. But note the similarity in the structures of sequence steps 2 and 5. It is usually possible to *clone* the logic and the tagname structure between steps, thus reducing the time impact. It does indeed use some processor memory, but the tradeoff is usually worth it. The next section deals with the device logic segments that mate to the sequential logic just described.

5. Device Control Detail Sheet (DCDS)

If a system is simple enough, then devices can be started or stopped and opened or closed within the sequence steps. But doing this takes the programmer only so far, as sooner or later a pump will need to be started from another sequence, or perhaps a throttling valve will need to be forced closed by an interrupt condition that was set after the step with the valve logic had stopped executing. The term *force* in this context implies that the normal conditions that govern the state of the bit or flag are being superseded, and an overriding condition is active.

Thus, the SSDS tells only part of the story. Most industrial applications require manual control as well as automated control. True manual control allows a device to be operated outside the purview of the automated control system. If the device is to be controlled in hand mode, then its safety interlocks need to be hardwired, as opposed to, or perhaps in addition to, being computerized.

Each device that is part of such an industrial application should have its logic described on a device logic detail sheet. Hardwired logic should be clearly delineated from software logic. If the device is controlled through software, its program logic should be segregated into an area of the program reserved for device logic. This section should be the last section executed in the program cycle.

Many devices likely have programmed interlocks that protect them from damage or prevent placing personnel at risk when controls are in automatic mode. These interlocks are active regardless of any other considerations. In our example, device interlock bits take the form XXXX_ILOK, where the XXXX is the device tagname.

What follows are descriptions of device logic for the key control elements associated with this project:

a. **Pump PP-10 Device Logic**

   Pump PP-10 must be able to function in either PLC-controlled mode or manual mode. Refer to Figure 3-61 for the following narrative:

   Motor M10 is a three-phase 480 VAC motor. Its 120 VAC control circuit is powered from the three-phase motor source via a 4:1 stepdown transformer (T10). The motor has thermal overloads that are hardwired into the control circuit and into the motor winding leads.

   LSL-47 is a hardwired interlock that shuts the motor off if it trips. The switch is configured for failsafe operation in which the normally open contacts are closed under normal operating conditions. These contacts open up under alarm conditions.

   The HOA switch HS-15B is shown in the Off position (centered). If the switch is rotated left to the Hand position, the contact sets move down and make (close) at the bottom (see commentary on switches in Chapter 5, Section A.2). This bypasses the PLC on/off command and starts the motor. If the switch is rotated right to the Auto position, the contact sets move up and make at the top. This enables the PLC output and provides status to the PLC (YL15B) that the switch is in Auto.

   HS-15A is a PLC digital output. This contact closes and opens based on the condition of a command flag that is controlled by the PLC program. The name of the command flag is PP10_CMD, and it is synonymous with HS-15A throughout the program.

   The M relay is the motor starter. When the M relay is energized, it closes contacts in the motor winding leads and closes its auxiliary run contacts to provide status feedback to the PLC (YL-15A). If YS-15A is true, then device status flag PP10_RNG is raised to indicate the motor is running.

   A DCDS provides a means for communicating intent and gaining customer approval for the device logic. It is also extremely valuable during the acceptance testing evolutions, providing a document that can be used in lieu of a checkout procedure. This sheet is similar in appearance to the

SSDS and operates under the same rules with respect to conditional cases and action time periods. The main difference is in the description of activities that need to be executed outside the control of the software program. Such hardwired controls are designated Zx, Yx, Xx, and so on.

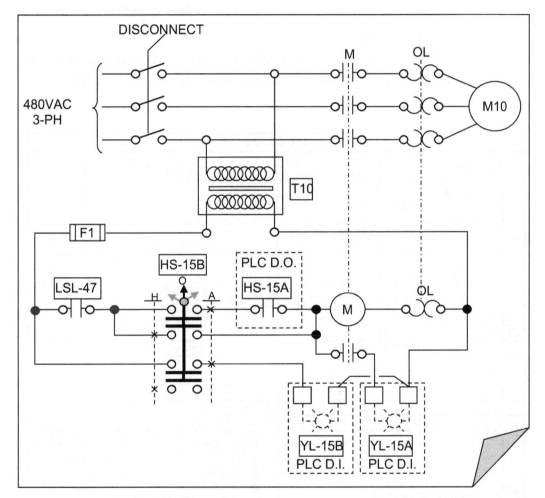

FIGURE 3-61. PUMP PP-10 MOTOR CONTROLS ELEMENTARY WIRING DIAGRAM

On the DCDS in Figure 3-62, the device logic for pump PP-10 is described.

1. Start/Open Conditions
   There are three conditions under which the pump may be started:
   - If S03_RUN is true, signifying the tracking mode is active, and if the pump is found to be running (PP10_RNG), and if PP10_CMD is false ("not"), then case A is satisfied, starting the pump in tracking mode.

- If S05_START is true, and the pump is not yet running [not(PP10_RNG)], then the "empty pump" step has just been activated, thus satisfying case B and starting the pump in "empty TK-10" mode.

- If HS-15A is placed in the Hand position, and the start button is pressed at the pump, then this is hardwired case Z.

2. Start/Open Actions
   Set PLC command bit PP10_CMD and start ride-through timer KC48. Or, if the hardwired start conditions were met, start the pump.

DEVICE CONTROL DETAIL SHEET				
DEVICE TAG:	PP-10 DISCHARGE PUMP		DEVICE INFO	
SERVICE:	TK-13 PRODUCT TANK		Device	PP-10
LOCATION:	Building 14, Bay 13		Name:	Discharge Pump
REFERENCE:			Descr.	Fixed Speed

START/OPEN CONDITIONS (PP10_SRT)		STOP/CLOSE CONDITIONS (PP10_STP)	
A1	S03_RUN (Tracking Mode)	A1	S05_PAUSE (Step 05 Paused)
A2	PP10_RNG (Pump is Running)	B1	S05_EXIT (Step 05 Exited)
A3	not(PP10_SRT)	C1	S03_RUN (Tracking Mode)
B1	S05_START (Empty Mode)	C2	not(PP10_RNG) (Pump is off)
B2	not(PP10_RNG) (Pump is not running)	D1	PP10_MM (Mismatch Alarm)
Z1	HS-15A in HAND position*	Z1	HS-15A in "OFF" position*
START/OPEN ACTIONS		**STOP/CLOSE ACTIONS**	
T1	Set PP10_CMD bit (Pump Start Command)	T1	Clear the PP10_CMD bit
T1	Start KC48 Ride-thru Timer		
T2	When YL15A sets, raise PP10_RNG flag	**INTERLOCKS (PP10_ILOK)**	
Z1	Start Pump*	A1	Not(LAL10)
RUN CONDITIONS (PP10_CMD)		A2	Not(PAH48T)
A1	PP10_RNG is True (Pump is running)	A3	Not(PAL48T) or (PAL48T and not(KC48))
		Z1	Motor Overloads*
		Z2	Motor Overtemp*
		Z3	Not(LAL-47)
Z1	HS-15A in HAND position*	**ALARMS**	
NOTES		1	Command Mismatch > 30-seconds
(*) Indicates a Hardwired condition.		2	
		REFERENCES	
		1	
		2	
		APPROVALS	
		DATE	SIGNATURE

FIGURE 3-62. PUMP PP-10 DEVICE CONTROL DETAIL SHEET

3. Run Conditions
   If the pump is running, then run conditions are satisfied, provided the interlocks are made, and the stop conditions are not met.

4. Stop/Close Conditions
   There are five cases in which the pump is stopped:

   - Case A: If S05_PAUSE is true
   - Case B: If S05_EXIT is true
   - Case C: If the S03_RUN command flag is true, but the pump remains off after a time period
   - Case D: If the PP-10 mismatch alarm activates
   - Case Z: The pump is stopped via hardwiring if HOA switch HS-15A is moved to the Off position.

5. Interlocks
   All interlock cases must be true to start or run. If the PLC interlocks are made (case A), then the PP10_ILOK flag is raised. Case A includes the following conditions:

   - Tank level not low (LAL10 false)
   - Pump discharge pressure not high (PAH48T false)
   - Pump discharge pressure not low (PAL48T false), or pump discharge pressure low and ride-through timer (KC48) timing

   If case Z is false, the pump is stopped through hardwired interlocks.

   Timer KC-48 was started in the start actions section. This timer allows the pump logic to ignore the fact that the pump discharge pressure is not above set point. As discussed earlier, this is necessary because the pump will shut down if the low-pressure alarm flag is raised, and the pressure will certainly be low until the pump gets up to speed. KC-48 is called a ride-through timer because it allows the system to ignore, or ride-through, a fault condition until the condition has had time to clear. When the pump gets up to speed, the discharge pressure rises, clearing the alarm condition before the timer trips. After the timer has timed out, the low-pressure alarm trip is unmasked and will shut down the pump if it trips or if it has not yet sealed in. The term sealed in this context is a relay term meaning the device (relay, motor, etc.) that is activated closes its own contact as a means of maintaining itself in the active state. Once it has sealed itself, then the original logic that caused it to activate can go away. The device will then stay in its active state until other (stop) conditions act to open the circuit elsewhere. This activity describes physical devices as well as software.

After the device control detail sheet is finalized, the information may be reduced to a series of logical statements as follows:

- PP10_ILOK == not(LAL-10) and not(LALL10T) and not(LALL47) and not(PAH-48T) and {not(PAL48T) or [(PAL48T and not(KC48)]}

which can be reduced to

- PP10_ILOK == not(LAL-10 or LALL10T or LALL47 or PAH48T) and {not(PAL48T) or [PAL48T and not(KC48)]}
- (**Note:** PAH48T is set if PAH48 is true for > 5 seconds)
- PP10_SRT == [S03_RUN and PP10_RNG and not(PP10_SRT)] or [S05_START and not(PP10_RNG)]
- PP10_STP == S05_PAUSE or S05_EXIT or [S03_RUN and not(PP10_RNG)] or PP10MM
- PP10_CMD == PP10_ILOK and not(PP10_STP) and [(PP10_SRT or PP10_RNG)]

So PP-10 is commanded to run if its interlocks are made, if a stop command is not received, and if the pump is either already running or a start bit cycles true.

b. **Valve HV-13 Device Logic**

Valve HV-13 functions only in PLC mode. There is no local hardwired control for this device. The DCDS for HV-13 is shown in Figure 3-63. This one is simpler than the pump's. HV-13 controls and may be resolved to a logic statement by using the device control detail sheet:

- HV13_CMD == HV13_ILOK and not(HV13_STP) and (HV13_SRT or HV13_RNG)

The structure of this statement looks remarkably like that of the pump. The syntax of the bit names was kept consistent even though start and stop could easily be changed to open and close.

6. Functional Logic Diagram

Another way to communicate functionality is through the use of logic diagrams. The information presented on a logic diagram is one step removed from the pro-

DEVICE CONTROL DETAIL SHEET

DEVICE TAG:	HV13 Fill Valve
SERVICE:	TK-13 PRODUCT TANK
LOCATION:	Building 14, Bay 13
REFERENCE:	

DEVICE INFO	
Device Name:	HV-13
Descr.:	Fill Valve

START/OPEN CONDITIONS (HV13_SRT)	
A1	S02_SRT (Automatic Fill Sequence Active)
B1	HV13_OPN (Operator Intervention)
C1	S02_RSM (Resume Filling Operation)

STOP/CLOSE CONDITIONS (HV13_STP)	
A1	S02_EXIT
B1	S02_PSD
	HV13_CLS (Operator Intervention)

STOP/CLOSE ACTIONS	
T1	Clear the HV13_CMD bit

START/OPEN ACTIONS	
T1	Raise the HV13_CMD Flag

INTERLOCKS (HV13_ILOK)	
A1	not(LAH10)
A2	not(LAHH10)
Z1	not(LSH-10)*

RUN CONDITIONS (HV13_CMD)	
A1	HV13_CMD True

ALARMS	
1	Command Mismatch for > 10 sec.

NOTES
(*) Denotes Hardwired Item

REFERENCES
1
2
3
4

APPROVALS	
DATE	SIGNATURE

FIGURE 3-63. HV-13 FILL VALVE DEVICE CONTROL DETAIL SHEET

gram, providing great functional detail. These diagrams describe the system graphically, showing all aspects of the control scheme broken into each logical element. Using the SFC and the sequence step detail sheet to gather the information greatly eases the process of creating the logic diagrams.

There are several logic diagram formats[4] with various symbol sets. For our purposes, we will use function blocks with the function acronym at the top of the block and the function tagname beneath (Figure 3-64).

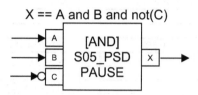

FIGURE 3-64. SAMPLE LOGIC DIAGRAM FORMAT

In this case, the output condition X is true if input conditions A and B are true, and condition C is false. As in the case shown above, signal flow in the logic diagrams presented here is generally from left to right. The logic is shown with the input signals originating in the field, or the requests originating in the HMI. Both enter the control system from the left, are processed in the controller, and then are shipped out to the field or back to the HMI to the right.

Referring to the SFC and the SSDSs makes it much easier to follow the logic narratives and diagrams that follow. Naming conventions should be parallel wherever possible. For example, where the run conditions are shown in the logic diagram, as described in the detail sheet, the function block should be named Run. A technique for simplifying the logic diagram is the use of rat holes, which are shortcuts that link signals on the same logic section. A rat hole (or "rathole") is represented by a labeled hexagon (Figure 3-65). If a signal enters a rat hole, it will assuredly emerge again at some other point in the logic. If the logic is complex, a method should be developed to keep track of the rat holes. Rat holes should only be used within the same program section. If the signal is passed to other sequences, then the entire tagname should be provided.

Speaking of tagnames, a list of acronyms should be developed and maintained, and the naming convention should be well documented. Figure 3-66 is a list of modifiers used in this project. The first portion of the tagname is either a step designation or an equipment designation. For example, S02_SRT is the start bit for sequence step 2.

The following are selected logic diagrams for the major sections of the TK-10 control scheme:

---

4. ISA-5.1.

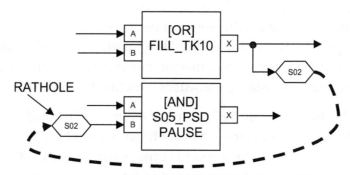

FIGURE 3-65. LOGIC DIAGRAM SHOWING RAT HOLES

NAMING CONVENTIONS	
MODIFIER	DESCRIPTION
_ILOK	INTERLOCK
_CMD	COMMAND
_REQ	REQUEST
_STP	STOP
_PSD	PAUSED
_EXIT	EXIT
_SRT	START
_RSM	RESUME
_ENA	ENABLE
_RDY	READY
_DN	DONE

FIGURE 3-66. NAMING CONVENTIONS FOR THIS PROJECT

a. **Tank TK-10 Control Sequence Step 02**

Sequence steps do not control field devices directly. Instead, they monitor process conditions and react by semaphore, raising and lowering flags to signal the completion of events and processes. Device logic monitors these flags and reacts by operating the equipment. Device logic is discussed later.

As stated, this method uses rat holes as shortcuts to link signals on the same logic section. If the signal leaves the original section and comes up elsewhere, this change is spelled out entirely in a box. Another convention is the function block names. The names listed in the box are the names of flags that are raised when the output of the function block is true. The outputs are denoted by an X box.

Sequence step 02 is the TK-10 fill step. Refer to Figure 3-57 if necessary. Note: The start parameters on the logic diagram exactly match the start parameters on the SSDS. Further, the exit parameters are much more complex than those listed on the detail sheet. That is because the exit conditions and the pause conditions must be accounted for. Also there are only two outputs to the outside world: "I'm running" and "I'm paused." The other tagnames are flags that may or may not be used elsewhere, but they are included to maintain a consistent structure between step logic sections.

The delay timer (Figure 3-67) is one of many used throughout the project. As indicated in the diagram, a delay timer does not react to a low-to-high transition on its input for a specific time period. It reacts immediately, however, when a negative transition is detected on its input. Delay timers are used primarily to debounce (condition) the signals. As a level reaches its high-level trip, for example, waves in the liquid may cause the level alarm to cycle on and off for a time until the liquid level rises far enough. The delay timer will ignore cyclic signals and will not activate until the signal is present for the time allotted. In the case of the timer (K02_RSM) shown in Figure 3-68, an output is not shown because it is used elsewhere. The K02_RSM flag rises after the timeout period, even though it is not shown.

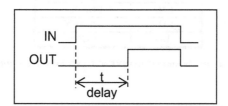

FIGURE 3-67. TIMING DIAGRAM FOR A DELAY TIMER

Some of the flags shown (such as ZLC13) are generated in logic sections that follow.

b. **Tank TK-10 Control Sequence Step 05**
Step 05 is the Empty Tank step. Refer to the SSDS and the SFC. Compare the logic diagram (Figure 3-69) to the detail sheet. Note the similar structures of the two sequence steps.

c. **Tank TK-10 Level Detect and Fill Control Logic**
The fill control device logic (Figure 3-70) exhibits a *continuous* processing requirement (i.e., the level detection section) and a *sequential* processing requirement (i.e., the fill valve open/close section).

1. TK-10 Level Detection (Continuous) Analog Alarm Logic

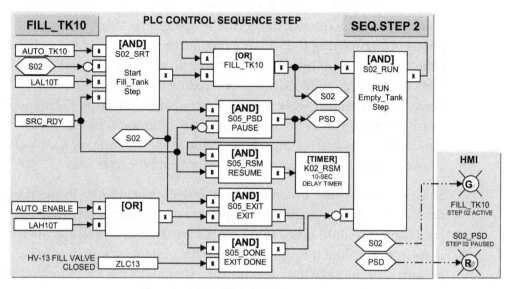

FIGURE 3-68. FILL_TK10 CONTROL LOGIC

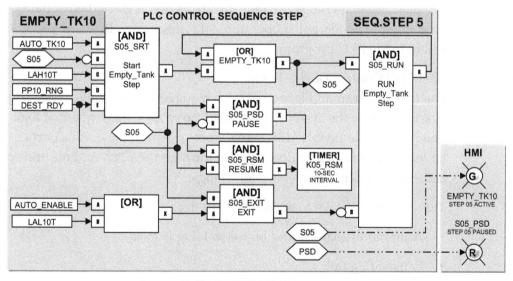

FIGURE 3-69. EMPTY_TK10 CONTROL LOGIC

The level transmitter (LT-10) sends level information to the PLC in the form of a 4–20 mA signal regardless of control mode. The signal hits an AI block, where it is converted from a 4–20 mA DC current value to an unsigned integer that ranges between 0 and 4095 increments (or counts). This integer is fed to a comparator block (CMP-10) and through the LI-10 rat hole to the HMI for display. (The integer value is converted to engineering units in the workstation.)

The comparator compares this process variable (PV) value to one or more alarm set points. At the HMI, the operator is able to enter set points in engineering units (EUs). Engineering units are degrees Fahrenheit, psi, or, as in this case, percentage fill of a tank. The HMI[5] converts the floating point EU value entered by the operator to a percentage of scale integer suitable for transmission to the PLC. These alarm set-point values are loaded into the comparator block, where they are used as trip settings. If the value at the PV input of the block exceeds a set point, the corresponding output bit cycles true.

Since the comparator is continuous, it reflects fluctuations induced in the PV, which may be due to an agitation or vibration, or perhaps a poor signal-to-noise ratio. Timer blocks are used to keep these fluctuations from rippling through the rest of the system. A signal must be present continuously for a specific time period for the PV excursion to be recognized by the system as a true alarm.

Each of the comparator's outputs feeds the alarm manager in the HMI. The alarm manager logs alarms, prioritizes them, gives them a date/time stamp, and alerts the operator. All outputs are also used elsewhere in the PLC program.

2. TK-10 Fill Control Device Logic

   The logic diagram in Figure 3-70 also manages the fill valve. The operator must place the system in automatic to engage the sequential logic. The Auto Mode Enable HMI signal does not originate on this particular logic diagram, but it does emerge through the ENA rat hole and so becomes a factor in the level control logic.

   There are three ways to open valve HV-13:

   - Auto mode enabled and high-level flag is false
   - Paused flag falls
   - Operator opens valve by pushing a button on the HMI

   These three signals are OR'd to satisfy the start (open) condition shown on the SCDS.

   The start (open) signal is fed to an interval timer that generates a timed pulse when its input goes true. This pulse is OR'd with the valve

---

5. The engineering unit (EU) conversion is usually done at the HMI, though it can be done in the PLC. The HMI is preferred because it minimizes the amount of information that needs to be shipped across the network. A floating-point number takes 32 bits, whereas an integer takes 16.

PART III – CHAPTER 10: CONTROL SYSTEMS INTEGRATION (CSI)

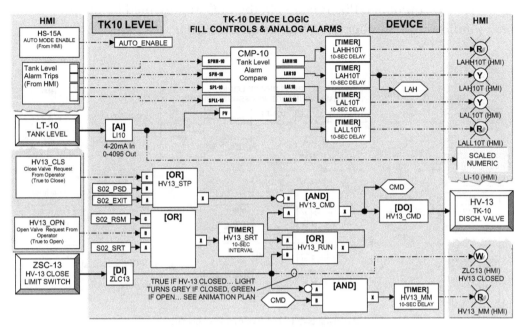

FIGURE 3-70. DEVICE LOGIC FOR TK-10 FILL CONTROLS AND ANALOG ALARMS

closed limit switch ZSC-10 to satisfy the run condition HV13_RUN. An Open command (HV13_CMD) is issued to the valve provided, a valid stop signal has not been received.

The stop (close) signal is generated if the sequence step exit or pause conditions are met or if the operator presses the stop button at the HMI.

d. PP-10 Pump Discharge Pressure Control Logic

Recalling our earlier discussion, the pump motor is a fixed speed motor, so when the pump is running, it moves the same amount of material at all times. However, the amount of flow that can be accommodated at the destination can vary, as downstream block and fill valves open/close as material is allowed to flow to various destinations. This presents a problem: How can you protect the pump from high-pressure spikes on its discharge? The answer is to provide an alternate path for the excess material so the destination receives what it needs, and the pump maintains a constant discharge pressure.

The alternate path provided (refer to the P&ID if necessary) is a recirculation line with a throttling valve (Figure 3-71). The throttling valve shunts off just enough material to protect the pump by recirculating the excess material back into the TK-10 vessel. Control of the throttling valve is based on the output of a PID function in the PLC, represented by the pressure indicating controller (PIC) bubble on the P&ID.

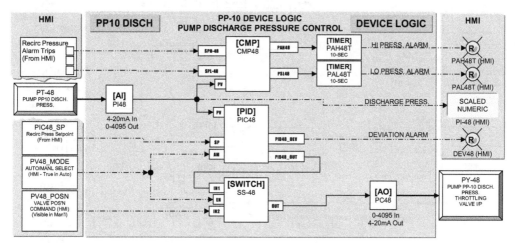

FIGURE 3-71. PP-10 DEVICE LOGIC

A pressure transmitter (PT) senses the discharge pressure and sends the signal to the PLC. The PLC applies this signal as the PV to a PID function block and to a comparator block (CMP). The PID block compares the process variable to an operator-entered set point (SP) and generates an output based on certain tuning parameters that are preloaded into the block.

The output of the PID block is fed to a switch that inhibits the automatic output when the system is in Manual mode, and passes a Manual position command to the valve from the operator. The PID block continues to execute, even when the system is in Manual. If the block itself is put in Manual mode, then its output value tracks the actual position of the valve based on the position requested by the operator. This provides a bumpless transfer to Automatic should the operator decide to do that. Also, the PID block is capable of sourcing several alarms. The only one shown here is the deviation alarm, which informs the operator of a large difference between the process variable and the set point.

Note: Bumpless is a controls term that indicates a condition that prevents a change in condition just because a control mode was switched. For example, a throttling valve is set manually to 25%, but the conditions are such that if it were in automatic, its position would be 50%. If the operator then switches to automatic mode, the valve would normally jump to 50%. Since such jumps are never good and can cause process upsets, program logic must be in place to cause the valve to integrate to the desired state over a time interval. In reverse, if the system is in automatic, and the operator changes it to manual, then the valve position will stay in manual mode but at the last position it was in while in auto. The operator may then take control of the valve and move it to the desired position.

The CMP monitors the process variable, which is pump discharge pressure, and tests for alarm conditions. In this case, the operator has preset two alarm settings: high pressure (PAH) and low pressure (PAL). Either will cause an alarm event to be recorded in the HMI.

e. **PP-10 Pump Control Logic**

The device logic sequence in Figure 3-71 depends heavily on the other sequences already discussed. Alarm flags developed previously are used here as permissives to allow the pump to run. The culmination of all this logic is the setting of the HS-15A command flag PP10_CMD. Assuming the conditions are right for automatic operation, the pump starts and runs based on high tank level (LAH) and pauses (stops) based on the low-tank-level alarm (LAL) (Figure 3-72). Like the other logic diagrams, this one uses the same nomenclature, naming the function blocks after the proper area in the sequence control detail sheet.

1. Interlocks

    There are five interlock conditions:

    - Low-level switch LSLL-47 normal.
    - High-pressure trip PAH48T normal.
    - Low-low level trip LALL10T normal.
    - Low-discharge-pressure trip PAL48T normal after ride-through.
    - Command mismatch alarm normal.

    If any of these are abnormal (i.e., in alarm), then the RUN_INHIBIT flag is raised, and a red alarm light is illuminated at the HMI. Also, each of these conditions has its own HMI status light. If none of these alarms is active, then the PP10_ILOK flag is raised, and the RUN_INHIBIT flag is lowered. The two flags are used for better readability in the PLC logic.

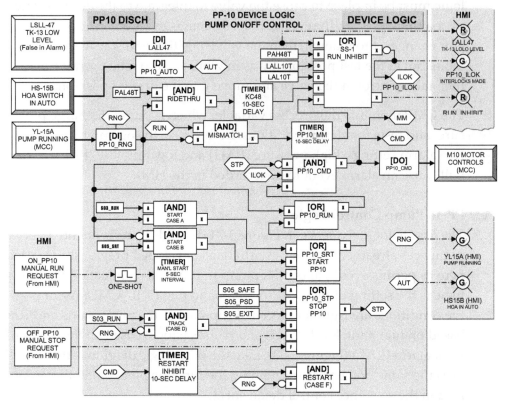

FIGURE 3-72. PP-10 DEVICE ON/OFF LOGIC

2. Start Conditions

   There are three start cases:

   - Step 3, tracking mode, is active (S03_RUN), and the pump is already running (PP10_RNG).

   - Step 5 (EMPTY_TK10) is active, and the pump is not already running.

   - The operator wants to start the pump and presses the PP-10 start request button at the HMI.

   Starting the pump causes the motor windings to heat up. Once it is running or off, they will cool again. If a pump start is detected, then a new start should be prevented until the windings have had time to cool. Therefore, the SRT_INHIBIT flag rises after a start and needs to be lowered for any future start request to take effect.

   The start of the pump is detected by use of a one-shot, rising (OSR) function block. The OSR looks for a leading edge on an input signal (Figure 3-73).

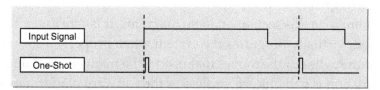

FIGURE 3-73. ONE-SHOT, RISING (OSR) EDGE FUNCTION BLOCK

The OSR function provides a short-duration pulse any time a rising edge is detected at its input. In our start-inhibit logic, this pulse is fed to a delay timer that detects this one-shot pulse and begins timing on the *trailing* edge of the pulse (Figure 3-74).

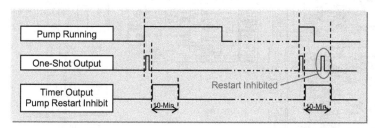

FIGURE 3-74. PUMP RESTART INHIBIT SIGNAL PROCESSING

The delay timer produces a precise output pulse that is, in this case, of 10 minutes duration. This output, when true, prevents a restart, thus giving the motor windings time to cool.

3. Run Conditions
   If the motor's auxiliary contacts close, then the pump will seal in, and the RUN flag will stay raised, even though the start pulse clears.

4. Stop Conditions
   Five conditions will stop the pump:

   - If the safe-state conditions are met in step 5 (S05_SAFE).
   - If the pause conditions are met in step 5 (S05_PSD).
   - If the exit conditions are met in step 5 (S05_EXIT).
   - If the tracking mode (S03_RUN) is active, and the pump is not running (PP10_RNG).
   - If the operator presses the pump stop button at the HMI.

f. **Logic Diagram Summary**

To summarize this section on logic diagrams, it is safe to say that the logic diagram virtually programs the system when properly done. The logic diagram spells out the moves that need to be made in great detail, and this information is valuable, regardless of the software platform in use. The end result is a drawing similar to Figure 3-75.

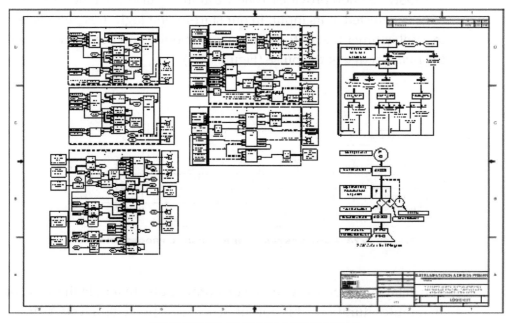

FIGURE 3-75. SAMPLE LOGIC DIAGRAM

7. Logic Diagram Standard ISA-5.1

Historically, communication of system logic was defined in SAMA[6] Standard PMC 22.1-1981.[4] This standard described a method of diagramming control logic that has seen wide use—and is still in vogue to a large degree, particularly in boiler control applications. However, the term SAMA no longer has any meaning, as that organization no longer exists. The ISA has incorporated key aspects of the SAMA standard into a new standard, ISA-5.1-2009, *Instrumentation Symbols and Identification*. Figure 3-76 shows some of the basic symbols.

As shown in Figure 3-76, the logic diagram addresses the sequential and the continuous functions of the control system. For sequential functions, diagramming the logic is similar to previously presented diagramming, except for some minor symbology differences. Continuous functions are a different matter however.

---

6. Scientific Apparatus Maker's Association.

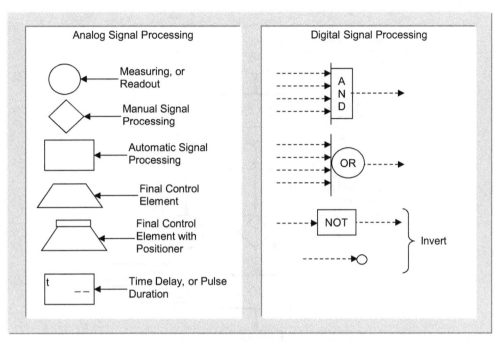

FIGURE 3-76. SELECTED SAMA SYMBOLS NOW INCORPORATED INTO ISA-5.1

The PIC-48 control loop is shown in Figure 3-77. The signal originates at PT-48. It is fed to an analog input point in the PLC (the I/O address is filled in at a later time). The control function is a proportional plus integral controller. The T symbol indicates that mode switching is available. The A indicates the mode may be switched to manual. The I represents a process variable indicator that, in this case, is actually a value on an HMI screen. The input signal is processed, and the resulting output is fed to an analog output function block (to be addressed later). The analog output function block passes the information to the final control element via the I/P transducer.

8. FEL2 Systems Integration Summary

The most important phase of the control logic production effort is the one that occurs before the programming even begins. If the control scheme is properly prepared, the actual writing of the PLC program should be a bit anticlimactic. Of course, there can always be "discoveries" that can mess up the best of plans. And the moves the programmer must make to effect the agreed-upon design may vary from one PLC platform to the next.

Today PLC programmers have more software tools at their disposal. From one software solution to the next, the functionality has improved to the point where most of them can accommodate most of the controls situations that may arise. The PLC program can be much more DCS-like, for example. While ladder

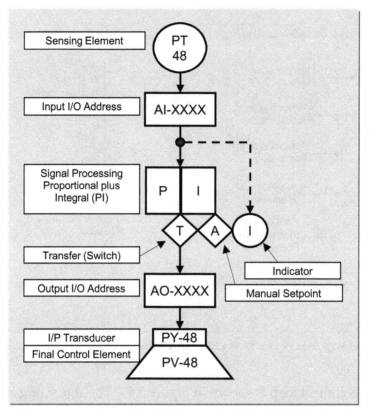

FIGURE 3-77. FUNCTIONAL CONTROL DIAGRAM (ISA-5.1)

logic is still prevalent, most of today's better PLC and PC-based software packages let the programmer choose from more than one programming structure.

In any case, the programming field is opening up to more people with a more varied sets of skills. A decade ago, PLC programmers were most likely controls technicians or engineers who had migrated into the systems integration field over a period of time. These individuals thoroughly understood relay logic and process controls, so over time they found they were writing software that looked a lot like relay logic. As a rule, they knew how the control system should behave since they had such a good grasp of the process. And they felt little need to document the software since its purpose was self-evident, and nobody else was likely to attempt to modify the program.

Today's new programmers have skipped most of that preparatory field work and, as a result, may have less knowledge of the "field craft" end of the controls business. On the other hand, they are probably better educated in the art of computer programming. If they can be made to understand the customer's needs at the process level, they just might be better equipped to implement the design. The key to success is good communications with those who understand the process and good by-the-numbers documentation techniques.

A technical description of the PLC program for this application will be left to others. Suffice it to say that, if the information presented in Section B is applied, the PLC programming and subsequent documentation task will be greatly simplified, and a quality product is more likely.

D. Operator Interface Specification Development – The HMI

When a control logic specification is devised, as described in the previous section, many of the core traits of the HMI are developed. In our example system, operators must be able to enable and disable automatic operation, to enter the set point for the recirculation pressure loop, and to enter alarm set points, among other actions.

However, simply defining the tasks the HMI must support is really a minor, though key, part of developing an HMI specification. Beyond providing functionality, the HMI design should put operators in their comfort zone. This includes color considerations, screen hierarchy, alarm management issues, and other esoteric considerations that would not be necessary to simply get the job done. Therefore, like the control logic specification process, the operator interface specification is a task that needs the active participation of all concerned.

For more general guidelines regarding HMI development, and for some background on its evolution, refer to the HMI section in Chapter 6, Section D. The present section provides specific examples of ways to communicate design intent for the HMI.

Human-Machine Interface is a broad term that encompasses several functions:

- *Control graphics* are graphic screens that let the operator view the process and quickly gain an understanding of its condition. These screens depict process equipment, using animated graphics that use color (or contrast) and shape to present information to the operator. They also give the operator the ability to take action by mouse click and/or via touch screen.

- The *action/animation database* is mated to the control graphics package. It provides a means for linking the graphics to the control system and ultimately to the process equipment. This database lets the programmer give the active graphic elements names that are useful for troubleshooting, define their action/animation characteristics, and define their links to the outside world.

- The *device driver* is a software utility mated to a hardware connectivity device that manages (translates) HMI data requests and transfers them onto the communication network. (Note: The HMI issues requests, the process controller issues commands…).

- The *alarm manager* is a software utility that monitors *alarm tags*. These tags are linked to process sensors and switches and to embedded PLC logic. The alarm manager logs alarm events, dates and time stamps them, and alerts the operator by audio or visual means.

- The *historian* is a software utility that collects and stores data for future use. Each data element is given a unique name called a *historian tag*. These tags can be individually configured for sample frequency and other considerations. The data is stored in archive files, which may be retrieved and used for analysis. The *report generator* is a software utility that generates process reports.

This section does not provide a finished HMI specification. What it does provide is a *generic approach* that is platform independent. It provides a set of tools that can be used to communicate design intent and used as a design aid regardless of PLC, DCS, or HMI platform.

1. Animation Plan

Before work even begins, some decisions should be made concerning animation. An animation plan details the color/contrast scheme and the anticipated situational animation. For example, what color indicates an open valve, a stopped pump, a full tank? When will animation tools, like flashing lights and audible alarms, be used? What about visibility issues and text messaging? Each of the animation issues should be discussed ahead of time as they might influence screen design.

The plan associates colors and symbols to keywords. Figure 3-78 is the animation plan to be submitted for our project.

a. **Colors**

In our animation plan, vibrant colors are used sparingly. The background color is a muted, opaque taupe; equipment items are shown in an off-white color. Safe or off conditions are indicated by a light grey, while running, or open is indicated by bright green. Simple status alert indicators cycle from light grey (off) to white (on). Red and yellow colors are reserved for alarms and warnings.

b. **Visibility**

The use of conditional visibility gives the designer great flexibility. In our simple case, the use of animation is limited to showing the pressed/released status of a pushbutton or the position of a rotary switch and to adding emphasis to emergency indicators (if the need for any such is identified). Conditional visibility simply means there are two or more pictures of the same graphic element superimposed one on the other. Each picture is linked to a different logical statement so that only one of the pictures will ever become visible at a time.

c. **Flash and Beep**
Use of the flash, or strobe, feature should be reserved for serious problems. A slow flash indicates a condition that needs to be looked at as soon as possible. A fast flash is used for a condition that requires immediate attention. Frequently, though not always, the flash feature is accompanied by an audible beep that sounds at the same frequency as the fastest flash in the system. Before using the audible alarm feature, the designer should investigate the capabilities of the built-in, alarm-manager utility, which usually manages the audible alarm feature.

d. **Messaging**
The use of messaging allows the system to automatically broadcast alert and alarm messages across the plant network or possibly to page supervision personnel using dialup modems. Again, this feature may be covered in the alarm manager utility.

2. Screen Diagrams

The person who said, "A picture is worth a thousand words," had to be a systems integrator! What better way can there be to talk about graphics with an operator or production manager than with a diagram? A moviemaker might call these diagrams storyboards, as they provide insight into the story line. Providing insight is the purpose of these diagrams, for the operator and the integrator. The purpose of a screen diagram is to describe the behind-the-scenes functionality of a graphic screen. Like the SFC and the logic diagram described earlier, the screen diagram is a multipurpose tool. Initially, screen diagrams are used to communicate the integrator's intent and his understanding of what the customer is asking for. At the end of the project, the diagrams can become the centerpiece of an O&M manual that will be very useful to project participants.

One might argue there is not time in the budget for such an effort as this. On the contrary, this time is invariably well spent because it avoids costly rework due to poor communications or misunderstandings. Like the SFC and the logic diagram, the screen diagram, or a similar tool, it is essential to a well managed integration project.

A screen diagram combines a screen graphic with a component schedule. Half of the diagram has the screen graphic, and the other half has the component schedule. Each active graphic component is "bubbled" with a number. This number is indexed to the component schedule.

ANIMATION PLAN

GROUP	KEYWORD	COLOR	VISIBILITY	FLASH	BEEP	MSG
BACKGND	-	TAUPE	-	-	-	-
EQUIPMEN	-	OFF-WHITE	-	-	-	-
MOTORS	RUNNING	GREEN	-	-	-	-
	OFF	LT GREY	-	-	-	-
VALVES	OPENED	GREEN	-	-	-	-
	CLOSED	LT GREY	-	-	-	-
PBUTTONS	PRESSED	LT GREY	DOWN	-	-	-
	RELEASED	LT GREY	UP	-	-	-
STATUS INDICATOR	OFF	LT GREY	-	-	-	-
	ALERT	WHITE	-	-	-	-
	WARNING	YELLOW	-	-	-	-
	ALARM	RED	-	SLOW	SLOW	-
	EMERGENCY	RED	LITTLE/BIG	FAST	FAST	YES

FIGURE 3-78. ANIMATION PLAN

The component schedule describes the functionality of each component. It is a chart with a component description, a description of the action (if any) that results if the mouse clicks on (or a finger selects) the item and a description of the animation (if any). Information about the source or destination of the animating signal should also be included.

The following is a more detailed orientation to the screen diagram:

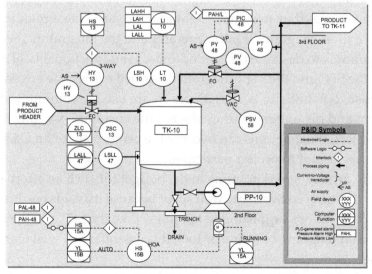

FIGURE 3-5 (REPRODUCED). P&ID PRESENTATION OF THE TK-10 SUBSYSTEM

The first step in developing a screen diagram is to duplicate the P&ID in the HMI software (Figure 3-5, reproduced below). The second step is to add graphic symbols that can be animated to provide alarm, data readout, and control functions (Figure 3-79). The third step is to add a component schedule in which all the data links are called out.

a. **Graphic Screen**

   To begin, usually all anyone has to work from is the P&ID. Happily, this makes a good place to start for a first-pass control graphics screen diagram. Simply copying and modifying the P&ID sometimes is enough to get the ball rolling.

   In Figure 3-79, all major instrumentation components are presented. Notice that the orientation of the components on our P&ID is physically realistic. This is important in helping the operator relate the screen to the process area. If the P&ID is not physically representative—and many times P&IDs are not—then the components should be rearranged to make the screen as representative as possible.

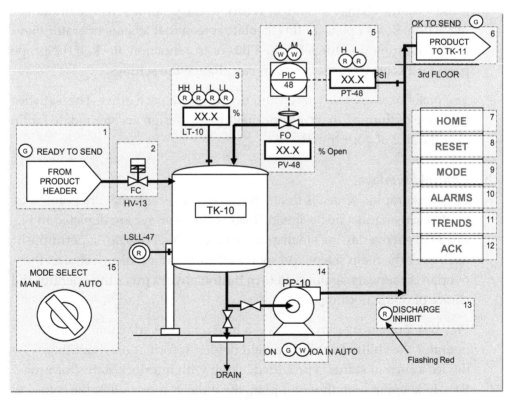

FIGURE 3-79. PRELIMINARY SCREEN GRAPHICS, TK-10 OVERVIEW SCREEN.

After the static background is set, the next step is to remove all the control system bubbles on the analog instrumentation and replace them with data readouts. Then, alarm lamps, control switches, or other animated devices and screen navigation buttons may be layered on.

Notice that the screen still looks remarkably similar to the P&ID. Any information not needed by the operator, such as the relief valve and the pneumatics, has been deleted. Numeric displays have replaced the analog bubbles, and indicator lamps provide visual annunciation. The control mode select switch has been added at the lower left.

The 15 numbered pick boxes (number in upper right of the box) indicate active areas on the screen that, if touched or clicked on, will initiate some kind of action. Under normal circumstances, the box outlines will not be visible, but are shown here for clarity. The buttons at the right (pick boxes 7–12) are navigation buttons that open up other screens. Likewise, pick boxes 1 and 6 take the operator to the existing graphic screens pointed to by the piping flags. Pick boxes 3 and 5 open data entry overlays that allow the operator to view and, if necessary, change the alarm set points. Boxes 2 and 14 open confirmation overlays that allow the operator to either continue with the change in state or return to the graphic screen without taking action. Box 4 opens a PID faceplate screen that lets the operator view alarm and tuning settings for the PID control function block. If the proper password is entered, the operator can change the settings.

The pick boxes are fairly large and do not touch each other. This satisfies the "rule of thumb" that calls for making the design accommodate fat fingers the size of your thumb.

b. **Control Overlays**

Once the graphic screen is done, the various overlays that are pointed to by that screen must be designed. Two of the overlays are depicted in Figure 3-80. An overlay is a small graphic screen that pops up, superimposing itself over the main screen. As soon as the operator makes a choice, the overlay disappears again. A Return button always gives the operator the choice of doing nothing.

Pick box 2 opens overlay 2. This is a true *control* overlay that gives the operator the ability to control a field device. Information about the device's current status is provided, along with interlock status information. In keeping with the color plan, the indicators are either the color indicated or a light grey. Pick box 15 opens an overlay that allows the operator to select the mode of operation for the system.

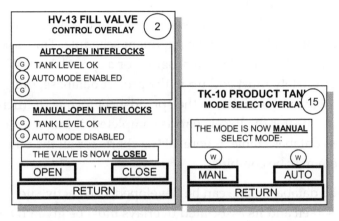

FIGURE 3-80. SAMPLE CONTROL OVERLAYS

Most HMI software packages support the kind of moves being presented here. It is important to note, for the purposes of the screen diagrams, overlays are considered to be a part of the graphic screen. Therefore, each active item on the screen, plus its overlays, eventually receives a unique ID number that ties it to the main graphic screen. Figure 3-81 presents the remaining overlays envisioned for this screen.

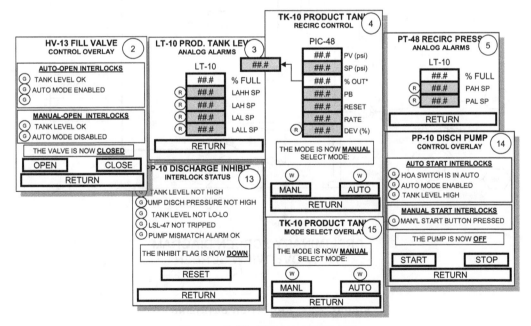

FIGURE 3-81. POP-UP OVERLAYS

c. Component Schedule

Okay, the graphics layout is now complete, and it is time to generate the database that will become the component schedule. This task can be done effectively in either a database program or a spreadsheet. While the information ultimately must end up in the project database, things probably work best, at least initially, in the spreadsheet format. If Microsoft software is used, after the Excel chart is completed, it may then be pulled into an Access database table, where all the charts for all the other screens on the system may be combined.

Before the database table can be developed, each active item on the graphic screen and its associated overlays must be given ID numbers (Figure 3-82).

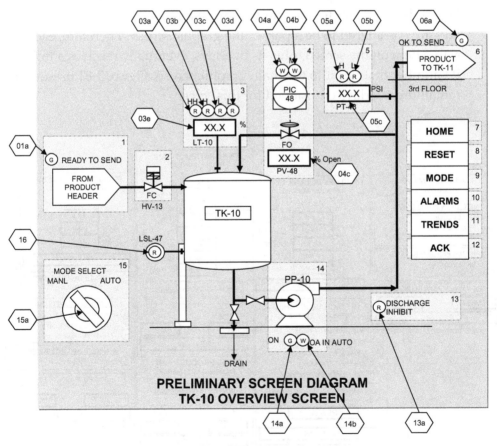

FIGURE 3-82. ANIMATION DETAILING

The item numbering scheme employed here is alphanumeric. This method groups all related items under the same number, with a letter suffix. For example, for pump PP-10, there are three items in the database: 14, 14a, and 14b. If additional items are shown on the overlay related to pick box 14, then they are numbered from 14c up (Figure 3-83). For example, the main graphic screen has only three items related to the pump. However, if the pick box is activated, an overlay appears that has additional related activity.

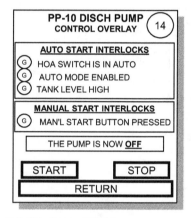

FIGURE 3-83. PUMP PP-10 CONTROL OVERLAY

Item 14b already appears on the main graphic. It is merely copied to the overlay, keeping the original data links and animation characteristics. On the other hand, item 14f is animated by the same PLC I/O point as item 14a, yet it gets its own letter suffix. Why? It is because it animates differently, with a conditional visibility change rather than with a color change. Figure 3-84 shows a section of the animation chart as it relates to item 14.

			Screen Diagram Query					
Node	Screen	Item	Description	Device	Action Tag	Action	Animation Tag	Animation
CR-1	TK10_OVIEW	14	PP10 Discharge Pump	Pick Box	TK10_OLAY14	Open Overlay #14	na	na
CR-1	TK10_OVIEW	14a	PP10 Running	Lamp	na	Na	TK10_RNG	Grey/Green
CR-1	TK10_OVIEW	14b	PP10 HOA in Auto	Lamp	na	Na	TK10_AUT	Grey/Green
CR-1	TK10_OVIEW	14c	Auto Mode Enabled	Lamp	na	Na	AUTO_ENABLE	Grey/Green
CR-1	TK10_OVIEW	14d	Interlocks OK	Lamp	na	Na	PP10_ILOK	Grey/Green
CR-1	TK10_OVIEW	14e	Restart Inhibit Time (decrements)	Numeric Disp.	na	Na	RESTART_INH	Value
CR-1	TK10_OVIEW	14f	Pump Status Message	Message	na	Na	PP10_RNG	"...Off" / "...On"
CR-1	TK10_OVIEW	14g	Manual Stop	Mom.PB	OFF_PP10	Stops Pump	na	na
CR-1	TK10_OVIEW	14h	Manual Start	Mom.PB	ON_PP10	Starts Pump	na	na
CR-1	TK10_OVIEW	14j	Return	Mom.PB	CLOSE_OLAY	Closes Overlay	na	na

FIGURE 3-84. ANIMATION CHART

Note that there are node names (e.g., CR-1, for control room node 1) and screen names (e.g., TK10_OVIEW), as well as the various items detailed above. The different elements of the screen, as designed above, should be pulled onto a final screen diagram that looks something like Figure 3-85.

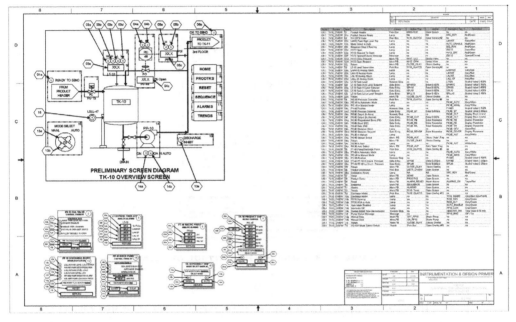

FIGURE 3-85. FINAL SCREEN DIAGRAM

The final diagram should then be submitted for review. This usually entails a formal presentation to the customer. After final approval of the conceptual design, the screens may be produced. Graphics may be improved in the process, but the underlying functionality should remain true to the diagram.

3. TAGNAME DATABASE, DEVICE DRIVER, AND I/O MAPPING

Before final production of the graphics screen, the database needs to be configured and mapped (Figure 3-86). Each element to be animated must be defined, named, and logged into the HMI database.

Certain animated elements on the graphics screen must be configured in the HMI tagname database. In the database, the animation item is defined, and, if the item is linked to the outside world, the database provides a connection between the screen and the I/O driver. More than one I/O driver may be running as part of an application. The tagname database maps the tag to the proper driver. The driver allows the designer to set up the method of polling (fixed rate or on exception), the memory locations to be polled, and the network node address of the source data location.

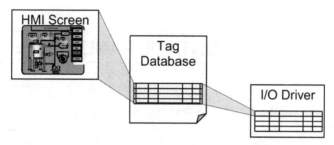

FIGURE 3-86. TYPICAL DATA PROGRESSION

4. FINISHED GRAPHICS SCREEN

The finished graphics screen as viewed by the operator is shown in Figure 3-87. The pump is running, and the tank is 80% full, with the high-level alarm active. Piping in which material is moving has been animated—indicated by the darker color here. Notice that the screen is a bit simplified from the original screen diagram. That is because some of the animation has been incorporated into the device graphics, eliminating the need for some of the lamps.

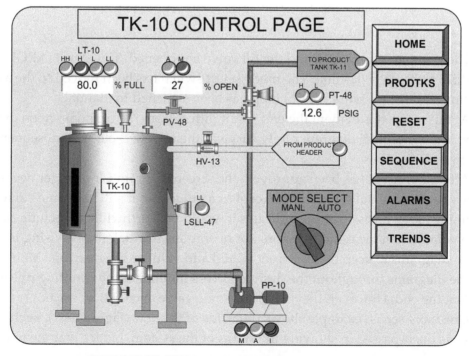

FIGURE 3-87. HMI SCREEN, PUMPING OUT IN MANUAL

In the second view of the same screen (Figure 3-88), the situation has changed. Instead of a high-level situation, the operator has placed the system in automatic, and tank is in a low-level alarm condition.

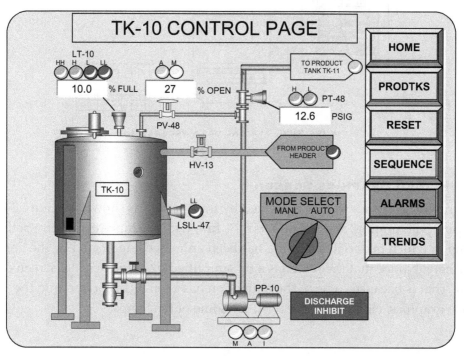

FIGURE 3-88. HMI SCREEN, FILLING IN AUTO

The pump has turned off and the fill valve has opened. The DISCHARGE INHIBIT message is flashing. The mode select switch has been moved to the auto position, and the pressure control loop has been switched to manual.

At this point, it is evident the end result differs from the original screen diagram. The conceptual diagram rarely survives the review cycle and subsequent design modifications.

After its concept has been approved, the screen diagram is no longer needed, from a developmental standpoint, except as a guide. But it still has some value as a training aid. If an assessment indicates it would be worthwhile to include each screen diagram in the training manual or operations manual, then revising the screen diagrams to keep them current would add value to the manuals. Maintaining the diagrams throughout the design process minimizes the resulting pulse of work at the end. However, they still have some value even if they are not revised.

One more screen example shows the value of the SFC (Figure 3-89). Such a screen is unsurpassed in showing the status of the system. Notice that the navigation buttons are in the same location on all three screens. The only change is in the button names, as one of those will change, depending on which screen is active.

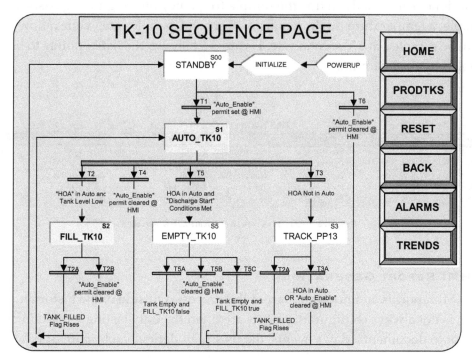

FIGURE 3-89. HMI SCREEN, SEQUENCE STATUS

5. Alarm Manager

Configuring alarms in some HMI packages is as simple as filling in a checkbox in the tagname database. It is pretty easy in any case. Unlike tags configured on the graphics screen, alarm tags are sampled, whether the graphic element is visible or not. So there is a cost in terms of network throughput. A sample alarm manager database for this application is shown in Figure 3-90.

HMI Alarm Manager Query							
Node	Screen	Item	Description	Device	Animation Tag	Animation	Alarm?
CR-1	TK10_OVIEW	02a	LAH10 Tank High Level Trip	Lamp	LAH10T	Green/Red	Yes
CR-1	TK10_OVIEW	03a	LAHH-10 Analog Alarm	Lamp	LAHH10T	Grey/Red	Yes
CR-1	TK10_OVIEW	03b	LAH-10 Analog Alarm	Lamp	LAH10T	Grey/Red	Yes
CR-1	TK10_OVIEW	03c	LAL-10 Analog Alarm	Lamp	LAL10T	Grey/Red	Yes
CR-1	TK10_OVIEW	03d	LALL-10 Analog Alarm	Lamp	LALL10T	Grey/Red	Yes
CR-1	TK10_OVIEW	13a	Discharge Inhibit	Lamp	TK10_INHIBIT	Grey/Red (SlowFlash)	Yes
CR-1	TK10_OVIEW	14d	Interlocks OK	Lamp	PP10_ILOK	Grey/Green	Yes
CR-1	TK10_OVIEW	16	LSLL-47 TK10 Low Level	Lamp	LALL47T	Grey/Red (SlowFlash)	Yes

FIGURE 3-90. SAMPLE ALARM MANAGER DATABASE

6. Historian

The historian is similar to the alarm manager in that it is simple to configure, and points defined there are always sampled. However, unlike the alarm manager, the designer can usually define the sample frequency on a tag-by-tag basis. For example, a temperature might require a 10-second sample rate, while a flow might be sampled at a 1-second rate. Figure 3-91 shows the list of points to be sampled by the historian.

HMI Historian Query							
Node	Screen	Item	Description	Device	Animation Tag	Animation	Historian?
CR-1	TK10_OVIEW	04c	PV-48 Position	Numeric Disp.	PV48S	Scaled Value 0-100%	Yes
CR-1	TK10_OVIEW	05c	PV-48 Position	Numeric Disp.	PV48S	Scaled Value 0-100%	Yes
CR-1	TK10_OVIEW	03e	LT-10 Tank Level	Numeric Disp.	LT10S	Scaled Value 0-100%	Yes
CR-1	TK10_OVIEW	04d	PIC48 Pressure Setpoint	Data Entry	PIC48_SP	Scaled Value 0-250psi	Yes
CR-1	TK10_OVIEW	04e	PIC48 Output (In Auto)	Numeric Disp.	PIC48_OUT	Display Pos'n Cmmd	Yes

FIGURE 3-91. HISTORIAN SAMPLING POINTS

7. HMI Report Generation

The HMI report is an important part of the process. Outside of the historian, which is not always configured, reports are often the only lasting record. In an attempt to document a day's events, the user may define batch, shift, or other reports.

Reports are often the weakest link in the HMI package. Sometimes the designer may export information into an Access database or another commercial software package, where reports are easily generated.

E. Network Single-Line Diagram Generation

The network single-line diagram is more than a single-line drawing. It does also show the network elements in a tiered single-line format. The basic design scheme of this type of drawing is similar to that shown in Figure 3-92.

The different network topologies are shown in their own tier. In our example, the pieces of equipment closest to the factory floor are shown at the bottom. The Remote I/O network is a network that is proprietary to the PLC vendor. It has been optimized for connectivity between different PLC components. Most PLC vendors have such a network.

In this case, the network requires a terminating resistor. Modbus-Plus is a protocol by Modicon® (Schneider Electric). Note the two terminating resistors on either end. The protocol is not a "token-ring," merely "token-passing." Modbus is another protocol based on RS232/485. Both of those have become de facto industry standards. Most external equipment purchased today will have Modbus and/

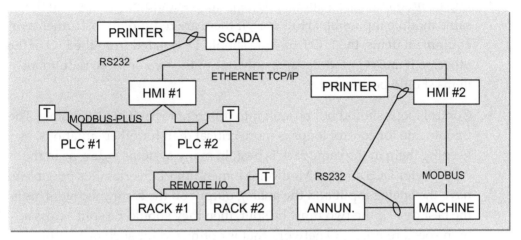

FIGURE 3-92. SIMPLE NETWORK SINGLE-LINE DIAGRAM

or Modbus-Plus communications options. Ethernet TCP/IP has been on the upper tiers for many years. It is now taking over the factory floor as well. This diagram provides just enough detail to indicate one-line connectivity.

F. OTHER SYSTEMS INTEGRATION TASKS

Generating DCS or PLC programs, designing HMI screens, and integrating the two are the primary responsibilities of the systems integrator. However, the task does not end there. The following are some of the other systems integrator's activities:

1. CONTROL SYSTEM CABINETRY DESIGN AND DELIVERY

Cabinet drawings are discussed elsewhere in this book, as the cabinet design is considered a basic engineering deliverable. However, the systems integrator sometimes performs this task. In fact, having the integrator do the PLC work, and include the cabinetry, makes a lot of sense. Doing this ties the control system task into a nice neat package. It reduces the interface requirements among groups and gives the integrator complete authority over the system, thus reducing the likelihood of finger-pointing later if things don't go as planned.

2. I/O ADDRESS ASSIGNMENT (PARTITIONING)

One key task for the systems integrator is I/O address assignment (partitioning). This can be done just as well by the design engineer, but for our present purposes, we will assume it is an CSI task. Regardless of who actually does the I/O configuration, the information will need to find its way into the project database.

Here are a few common-sense considerations for partitioning the I/O:

- I/O common to a particular equipment item should be grouped on the same module if possible. However, if there are redundant instruments or equipment items, the I/O from one should be isolated from the I/O of the other, so if an I/O module fails, only one of the two or more redundant items should be affected.

- Control loops should not be split into different racks if at all possible. The update rate for control loops is more critical than for other situations, so keeping them in the same rack is best. In many systems today, as in the ABB-Taylor DCS and the Modicon Momentum PLC series, it is possible to have the analog input and the analog output on the same module. This is the best case. If the two must be split, then they should be split across a high speed remote I/O network that is optimized for such tasks, not across a LAN that is doing other things.

- Getting back to the task at hand, if we refer to the network schematic begun earlier (refer to Figure 3-11), we see the new rack is attached to an existing remote I/O LAN as drop 03. There is no network on the system level, so this will be node 01.

a. **Hardware (HW) Address**

   Each I/O point should be given an address, called the Hardware Address, so the point may be easily located by a field technician. A good format for a hardware address is as follows:

   Nomenclature: DxxNxxRxxSxxPxx

   where Dxx = LAN drop number
   Nxx = Remote I/O node number
   Rxx = Rack number
   Sxx = Slot number
   Pxx = Point number

   Since we only need one rack, the first part of our HW address is D01N03R01 (Drop/Node/Rack). The slot and point numbers are assigned in the database. Figure 3-93 shows the resulting data. Because we have a 10-slot rack, there is enough room to spread the modules out. Leaving blanks now allows us to add modules later, if necessary, keeping like modules grouped in adjacent slots. This is important in systems with mixed power supplies (AC/DC or 24VDC/250 VDC). In our example, all I/O are 24 VDC, but we are spreading them out anyway because it is a good habit to maintain.

Tagname	I/O Type	HW Address	SW Address	Equip
PT-TK10-48	AI	D01N03R01S04P01		PP-10 F
LT-TK10-10	AI	D01N03R01S04P02		TK-10 F
PY-TK10-48	AO	D01N03R01S06P01		TK-10 F
ZSC-TK10-13	DI	D01N03R01S08P01		TK-10 F
YS-TK10-15A	DI	D01N03R01S08P02		PP-10 F
YS-TK10-15B	DI	D01N03R01S08P03		PP-10 F
LSLL-TK10-47	DI	D01N03R01S08P04		TK-10 F
HY-TK10-13	DO	D01N03R01S10P01		TK-10 F
HS-TK10-15B	DOI	D01N03R01S10P02		PP-10 F
PSV-TK10-58	NA	NA		TK-10 F
PV-TK10-48	NA	NA		TK-10 F
LSH-TK10-10	HW	NA		TK-10 F
HV-TK13-13	NA	NA		TK-10 F

FIGURE 3-93. HARDWARE ADDRESSES

b. **Software (SW) Address**

The software address is next. This task is not quite as easy since software addresses need to be unique and, depending on the platform, it is sometimes not possible to assign them by physical position in the system.

In our example, the customer has allocated a block of numbers between 100 and 150 for each type of I/O. Given this criteria, our result is shown in Figure 3-94.

The specific nomenclature is platform dependent. We will assume the PLC address nomenclature consists of the I/O type, followed by a sequential number. For example, the first analog point is AI0100 for analog input point 100. Given this I/O nomenclature specification, our system configuration is modified as shown in Figure 3-95.

3. Factory (or Functional) Acceptance Test (FAT)

The FAT is usually a customer requirement, but is a good process to follow in any case. Sometimes this test occurs after the equipment ships to the site (functional), but most of the time it happens while the equipment is still set up at the system integrator's location (factory). The purpose of the FAT is to uncover problems while in a shop environment, before the control system gets to the plant. If problems are discovered during the FAT, the programming team can address them more efficiently in their own environment.

A FAT procedure should be generated ahead of time and should be approved by the customer. The procedure should be as thorough as possible within the

Tagname	I/O Type	HW Address	SW Address	
PT-TK10-48	AI	D01N03R01S04P01	AI0100	P
LT-TK10-10	AI	D01N03R01S04P02	AI0101	T
PY-TK10-48	AO	D01N03R01S06P01	AO0100	T
ZSC-TK10-13	DI	D01N03R01S08P01	DI0100	T
YS-TK10-15A	DI	D01N03R01S08P02	DI0101	P
YS-TK10-15B	DI	D01N03R01S08P03	DI0102	P
LSLL-TK10-47	DI	D01N03R01S08P04	DI0104	T
HY-TK10-13	DO	D01N03R01S10P01	DO0101	T
HS-TK10-15B	DOI	D01N03R01S10P02	DO0102	P
PSV-TK10-58	NA	NA		T
PV-TK10-48	NA	NA		T
LSH-TK10-10	HW	NA		T
HV-TK13-13	NA	NA		T

FIGURE 3-94. SOFTWARE ADDRESSES

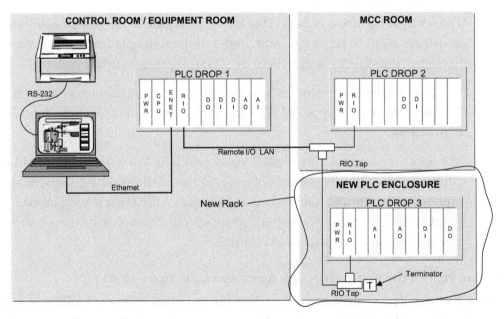

FIGURE 3-95. ADDING THE I/O MODULES

limits of the budget. If the control detail sheets described earlier were developed, then the FAT procedure could be as simple as writing a test fragment around each detail sheet type and following that procedure for each similar control detail. So it is possible to build a FAT procedure for, say, a hundred devices using only a dozen or so procedures. Also, building simulators and simulator logic takes time and adds cost, so it is wise to discuss the depth of simulation expected during the estimate; then, the procedure can be written to conform to the budget.

The FAT is really the first step in system checkout, and it feeds right into an onsite checkout procedure and eventually into the Site Acceptance Test and Commissioning. So, while the FAT can be time consuming, much of that time is recouped later if it is done right—not to mention providing the added value of a higher quality product. Also, if the programming team keeps a database and builds the detail sheets described earlier, the FAT procedure can be generated quickly and accurately.

The FAT should be well organized, with tests clearly defined and well considered. Each test procedure should have a place for the customer's sign-off upon successful completion of the test. Setup and prerequisites, if any, should be clearly defined before each test begins. And the test should be as inclusive as possible. For example, signals should be injected at the I/O termination bay if possible.

If physically injecting signals is impractical, then simulation software should be written to provide loop-back simulation of the field devices.

FAT setup should include the core elements of the control system—the controllers involved and the HMI at minimum. Also, a programmer's tool, such as a laptop, should be available and online to force bits and to enter simulation values. Several types of tests should be executed:

- *Analog Alarms:* Simulate each analog signal by disabling the I/O interface and entering dummy values—preferably raw, unscaled values—as near to the analog input block as possible. Vary the values and confirm that the scaling is correct at the HMI and that the alarms trip at the proper points. If an alarm is used as an interlock, activate the device logic of interest and demonstrate that the interlock will deactivate the device. Do not forget to restore the device logic to its original condition after this test.

- *Sequencing:* Demonstrate the program flow of control. From system startup, take the customer through each step in the process. Allow the customer to "fly" the system as the operator, pressing the buttons at the HMI and monitoring the screens.

- *Device Logic:* Demonstrate the proper operation of each field device. If a device has six interlocks, get the device logic into simulated run status and exercise the device by simulating the loss of all interlocks. This simulation should be done as near to the I/O interface logic as possible. For example, if an interlock is the result of a logical combination, then each element in the combination should be verified.

- *Reporting, Alarm Logging, and Historian:* Demonstrate proper operation of each of these. Report structure is frequently a major topic of discussion.

The execution of the FAT should be a fun event, but it should be well regulated. If the procedure is well written, and if the staff is disciplined in its execution, then the stress of the event, from the customer's point of view, will be reduced greatly. Sometimes these events last for several days, with the customer (in the case of a FAT) living in a hotel, so long hours are the norm.

4. Site Acceptance Test (SAT)

The SAT is sometimes called a checkout (Chapter 18, section A.3). Whatever the name, a properly executed SAT greatly improves the chances of a smooth commissioning process, just as a well executed FAT eases the SAT. Again, it usually falls to the systems integrator to generate this procedure. The SAT procedure falls into the Cold Commissioning Test aspect of the Deployment Phase (Phase 3).

If a well executed FAT has occurred previously, the SAT will be greatly simplified. In fact, the SAT procedure can be based on the FAT procedure, with the exception that the actual field devices will be exercised instead of simulated. The SAT could be as simple as a database listing coupled with a set of generic procedures, one procedure for each device type described in the FAT section. The database listing should be grouped by I/O type, as that determines the type of test to be performed.

- *Analog Alarms:* An analog alarm procedure can be generated by replacing the phrase "simulate by entering bogus data" in the analog alarm section of the FAT with the phrase "verify proper ambient reading" or "simulate by injecting signal." Wherever possible, the actual signal from the transmitter should be used, as that saves time. For example, if the ambient temperature is 78°F and the resistance temperature detector (RTD) signal is 150° or zero, then there could be a problem. However, if the signal is 78, then it is likely the RTD is wired correctly. It is advisable to have an electrician check to make sure the RTD is the correct one. If a proper FAT was performed, checking all the PLC alarm trips and interlocks can be avoided. However, if there are any external devices, like recorders, totalizers, and annunciators, these checks should be made.

- *Discrete Inputs:* All externally generated discrete inputs should be checked. As before, if the FAT was properly executed, this task is made easier by having two electricians, one in the field forcing or simulating contact closures and one watching the lights at the I/O module. Ideally, each discrete signal should be verified at the PLC program level as well as at the HMI.

- *Analog Outputs:* The valves should be stroked by placing them in manual mode at the HMI and forcing them into various positions. Ensure they respond as desired. If variable-speed motors can be operated, do so. Otherwise, disconnect them, and check the output of the variable-frequency drive (VFD) to make sure it is operating correctly.

- *Digital Outputs:* Motors should be bumped by placing them in manual mode at the HMI and issuing start commands. If a motor should not be turned via manual mode at the HMI, the windings should be disconnected. Such motors should be bumped locally to verify proper direction of rotation. Valves should be opened and closed from the HMI, and all alarm sirens, horns, beacons, and so on should be activated by forcing them on from the controls computer.

As with the FAT, the customer should monitor this test. Space should be reserved for the customer's initials next to each line item on the list; the customer should initial each line item as its proper operation is verified.

5. Commissioning

The commissioning procedure may or may not be a CSI product. Usually, this procedure is provided by the plant operations department. The major players at this point are the construction team and the systems integrator, who are still in the line of fire. Invariably, unforeseen problems arise and the integrator must be able to respond quickly and decisively. And no matter how well the FAT and SAT were conceived and executed, program bugs will be found.

Commissioning ends when product is generated that is within the tolerance described by the specification. The customer will sometimes add stipulations to the RFP, such as "must operate without difficulties for a period of one month," or other similar requirements, but generating product up to specification is certainly a major milestone. Technically, the project is not over until the warranty expires. Usually, if a warranty is purchased, it lasts for one year. The warranty does not cover extra-scope work, but sometimes it is not possible to fully exercise a system when first making product. A process or system may be sufficiently complex that a specific instance may not occur until some time has passed; if it does occur, the customer will want some assurance that ongoing support is available. The integrator should assume some such instances will occur and build that expectation into the budget.

6. Operations and Maintenance (O&M) Manual

The customer needs to know how to operate and maintain the new control system. Sometimes this information is combined into one book, and sometimes it is split into two, depending upon customer preference. The following are features that should be included in each section:

a. **Operations**

The operations manual should have an overview section in which the basic production process is explained. Material handling issues should be discussed here as well. The plant will need to provide some of the material for the overview. The operations section should also include the following:

- Screen diagrams and a narrative for each, written from the point of view of the operator.

- An alarm response section so that the operator can quickly find the proper actions to take in the event of an alarm.

- Any SFCs, CFCs, sequence step detail sheets, and control narratives that were developed earlier.

- A procedures section into which the operations department may insert operating procedures as they are developed and modified as a result of the new control scheme.

b. **Maintenance**

The same overview written for the operations manual is useful in the maintenance manual. In addition, this manual should contain:

- Catalog cuts from all the control equipment used

- Screen diagrams (a narrative is probably not needed)

- Troubleshooting procedures and tips

- A contact list of equipment vendors and support organizations

- A hard copy of the control logic

7. ONSITE TRAINING

Training is almost always a deliverable of systems integrators. They know the control system, and after writing the software, they know quite a bit about the operation of the plant. So the systems integrator is the logical choice to lead plant operator and maintenance technician training.

This should really be done before startup because a tremendous side benefit of this process is feedback from the operators and maintenance personnel. Many times, the trainer becomes the trained. If this is done early enough, the lessons learned may be plowed back into the system before it "hits the street," thereby improving the product.

As suggested, the FAT is a great tool that can be used as the basis for training. It has sections that appeal to operators and maintenance personnel: the sequential and alarming section for the operators and the simulated inputs and outputs section for the maintenance folks, who will also want to discuss networking and other areas of concern to them.

G. Systems Integration Summary

The CSI aspect of the process controls field is varied and challenging. Unlike some of the other design disciplines, systems integrators have the satisfaction of seeing the results of their labor. Their work affects almost every aspect of the automation project, and, as a key component, the CSI team must function at peak performance to achieve success.

In closing, the customer needs to be confident that his facility will operate safely and in the most productive manner possible. That confidence is greatly increased if customer personnel were involved in the process of developing their own control system. Involving the customer in the decision process from the very beginning, providing an organized approach to the work, and executing meaningful testing plans that provide demonstrable quality assurance is the key to boosting that confidence. The integrator will likely complete the project with better customer rapport, and with a better sense of accomplishment, having completed a quality job. With that result, future projects are more likely.

PART III – CHAPTER 11: INFORMATION MANAGEMENT

Chapter 9 provides an overview about how a database can be used as a tool to manage information. But what kind of information needs to be managed? Regardless of the tool used, certain types of information need to be controlled. The methods of collecting information, as well as the maintenance and distribution of information, are critical to the success of the project. This chapter steers clear of the "how" of data collection, and focuses more on "what" data to collect, and who should be in the loop.

As discussed in Chapter 9, two primary types of information must be managed by the project database: document and instrument information.

A. DOCUMENT CONTROL

Proper document management is an undervalued component in the efficient execution of a design project. In most cases, the exchange of documents is the first contact between the customer and the engineering service provider, so that the exchange of information and documents can set a tone for the entire relationship. The larger the project, the more important this topic becomes. The design team needs to have a way to track documents from the time they are generated until they are issued for record.

Chapter 9 provides a database-centered approach to document control, whereby documents may be logged into a database upon receipt and tracked through the life of the project by entering dates in various fields to indicate achievement of milestones. Regardless of how the information is captured and used, it needs to be managed through the various stages of the project.

If the project is a retrofit, like our test case, the likelihood is high that the customer will have existing drawings that will need to be modified. Determining which drawings to request frequently involves walking the facility down and investigating the document set. The customer is usually heavily involved in this, as he will have the best perspective on which documents are available and which are involved. A list of documents required for the project will be created.

Once a list is created, the service provider must make a request to "check out" the drawing originals. Back in *the day*, drawings were physical "board" drawings, and the original vellum would have been pulled from a drawing vault and handed over. Today, the process is more virtual, as most documents are stored digitally.

All engineering documents have an assigned document number and associated revision number. Care should be taken to understand the difference between the *official revision* of record and *active* revisions. The official revision of record depicts the "As-Built" state of the facility. It represents the facility as it is in its present un-modified, or as-built state, prior to the advent of the current project. The active revision, also sometimes referred to as the *pending* revision, is the version of the document that is undergoing design modifications for the project. It represents what will be at some point in the future. The active revision provides a view of the facility that is preliminary and perhaps un-checked, and should not be used for troubleshooting purposes.

For example, "Drawing X" is sitting in the customer's vault at revision level 4 (Rev 4). Rev 4 represents the as-built condition of the plant. When the drawing is ordered by the engineer, the drawing's active revision becomes Rev 5. While the project is ongoing, the engineer works the drawing as a Rev 5 drawing, while the plant continues to operate under the Rev 4 version.

The active revision can be in one of two states through the life of the project: Preliminary or Approved for Construction. If the information on the document is preliminary, unchecked, or otherwise not complete, the active revision number will generally have a letter modifier. There can be several different levels of pending revisions, which is why pending revisions are differentiated from as-built revisions by use of a letter-designator. Generally, if a document is sent to the customer for review or comment while in a preliminary state, the letter designator is incremented to reduce the likelihood of confusion later. Once the document is completed and checked, and approved by the customer, the revision level on the document is incremented to the active revision with no letter modifier. The Active revision will be the revision used by the constructor to build the facility. If modifications are made to the document during construction, then the modification notes and sketches are maintained until the completion of the project, at which time the changes are made and the revision level raised again. In all cases, at the conclusion of the project, all documents will be returned to the customer in an as-built state.

B. Instrument and I/O List

Tracking instruments and managing the I/O list are major project requirements. These tasks can be accomplished in a number of ways, but the database is one of the best because of the multiple uses the data can be made to serve.

1. INSTRUMENT AND I/O LIST TABLE

The instrument and I/O list table contains all the instrumentation and end devices associated with a project, and all information pertinent to a particular instrument item. An instrument may generate more than one signal, in which case it will have more than one data record.

This table is more than an I/O list, however. Every instrument should be listed. If there is a mix of existing and new instruments, a new field should be added to capture that information. In our case, all items are new, so that field is not present.

Also, this data table contains all drawings on which an instrument may be found; it contains the instrument's cabling requirements and conduit routing number, and it provides a means for managing the I/O list.

While these tasks can be accomplished any number of ways, the advantage of using a comprehensive database is that the tasks can be generated without reentering the data. The database, as shown in Figure 3-96, has fields that cover many aspects of the design process.

As the project progresses, a lot of data can be amassed for each item (Figure 3-97). Once collected, this data can be used for multiple tasks. The following are some of the queries and reports that can be built.

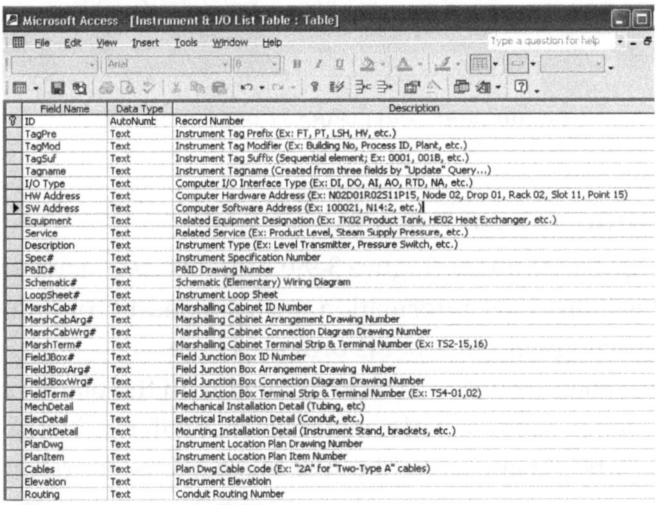

Field Name	Data Type	Description
ID	AutoNumk	Record Number
TagPre	Text	Instrument Tag Prefix (Ex: FT, PT, LSH, HV, etc.)
TagMod	Text	Instrument Tag Modifier (Ex: Building No, Process ID, Plant, etc.)
TagSuf	Text	Instrument Tag Suffix (Sequential element; Ex: 0001, 001B, etc.)
Tagname	Text	Instrument Tagname (Created from three fields by "Update" Query...)
I/O Type	Text	Computer I/O Interface Type (Ex: DI, DO, AI, AO, RTD, NA, etc.)
HW Address	Text	Computer Hardware Address (Ex: N02D01R02S11P15, Node 02, Drop 01, Rack 02, Slot 11, Point 15)
SW Address	Text	Computer Software Address (Ex: 100021, N14:2, etc.)
Equipment	Text	Related Equipment Designation (Ex: TK02 Product Tank, HE02 Heat Exchanger, etc.)
Service	Text	Related Service (Ex: Product Level, Steam Supply Pressure, etc.)
Description	Text	Instrument Type (Ex: Level Transmitter, Pressure Switch, etc.)
Spec#	Text	Instrument Specification Number
P&ID#	Text	P&ID Drawing Number
Schematic#	Text	Schematic (Elementary) Wiring Diagram
LoopSheet#	Text	Instrument Loop Sheet
MarshCab#	Text	Marshalling Cabinet ID Number
MarshCabArg#	Text	Marshalling Cabinet Arrangement Drawing Number
MarshCabWrg#	Text	Marshalling Cabinet Connection Diagram Drawing Number
MarshTerm#	Text	Marshalling Cabinet Terminal Strip & Terminal Number (Ex: TS2-15,16)
FieldJBox#	Text	Field Junction Box ID Number
FieldJBoxArg#	Text	Field Junction Box Arrangement Drawing Number
FieldJBoxWrg#	Text	Field Junction Box Connection Diagram Drawing Number
FieldTerm#	Text	Field Junction Box Terminal Strip & Terminal Number (Ex: TS4-01,02)
MechDetail	Text	Mechanical Installation Detail (Tubing, etc)
ElecDetail	Text	Electrical Installation Detail (Conduit, etc.)
MountDetail	Text	Mounting Installation Detail (Instrument Stand, brackets, etc.)
PlanDwg	Text	Instrument Location Plan Drawing Number
PlanItem	Text	Instrument Location Plan Item Number
Cables	Text	Plan Dwg Cable Code (Ex: "2A" for "Two-Type A" cables)
Elevation	Text	Instrument Elevatioin
Routing	Text	Conduit Routing Number

FIGURE 3-96. INSTRUMENT AND I/O LIST TABLE, DESIGN VIEW

Tagname	I/O Type	HW Address	SW Address	Equipment	Service	Description	Spec#
HV-TK13-13	NA	NA	NA	TK-10 Product Feed Tank	Product Supply	Feed Valve	
HY-TK10-13	DOI	N01D03R01S04P01	DO0101	TK-10 Product Feed Tank	Product Supply	Solonoid, 3-Way	NA
LT-TK10-10	AI	N01D03R01S08P02	AI0101	TK-10 Product Feed Tank	Product Level	Level Transmitter	
LSH-TK10-10	HW	NA	NA	TK-10 Product Feed Tank	Product Level	Level Switch	NA
PT-TK10-48	AI	N01D03R01S08P01	AI0100	PP-10 Product Pump	Discharge Press	Pressure Transmitter	
PY-TK10-48	AO	N01D03R01S10P01	AO0100	TK-10 Product Feed Tank	Recirc Pressure	I/P Transducer	NA
PV-TK10-48	NA	NA	NA	TK-10 Product Feed Tank	Recirc Pressure	Throttling Valve	
PSV-TK10-58	NA	NA	NA	TK-10 Product Feed Tank	Pressure	Relief Valve	
LSLL-TK10-47	DI	N01D03R01S06P05	DI0104	TK-10 Product Feed Tank	Low Level	Level Switch	
HS-TK10-15B	DOI	N01D03R01S04P02	DO0102	PP-10 Product Pump	Motor Controls	Start Cmmd	NA
YS-TK10-15B	DI	N01D03R01S06P03	DI0102	PP-10 Product Pump	Motor Controls	Run Status	NA
YS-TK10-15A	DI	N01D03R01S06P02	DI0101	PP-10 Product Pump	Motor Controls	Switch in Auto	NA
ZSC-TK10-13	DI	N01D03R01S06P01	DI0100	TK-10 Product Feed Tank	Product Supply	Valve Closed Status	NA

Tagname	P&ID#	Schematic#	LoopSheet#	MarshCab#	MarshCabArg	MarshCabWrg	FieldJBox#	FieldJBoxArg	FieldJBoxWrg	FieldTerm#	MechDetail	ElecDetail	MountDetail	PlanDwg	PlanItem	Cables	Elevation	Routing
HV-TK13-13	PID001	NA	NA	NA		NA	NA	NA	NA		NA	NA	In-Line	PLAN001	04	NA	2H	NA
HY-TK10-13	PID001	SCH-003	NA	TC-2	ARR-001	WRG-001	FJB-TK10-01	ARR-002	WRG-002		MECH-001	NA	Integral	PLAN001	04	2A	2H	A1-PLAN001
LT-TK10-10	PID001	NA	LOOP-LT-10	TC-2	ARR-001	WRG-001	NA	NA	NA		NA	ELEC-001	MOUNT-001	PLAN001	01	1B,3A	2H	D1-PLAN001,A1
LSH-TK10-10	PID001	NA	NA	NA		NA	FJB-TK10-01	ARR-002	WRG-002		NA	ELEC-001	NA	PLAN001	01	2A	2H	A1-PLAN001
PT-TK10-48	PID001	NA	LOOP-PIC48	TC-2	ARR-001	WRG-001	NA	NA	NA		NA	ELEC-001	MOUNT-001	PLAN001	03	1B	2H	D1-PLAN001
PY-TK10-48	PID001	NA	LOOP-PIC48	TC-2	ARR-001	WRG-001	NA	NA	NA		MECH-002	ELEC-001	Integral	PLAN001	02	1B	2H	D1-PLAN001
PV-TK10-48	PID001	NA	NA	NA		NA	NA	NA	NA		NA	NA	In-Line	PLAN001	02	NA	2H	NA
PSV-TK10-58	PID001	NA	NA	NA		NA	NA	NA	NA		NA	NA	MOUNT-001	PLAN001	05	NA	2H	NA
LSLL-TK10-47	PID001	SCH-003	NA	TC-2	ARR-001	WRG-001	FJB-TK10-01	ARR-002	WRG-002		NA	ELEC-001	MOUNT-001	PLAN001	07	5A	2M	A1-PLAN001
HS-TK10-15B	PID001	SCH-003	NA	TC-2	ARR-001	WRG-001	MCC							MCC001				
YS-TK10-15B	PID001	SCH-003	NA	TC-2	ARR-001	WRG-001	MCC							MCC001				
YS-TK10-15A	PID001	SCH-003	NA	TC-2	ARR-001	WRG-001	FJB-TK10-01	ARR-002	WRG-002		NA	NA	MOUNT-002	PLAN001	06	2A	2L	A6-PLAN001
ZSC-TK10-13	PID001	SCH-003	NA	TC-2	ARR-001	WRG-001	FJB-TK10-01	ARR-002	WRG-002		NA	ELEC-001	Integral	PLAN001	04	3A	2H	A1-PLAN001

FIGURE 3-97. INSTRUMENT AND I/O LIST DATABASE, DATASHEET VIEW

2. PRELIMINARY DESIGN QUERY

The preliminary design query (Figure 3-98) data fields contain data collected from preliminary engineering documents or developed during Phase 1 of the project.

Microsoft Access - [Preliminary Design Query : Select Query]

P&ID#	TagPre	TagSuf	Tagname	I/O Type	HW Address	TagMod	SW Address	Equipment	Service	Description	Spec#
PID001	PT	48	PT-TK10-48	AI	N01D03R01S08P01	TK10	AI0100	PP-10 Product Pump	Discharge Press	Pressure Transmitter	
PID001	LT	10	LT-TK10-10	AI	N01D03R01S08P02	TK10	AI0101	TK-10 Product Feed Tank	Product Level	Level Transmitter	
PID001	PY	48	PY-TK10-48	AO	N01D03R01S10P01	TK10	AO0100	TK-10 Product Feed Tank	Recirc Pressure	I/P Transducer	NA
PID001	ZSC	13	ZSC-TK10-13	DI	N01D03R01S06P01	TK10	DI0100	TK-10 Product Feed Tank	Product Supply	Valve Closed Status	NA
PID001	YS	15B	YS-TK10-15A	DI	N01D03R01S06P02	TK10	DI0101	PP-10 Product Pump	Motor Controls	Switch in Auto	NA
PID001	YS	15A	YS-TK10-15B	DI	N01D03R01S06P03	TK10	DI0102	PP-10 Product Pump	Motor Controls	Run Status	NA
PID001	LSLL	47	LSLL-TK10-47	DI	N01D03R01S06P05	TK10	DI0104	TK-10 Product Feed Tank	Low Level	Level Switch	
PID001	HS	15B	HS-TK10-15B	DOI	N01D03R01S04P02	TK10	DO0102	PP-10 Product Pump	Motor Controls	Start Cmmd	NA
PID001	HY	13	HY-TK10-13	DOI	N01D03R01S04P01	TK10	DO0101	TK-10 Product Feed Tank	Product Supply	Solonoid, 3-Way	NA
PID001	LSH	10	LSH-TK10-10	HW	NA	TK10	NA	TK-10 Product Feed Tank	Product Level	Level Switch	NA
PID001	PSV	58	PSV-TK10-58	NA	NA	TK10	NA	TK-10 Product Feed Tank	Pressure	Relief Valve	
PID001	PV	48	PV-TK10-48	NA	NA	TK10	NA	TK-10 Product Feed Tank	Recirc Pressure	Throttling Valve	
PID001	HV	13	HV-TK13-13	NA	NA	TK13	NA	TK-10 Product Feed Tank	Product Supply	Feed Valve	

FIGURE 3-98. PRELIMINARY DESIGN QUERY

3. PLAN DRAWING TAKEOFF QUERY

The plan drawing takeoff query (Figure 3-99) contains data fields that pertain to the instrument location plans. Information, such as conduit routing number, cable type, and plan drawing item number is provided here.

A component schedule needs to be associated with the plan drawing in order to provide information relative to the materials that will need to be procured. That schedule may be generated right out of this database by running Plan Dwg Takeoff Query, hiding a few of the fields that are not needed for the schedule and copying the data and pasting it into an Excel spreadsheet (Figure 3-99).

This information may then be placed directly on the drawing either as an embedded Excel object or as CADD data by copying and pasting the individual data items. In either case, the data entry task is avoided, thus eliminating the risk of injecting typographical errors.

Part III – Chapter 11: Information Management

Figure 3-99. Plan drawing takeoff query

4. Plan Dwg Takeoff Query Report

A report can be generated to show all data, organized by instrument location plan drawing and sorted by plan item number (Figure 3-100).

Figure 3-100. PlanDwgTakeoffQuery report

Tagname	Equipment	Service	Description	P&ID#	PlanDwg	PlanItem	Elevation	Cables	Routing
YS-TK10-15B	PP-10 Product Pump	Motor Controls	Run Status	PID001	MCC001				
HS-TK10-15B	PP-10 Product Pump	Motor Controls	Start Cmmd	PID001	MCC001				
LSH-TK10-10	TK-10 Product Feed Tank	Product Level	Level Switch	PID001	PLAN001	01	2H	2A	A1-PLAN001
LT-TK10-10	TK-10 Product Feed Tank	Product Level	Level Transmitter	PID001	PLAN001	01	2H	1B,3A	D1-PLAN001, A1-PLAN001
PV-TK10-48	TK-10 Product Feed Tank	Recirc Pressure	Throttling Valve	PID001	PLAN001	02	2H	NA	NA
PY-TK10-48	TK-10 Product Feed Tank	Recirc Pressure	I/P Transducer	PID001	PLAN001	02	2H	1B	D1-PLAN001
PT-TK10-48	PP-10 Product Pump	Discharge Press	Pressure Transmitter	PID001	PLAN001	03	2H	1B	D1-PLAN001
ZSC-TK10-13	TK-10 Product Feed Tank	Product Supply	Valve Closed Status	PID001	PLAN001	04	2H	3A	A1-PLAN001
HY-TK10-13	TK-10 Product Feed Tank	Product Supply	Solonoid, 3-Way	PID001	PLAN001	04	2H	2A	A1-PLAN001
HV-TK13-13	TK-10 Product Feed Tank	Product Supply	Feed Valve	PID001	PLAN001	04	2H	NA	NA
PSV-TK10-58	TK-10 Product Feed Tank	Pressure	Relief Valve	PID001	PLAN001	05	2H	NA	NA
YS-TK10-15A	PP-10 Product Pump	Motor Controls	Switch in Auto	PID001	PLAN001	06	2L	2A	A6-PLAN001
LSLL-TK10-47	TK-10 Product Feed Tank	Low Level	Level Switch	PID001	PLAN001	07	2M	5A	A1-PLAN001

FIGURE 3-101. PLAN DRAWING COMPONENT SCHEDULE (MICROSOFT® ACCESS TO MICROSOFT® EXCEL)

5. X-REF DOCUMENT CROSS-REFERENCE QUERY

The X-ref query (Figure 3-102) provides a list that cross-references the instrument list table with all of the major documents on which it appears. As designers progress through the design process, they should log the event each time they place an instrument or its wiring on a drawing. If this is done, the customer will have a useful maintenance tool. As a side benefit, the design check should be much easier.

Tagname	Spec	P&ID#	Schematic	LoopSheet#	MarshCab	FieldJBoxWrg#	MechDetail	ElecDetail	MountDetail	PlanDwg
HV-TK13-13		PID001	NA	NA	NA	NA	NA	NA	In-Line	PLAN001
HY-TK10-13	NA	PID001	SCH-003	NA	WRG-001	WRG-002	MECH-001	NA	Integral	PLAN001
LT-TK10-10		PID001	NA	LOOP-LT-10	WRG-001	NA	NA	ELEC-001	MOUNT-001	PLAN001
LSH-TK10-10	NA	PID001	NA	NA	NA	WRG-002	NA	ELEC-001	NA	PLAN001
PT-TK10-48		PID001	NA	LOOP-PIC48	WRG-001	NA	NA	ELEC-001	MOUNT-001	PLAN001
PY-TK10-48	NA	PID001	NA	LOOP-PIC48	WRG-001	NA	MECH-002	ELEC-001	Integral	PLAN001
PV-TK10-48		PID001	NA	NA	NA	NA	NA	NA	In-Line	PLAN001
PSV-TK10-58		PID001	NA	NA	NA	NA	NA	NA	MOUNT-001	PLAN001
LSLL-TK10-47		PID001	SCH-003	NA	WRG-001	WRG-002	NA	ELEC-001	MOUNT-001	PLAN001
HS-TK10-15B	NA	PID001	SCH-003	NA	WRG-001					MCC001
YS-TK10-15B	NA	PID001	SCH-003	NA	WRG-001					MCC001
YS-TK10-15A	NA	PID001	SCH-003	NA	WRG-001	WRG-002	NA	NA	MOUNT-002	PLAN001
ZSC-TK10-13	NA	PID001	SCH-003	NA	WRG-001	WRG-002	NA	ELEC-001	Integral	PLAN001

FIGURE 3-102. DOCUMENT CROSS-REFERENCE (X-REF) QUERY

6. X-REF DOCUMENT CROSS-REFERENCE REPORT

The X-ref query report is linked to the X-ref query. It formats the information collected by the query and provides a listing that crosses the instrument record to all the major related documents (Figure 3-103).

Cross-Reference Report

Tagname	Spec#	PID#	Schematic#	LoopSheet#	MarshCabW	FieldJBoxWr	MechDet	ElecDeta	MountDet	PlanDwg
HS-TK10-15B	NA	PID001	SCH-003	NA	WRG-001					MCC001
HV-TK13-13		PID001	NA	NA	NA	NA	NA	NA	In-Line	PLAN001
HY-TK10-13	NA	PID001	SCH-003	NA	WRG-001	WRG-002	MECH-001	NA	Integral	PLAN001
LSH-TK10-10	NA	PID001	NA	NA	NA	WRG-002	NA	ELEC-001	NA	PLAN001
LSLL-TK10-47		PID001	SCH-003	NA	WRG-001	WRG-002	NA	ELEC-001	MOUNT-001	PLAN001
LT-TK10-10		PID001	NA	LOOP-LT-10	WRG-001	NA	NA	ELEC-001	MOUNT-001	PLAN001
PSV-TK10-38		PID001	NA	NA	NA	NA	NA	NA	MOUNT-001	PLAN001
PT-TK10-48		PID001	NA	LOOP-PT-48	WRG-001	NA	NA	ELEC-001	MOUNT-001	PLAN001
PV-TK10-48		PID001	NA	NA	NA	NA	NA	NA	In-Line	PLAN001
PY-TK10-48	NA	PID001	NA	LOOP-PT-48	WRG-001	NA	MECH-002	ELEC-001	Integral	PLAN001
YS-TK10-15A	NA	PID001	SCH-003	NA	WRG-001	WRG-002	NA	NA	MOUNT-002	PLAN001
YS-TK10-15B	NA	PID001	SCH-003	NA	WRG-001					MCC001
ZSC-TK10-13	NA	PID001	SCH-003	NA	WRG-001	WRG-002	NA	ELEC-001	Integral	PLAN001

Friday, June 28, 2002

FIGURE 3-103. DOCUMENT CROSS-REFERENCE REPORT

C. DATABASE SUMMARY

The Project Control Database, as presented in this chapter, is a single comprehensive database that will manage document control as well as engineering and design functions. This database is a tool for the designer, much like a hammer is for a carpenter. Properly wielded, it can play a huge role in turning out a quality product. It is a design tool to be used during design development and a quality management tool to be used during the design check. It is a construction management resource from which a checkout checklist can be built. It then becomes a maintenance tool. For versatility and usefulness, this database is unsurpassed.

Part III – Chapter 12: Instrument Specifications

An *instrument* can be either a sensing element (switch, transmitter, etc.), or a final control element (valve, actuator, solenoid, horn, etc.). Instruments do not have to be wired. For example, a site glass on a tank, a gauge, or a self-actuating valve, such as a relief valve, are also classified as instruments. Specification data sheets (also called instrument spec sheets, instrument data sheets, mil sheets) are used to describe an instrument to a degree that leaves no doubt as to what will be supplied. The format of the instrument specification can change, depending on the instrument type. This is because properly specifying a particular instrument depends on process parameters, such as temperature, pressure, flow rate, specific gravity, and viscosity that are specific to that type of instrument. Other important information includes manufacturer, model number, materials of construction, face-to-face dimensions, process connections, power requirements, calibrated range, output, conduit connection size, area classification, enclosure type, mounting type, and set points.

ISA provides typical data sheet forms in ISA-20-1981, *Specification Forms for Process Measurement and Control Instruments, Primary Elements, and Control Valves*.

An instrument spec sheet should be generated for every instrument in a project. A generic data sheet is usually created for identical applications, given a unique number, and assigned to multiple instruments. For example, if 10 identical pressure indicators measuring the same process information are to be purchased, one data sheet could be generated with 10 unique tag numbers for the indicators listed on it. This technique applies only to the most basic type of instrument, such as indicators or end devices, such as on/off valves. Instruments that handle multiple parameters cannot be easily combined. However, if multiple process trains (identical process lines) are included in the project, generic instrument spec sheets are advisable for all instrument types.

Each spec sheet gets a unique spec number that should be listed in the database against that instrument. A good design process generates a bi-directional flow of information. Since instrument tag numbers are listed on the instrument spec sheet, the specification numbers should be entered for the instruments in the instrument database. This allows rapid cross-referencing between instrument and spec and back, which will be greatly appreciated by the customer. Aside from providing assistance to the customer, this practice proves valuable during the design phase as well. The database can then be used to track the status of the instrument spec sheets in the normal document control process. For instance, when an engineer begins a specification for an instrument, the first portion of the instrument spec number (usually the job number) can be listed in the database under the instrument spec field for that instrument. After the specification has been completed, the entire instrument spec number can be added to the database. When the instrument spec sheet is issued for review, the date can be added to the database in the issue-for-review field.

When an instrument is purchased, the purchase order number can be listed in the P.O. number field for that device. At any given time, queries or reports can be created to indicate how many devices have specs in process, how many specs are complete, how many specs have been issued for review, and how many instruments have been purchased. A little bit of data entered during the specification process will save an immense amount of time when trying to determine the status of the procurement process.

It is of great value to the instrument designer to have the instrument spec sheets available at the beginning of a project. However, like most deliverables, they are a "design in process" until the close of the project. Therefore, many parameters for particular instruments must be determined on an as-needed basis. Sometimes others need the information to complete their work before the specs have been completed. Guidelines can be agreed upon early in the job. For instance, it might be decided that all on/off valve data sheets will provide piping in face-to-face dimensions, or that all analog devices will be two-wire loop powered wherever possible. The specifying engineer would then notify the E/I&C designer of any exceptions.

A. Purpose

The purpose of the instrument specification data sheet is to catalog the various parameters that define the instrument as an entity in the process. Other engineering disciplines rely on this data during the design and construction phases of a project. Even after these phases are complete, the instrument specification sheet continues to be valuable. It is a living document, which means it is updated, as required, to show any changes the device undergoes throughout its lifetime. Instrument and end device specs should be readily available for every such item installed in a plant.

The following are some of the groups that use this versatile and valuable product of the I&C engineer:

1. Mechanical Designers

Mechanical designers, or pipers, are interested in the face-to-face dimensions of all inline instruments. They need to show these instruments, with the correct dimensions, in the piping in the 3-D model, which is used to extract isometric drawings, or "spools." These spools are either sent to a shop to fabricate the piping, or field installers will use them to fabricate the piping on site. Either way, the spool fabricator must know the face-to-face dimensions of every inline instrument.

The pipers also need to know if they will have to supply reducers for control valves or other inline instruments. More often than not, a control valve is one line size (pipe diameter) smaller than the pipe. If reducers are required, the face-to-face dimensions of the reducers must be accounted for in the model.

Another point of interest for pipers is the envelope (in space) that any instrument occupies. For example, an on/off valve is typically made up of the valve and the "topworks," which includes an actuator, a solenoid valve, limit switches, and sometimes an air accumulator tank. The pipers must ensure they leave enough room for the entire assembly. Otherwise, a pipe may be routed through the area in space where the topworks resides, resulting in the necessity for field rerouting of pipe.

The pipers also need to know what process connection is specified for each instrument. For example, an RTD's process connection could be specified as a 1½ inch or 2 inch 150#[7] flange. The instrument department and pipers must work closely together to ensure that all instruments and piping are perfectly mated. Sometimes, it is necessary for additional materials, such as reducer bushings, to be provided.

2. Instrument Designers

Instrument designers need instrument specs to determine how to properly wire each instrument. The instrument engineer should supply the manufacturer's cut sheets and wiring diagrams to the instrument designer for every wired instrument. There is no reason for more than one person to compile information for an instrument. The following are some of the parameters specified on an instrument spec that are needed for the designer to properly wire an instrument:

- Is an analog instrument a two-wire or a four-wire device? Four-wire devices require an external power source, whereas two-wire devices can be powered through the PLC or DCS.

- What is the contact arrangement of a switch? Double-pole double-throw or single-pole double-throw, etc.?

---

7. 150# is shorthand for 150 pounds per square inch.

- Does the instrument require the use of intrinsically safe wiring practices? If so, the manufacturer must provide the certified Factory Mutual (FM) approval wiring drawing that shows the complete wiring diagram.

- What are the power requirements: 120 VAC or 24 VDC?

- What is the maximum current for which the instrument is rated? The fuse will need to be sized for a value that will protect the instrument.

- What are the terminal identification numbers on the instruments? These identification numbers are not shown on the instrument specs but can be found in the manufacturer's literature, given the model number, as shown on the instrument spec.

The instrument designer also depends on the instrument spec to define the instrument's mechanical installation details and, subsequently, the installation bills of materials. The designer specifies any additional material necessary for an instrument to be mated to a pipe or vessel. For instance, local pressure indicators are typically specified with ½ inch National Pipe Thread (NPT) connections. If the piping spec calls for 1½ inch, 150# flanged isolation valves, the designer calls for a 1½ inch, 150# blind flange to be drilled and tapped for ½ inch NPT. The blind flange is bolted to the isolation valve, and the indicator is screwed into the ½ inch hole that was drilled and tapped. Doing this is a lot more economical than trying to purchase the indicator with a flange.

When the instrument designer is preparing the electrical installation details, he consults the instrument spec to determine the conduit connection size for a particular instrument. Another point when preparing electrical details arises when working on an on/off valve package. Some limit switches can be specified with additional terminals inside the enclosure for the solenoid valve. This way, only one run of conduit and one seal have to be provided between the junction box and the on/off valve.

The instrument designer must know how to mount each instrument when preparing the mounting details. The instrument spec defines which mounting technique should be used. Any mounting apparatus supplied with the instrument can be determined via the model number. The instrument designer specifies for purchase any additional material required on a mounting detail.

To create power distribution drawings and properly size the power supplies and fuses, the instrument designer must also know how much power is required for each instrument. Also, some instruments require a special cable that is sometimes supplied by the manufacturer. The instrument spec sheet indicates how much of what type of cable will be supplied by the manufacturer, if any.

3. Other Users

- *Structural designer:* The only purpose for which the structural designer needs the instrument spec is designing the steel structures around vessels that require load cells for weighing. The bolt hole patterns for the load cells and the tank legs or lugs (if the tank hangs through the floor) must match.

- *Construction:* When an instrument is received at the plant site, a construction representative must inspect it and verify that it is the instrument described on the instrument specification sheet. Construction personnel also refer to the instrument spec when they install the instrument and the wiring for a wired instrument.

- *Maintenance:* After installation, calibrated instruments must be recalibrated on a periodic basis. Maintenance personnel use the instrument spec to obtain calibration parameters. For instance, a 4–20 mA output flow transmitter with a calibrated range of 0–50 gpm (gallons per minute) must be calibrated so that when the flow is 0 gpm, the transmitter outputs 4 mA, and when the flow is 50 gpm, the transmitter outputs 20 mA.

- *Computer programmers* need the instrument specs to obtain the calibration ranges and proper units for their HMI screens and for the configuration and programming for the DCS or PLC.

- *Purchasing:* When the instrument spec has been reviewed and approved by the customer, it is sent to the purchasing agent to be used to procure the instrument. When the purchasing agent receives the instrument spec, a purchase order number may be assigned.

- *Vendors* usually represent many different manufacturers, who are referred to as the vendor's "principals." The purchasing agent forwards the instrument spec to the vendor who represents that particular manufacturer. The vendor sends the spec to the manufacturer for quoting prices and lead time. That information is relayed back to the instrument engineer. If the price and delivery date are within the budget and schedule, the manufacturer supplies the instrument as specified on the instrument spec.

B. Interfaces

Predecessors to the instrument specs are HMBs, or adequate process information and P&IDs. Pump specs are needed for control valves, flowmeters, pressure instruments, and restriction orifices in pumped lines. Successors to the instrument specs are wiring drawings, mechanical installation details, mounting details, electrical details, and piping isometrics.

C. EXAMPLES

The following project, as described previously, is to install a product tank, a pump, and associated instrumentation and controls. This chapter looks at two of the instruments: level transmitter LT-10 and pressure control valve PV-48. Refer to Chapter 7 for more information on the P&ID (Figure 3-5, reproduced below).

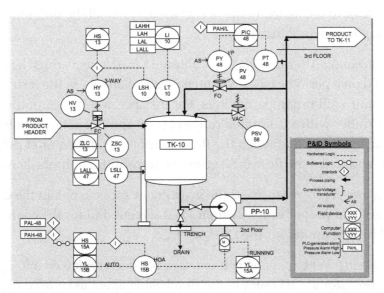

FIGURE 3-5 (REPRODUCED). P&ID PRESENTATION OF THE TK-10 SUBSYSTEM

1. LT/LSH-10

An ultrasonic level transmitter was chosen for level detection in TK-10. An ultrasonic transmitter works for many liquid applications. Sometimes batch processes are used for more than one end product, requiring different intermediate products in specific tanks. By choosing an ultrasonic transmitter for our tank, the intermediate product can be changed without having to recalibrate the level transmitter.

The operation of an ultrasonic transmitter is based on ultrasonic sound waves (Figure 3-104). A non-contacting ultrasonic transducer is mounted in the top of the tank. Electrical signals are converted to ultrasonic pulses, which are transmitted to the surface of the liquid and reflect back to the transducer. The time required for the pulses to travel to the surface and echo back to the transducer is converted into a 4–20 mA signal that is proportional to the liquid level in the tank.

Since the speed of travel of ultrasonic waves increases as temperature rises, most transmitters are equipped with temperature compensation.

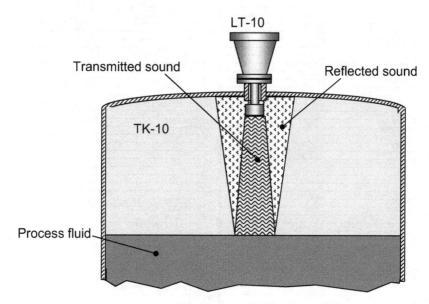

FIGURE 3-104. ULTRASONIC LEVEL TRANSMITTER

An ultrasonic level transmitter's level reading is only as good as the echo it receives. The echo can be weak due to dispersion (which reduces sound intensity by the square of distance) and absorption (which in dry air reduces its energy level by 1 to 3 decibels/meter).[5] Since our tank is only 10 ft tall, there are no concerns about dispersion. The vapor space above our product should not pose an absorption problem.

Relays with Form-C contacts, to trip for low and high levels, are provided with the ultrasonic level transmitter. This saves the cost of a separate level switch. See Figure 3-105 for the instrument spec (device specification) sheet for LT/LSH/LSL-10. Notice the "/" between the tag prefixes. This indicates there is only one instrument, but it performs multiple functions.

Several items shown on the instrument spec sheet are of particular interest to the instrument designer:

- *Type* shows this instrument is a transmitter and a switch.
- *Area Classification* shows the transducer is located in a hazardous area, and the electronics are located in a non-hazardous area.

#				#		
1	RESPONSIBLE ORGANIZATION	ULTRASONIC CONTACT-TYPE		6	SPECIFICATION IDENTIFICATIONS	
2	MESA ASSOC., INC.	LEVEL SWITCH		7	Document no	2704-08
3	10604 MURDOCK	Device Specification		8	Latest revision	1 Date 5/29/2002
4	KNOXVILLE, TN			9	Issue status	IFC
5	www.mesainc.com			10		
11	PROCESS CONNECTION			60	SWITCH MECHANISM continued	
12	Body/Fitting type			61	Enclosure material	NEMA-4X
13	Process conn nominal size	3" Rating		62	Exterior coating material	MANUF. STANDARD
14	Process conn term type	Style		63	Area Classification	NON-HAZARDOUS
15	Wetted material	GLYCOL		64	Mounting	PANEL-MOUNT
16	Flange/Fitting material	3" 150# FLUSH		65	Type	Qty = 4 FORM "C" CONTACTS
17	Seal/O ring material			66	PERFORMANCE CHARACTERISTICS	
18				67	Max press at design temp	50 PSIG At 22 DEGF
19				68	Min working temperature	Max
20	SENSING ELEMENT			69	Repeatability/Sensitivity	
21	Configuration type	KYNAR		70	Min ambient working temp	-40 DEGF Max 203 DEGF
22	Nominal temp rating	30 DEGF		71	Contacts ac rating	120VAC,5A At Max
23	Number of sensors	1		72	Contacts dc rating	At Max
24	Probe diameter	2.5"		73	Max sensor to receiver lg.	
25	Insertion length			74	Level Range (Transducer)	1-33 FT
26	Integral cable length	3FT		75	Level Range (Electronics)	1-200 FT
27	Self test type			76	Accuracy	0.25% OF TARGET RANGE
28	Cert/Approval type			77	Calibrated Span	0-10 FT
29	Area Classification	HAZ., Class 1, Grp C, Div 1		78		
30	Orientation	VERTICAL		79		
31	Beam Angle	12 DEGREES		80		
32	CONNECTION HEAD OR PREAMPLIFIER			81		
33	Type	ANALOG		82		
34	Output signal type	4-20mA		83		
35	Enclosure type no/class	HAZ., Class 1, Grp C, Div 1		84		
36	Signal power source	120VAC,5A		85	ACCESSORIES	
37	Gain type			86	Connecting cables length	RG62U COAX, 150 FT
38	Cert/Approval type			87	Mounting hardware	
39	Mounting location/type	MANUF. STANDARD		88	External relay	
40	Failure/Diagnostic action	UPSCALE		89	Intrinsic safety barrier	
41	Enclosure material	MANUF. STANDARD		90	Conduit Connection	1" MFNPT
42				91		
43				92		
44	SWITCH MECHANISM			93	SPECIAL REQUIREMENTS	
45	Housing type	MANUF. STANDARD		94	Custom tag	
46	Element style			95	Reference specification	
47	Input signal type	115VAC @ 60HZ		96	Compliance standard	
48	Output signal type	115VAC @ 60HZ		97	Integral Temp. Compensa.	-58-203 DEGF
49	Enclosure type no/class	NEMA-4X		98		
50	Control logic			99		
51	Gain type			100	PHYSICAL DATA	
52	Signal power source			101	Estimated weight	
53	Contacts arrangement	FORM C Quantity 4		102	Overall height	
54	Failsafe style	OPEN ON ALARM		103	Removal clearance	
55	Integral Indicator style	DIGITAL. LCD		104	Signal conn nominal size	Style
56	Cert/Approval type	UL		105	Mfr reference dwg.	
57	Mounting location/type	REMOTE. GENL PURPOSE		106		
58	Failure/Diagnostic action	OPEN ON FAIL		107		
59	Switch time delay	0.5-5SEC		108		

#	CALIBRATIONS AND TEST		SETPOINT		OUTPUT	
110	TAG NO/FUNCTIONAL IDENT	MEAS/SIGNAL/TEST	LRV	URV ACTION	LRV	URV
111	LSH-10	Lg. conn to sp 1-Output				
112	LSL-10	Lg. conn to sp 2-Output				
113		Lg. conn to sp 3-Output				
114		Lg. conn to sp 4-Output				
115	LT-10	Lg. conn to Sig-Output				
116						
117						

#	COMPONENT IDENTIFICATIONS		
118	COMPONENT TYPE	MANUFACTURER	MODEL NUMBER
119	LT/LSH/LSL-10	MILLTRONICS	XPS-10/AIR RANGER SPL

Rev	Date	Revision Description	By	Appv1	Appv2	Appv3	REMARKS

FIGURE 3-105. INSTRUMENT SPECIFICATION FOR LT/LSH/LSL-10

- *Equipment Category* shows the electronics are in a NEMA-4X enclosure, and the transducer is hermetically sealed and rated for a Class I, Division 1, Group C&D area.

- *Power Supply* shows 120 VAC is the power requirement.

- *Output Signal* shows the output is 4–20 mA.

- *Switch Type and Rating* shows there are four SPDT (single-pole, double-throw) relays rated for 5 Amps.

- *Electronics* shows the electronics are remote in relation to the transducer.

- *Elect. Mounting* shows the electronics are panel mounted.

- *Conduit Connections* shows the transducer has a 1 inch MNPT conduit connection.

- *Note* shows the vendor will supply 150 ft of RG-62U coaxial cable that must be run in a separate conduit.

2. PV-48

A control valve may be used to control process flow, pressure and temperature. Unlike an on/off valve, which is discrete (fully open or fully closed), a control valve modulates, varying its percentage of opening to any position between open and closed. The percentage of opening is based on an independent analog signal, usually a 4–20 mA signal, generated from a DCS or PLC. An I/P transducer converts the 4–20 mA signal into a 3–15 psig pneumatic signal, which is applied to the control valve actuator, which, in turn, varies the valve position. A positioner can be used for position feedback. A globe valve, which is one of the most commonly used control valves, was chosen for our application (Figure 3-106). The main advantages of the traditional globe design include the simplicity of the spring/diaphragm pneumatic actuator, the availability of a wide range of valve characteristics, the relatively low likelihood of cavitation and noise, the availability of a wide variety of specialized designs for corrosive, abrasive, and high-temperature or high-pressure applications, the linear relationship between the control signal and valve-stem movement, and the relatively small amounts of dead band and hysteresis.[6]

Many different parameters are required to properly size a control valve, but they go beyond the scope of this book. The process parameters include the process pressure upstream and downstream of the valve and the process fluid's flow rate, specific gravity at flowing temperature, critical pressure and vapor pressure.

C_V is the primary calculated (unitless) value used to determine appropriate valve size and trim size. One C_V equals the flow of one U.S. gallon per minute of water at 60°F and under a pressure drop of 1 psi.[7] For a liquid service with a subcritical flow, the volumetric C_V equation is

$$C_V = Q \times (S.G./\Delta P)^{1/2}$$

where Q = liquid flow rate, in U.S. gallons per minute
 S.G. = specific gravity at flowing temperature
 ΔP = upstream pressure (P_1) – downstream pressure (P_2).

CONTROL VALVE DATA SHEET

PROJECT: Successful Instrumentation & Controls Design	DATA SHEET 2 OF 2
UNIT:	SPEC: 2704-83
P.O.:	TAG: PV-48
ITEM:	DWG: P&ID-001
CONTRACT:	SERVICE: TF-10 Feed Tank
MFR. SERIAL*:	

SERVICE CONDITIONS

1. Fluid: GLYCOL — Crit Press PC: ___

#		Units	Max Flow	Norm Flow	Min Flow	Shut-Off
2	Flow Rate	GPM	20	10	12	—
3	Inlet Pressure	PSIG	45			
4	Outlet Pressure	PSIG	32			
5	Inlet Temperature	DEGF	22			
6	Spec Wt. / Spec Grav / Mol Wt.	SG		1.138		—
7	Viscosity / Spec Heats Ratio	Cp /	1.13 /			—
8	Vapor Pressure P_v	Pv / Pc (psia)	1000 / 116.5			—
9	Required C_v*	Cv	6.054	3.579	2.975	—
10	Travel*	%	100		0	0
11	Allowable / Predicted SPL*	dBA				—

LINE
- 13. Pipe Line Size In: ___
- 14. and Schedule Out: ___
- 15. Pipe Line Insulation: ___

VALVE BODY / BONNET
- 16. Type*: GLOBE
- 17. Size*: 3/4" ANSI Class 150#
- 18. Max Press/Temp: 150 PSIG / 350 DEGF
- 19. Mfr & Model: VALTEK / MARK ONE
- 20. Body / Bonnet Matl*: CARBON STEEL
- 21. Liner Material / ID*:
- 22. End In:
- 23. Connection Out:
- 24. Flg Face Finish: RAISED FACE, FACE-TO-FACE: 7.25"
- 25. End Ext / Matl:
- 26. Flow Direction*: FLOW UNDER PLUG
- 27. Type of Bonnet*: STANDARD
- 28. Lub & Iso Valve: ___ Lube: ___
- 29. Packing Material*: TFE
- 30. Packing Type*: STANDARD

TRIM
- 32. Type*: STANDARD
- 33. Size*: 0.62 Rated Travel: 100%
- 34. Characteristic*: EQUAL PERCENTAGE
- 35. Balanced / Unbalanced*: BALANCED
- 36. Rated* C_v: 8.6 F_L: 0.8 X_T: 0.7
- 37. Plug / Ball / Disk Material*: 316 S.S.
- 38. Seat Material*: TFE
- 39. Cage / Guide Material*: GLASS-FILLED TEFLON
- 40. Stem Material*: 316 S.S.

SPECIALS / ACCESSORIES
- 43. NEC Class: GP Group: ___ Div: ___
- 44. Vendor to permanently attach stainless steel tag with above tagname.
- 45. Class IV shutoff required.

ACTUATOR
- 53. Type*: DOUBLE-ACTING CYLINDER
- 54. Mfr & Model*: VALTEK 25
- 55. Size*: ___ Eff Area: 25 SQUARE INCHES
- 56. On / Off: ___ Modulating: MODULATINIG
- 57. Spring Action Open / Close: OPEN
- 58. Max Allowable Pressure*: 150 P.S.I.G.
- 59. Min Required Pressure*: 3 P.S.I.G.
- 60. Available Air Supply Pressure:
- 61. Max: 60 P.S.I.G. Min: ___
- 62. Bench Range*: 3-15 P.S.I.G. / ___
- 63. Actuator Orientation: VERTICAL
- 64. Handwheel Type: N/A
- 65. Air Failure Valve: N/A Set At: ___

POSITIONER
- 67. Input Signal: 4-20 Ma
- 68. Type*: I/P W/ POSITIONER
- 69. Mfr & Model*: VALTEK NT3000 / XL SERIES
- 70. On Incr Signal Ouptut Incr / Decr*: INCR
- 71. Gauges: YES By-Pass: YES
- 72. Cam Characteristic*: ___

SWITCHES
- 74. Type: ___ Quantity: ___
- 75. Mfr & Model: ___
- 76. Contacts / Rating: ___
- 77. Actuation Points: ___

AIRSET
- 79. Mfr & Model*: ___
- 80. Set Pressure*: ___
- 81. Filter: ___ Gauge: ___

TESTS
- 83. Hydro Pressure*: ___
- 84. ANSI / FCI Leakage Class: ___

Rev	Date	Revision	Orig	App
0	6/29/2002	ORIGINAL	LGT	MDW

Information supplied by manufacturer unless already specified

© 1981 ISA — Second Printing — ISA FORM S20.50, Rev. 1

FIGURE 3-106. INSTRUMENT SPECIFICATION FOR CONTROL VALVE PV-48

C_V tables can be found in the manufacturer's literature. They list the CV values for a particular valve at different percentage openings (e.g., 10%, 20%). The chosen valve usually has a calculated C_V equal to the valve's C_V at approximately 70% to 80% open, based on the C_V table.

Many additional factors are used to properly size control valves, including cavitation, liquid pressure recovery factor (F_L), liquid critical pressure ratio factor (F_f), flashing, choked flow, Reynolds number factor (F_R), piping geometry factor

(F_P), valve exit velocity, and noise. For more information on sizing control valves, refer to ANSI/ISA-75.01-1985, *Flow Equations for Sizing Control Valves*.[8]

D. SUMMARY

Instrument specification sheets lay the groundwork for virtually all deliverables produced by the instrumentation design group. They are vital to the instrument designer for designing the wiring diagrams, the power distribution drawings, and the instrument mechanical, electrical and mounting installation details. Mechanical pipers, construction and maintenance personnel, programmers, purchasing personnel and vendors also find them useful.

Part III – Chapter 13: Physical Drawings

Physical drawings put the "A" in "A&E," at least from the perspective of the I&C group. I&C folks might find it hard to believe, but it is impossible to build an industrial facility with just E&I drawings and software. There must be a point at which the rubber meets the road. That point is the set of physical drawings. These drawings consist of control room arrangements, termination room arrangements, electrical room arrangements, process area plans, installation details, and so on. Since the addition is small in our scenario, the control room and termination room already exist. As a result, this chapter deals mostly with the process area. However, the control and termination rooms are given a cursory examination.

A. Control Room

The control room is the nerve center of the facility. Plant owners are usually motivated to provide operators with a comfortable environment from which to manage the plant, but designing a modern control room is not that different from designing any other office space. Industry has learned over the years that the human factor in engineering plays a key role in reducing operator-induced process problems. Designing an environment that does not induce fatigue or frustration must be the primary goal of the designer when taking on this task. An operator's activities require, for example, proper lighting, a comfortable temperature, and a logically arranged room. The following is a list of some key considerations and some suggested methods for designing a good control room environment:

1. Environmental Issues

a. **Heating, Ventilation and Air Conditioning (HVAC)[9]**
   Dry-bulb air temperature should be approximately 76°F. Relative humidity should be between 20% and 60%. Air velocity should be less than 0.23 meters/second (m/s), or 45 feet per minute (fpm). Also, temperature at the operator's feet and head should be about the same. Ventilation should be at least 15 cubic feet per minute (cfm) per occupant.

b. **Noise Levels**

The control room should be isolated from the process area in terms of noise level. Noise should be limited to about 55 decibels (dB) to permit normal voice communication at a distance of 10 feet.

c. **Lighting**

Lights should be situated to reduce or eliminate shadows and glare. The two can be difficult to manage, as correcting one may make the other worse. In general, it is better to have many lights. They can be dim, or at least have a dimmer switch to allow the operator to control the light level to his needs. They can also have special diffusers or filters. In areas where light may be too diffuse, supplemental lighting should be provided. Flat paints and textured finishes also help to reduce glare. If monitor glare is still a problem, monitors may be fitted with non-glare filters.

2. Physical Arrangement

The physical arrangement includes office furniture, a computer or two, printers, file cabinets, closed-circuit television monitors, annunciators, and other equipment items. It is up to the designer to put these elements together into a room design that complements the set of tasks the operator must execute. The following are a few considerations for the physical arrangement of the control room:

- Furniture should be arranged such that operators have an unobstructed view of the control room equipment. Items that require their attention on a regular basis should be most convenient to them. The room should be organized with operator interaction in mind.

- Traffic flow through the room should be routed away from the operator's normal work area. Consideration should be given to access paths for maintenance teams and their equipment. Also, if emergency procedures require special access, this should be accommodated. Unnecessary traffic should be minimized, perhaps by the use of signs on the doors.

- Operators should be able to maneuver within their work area without coming in contact with "knee knockers" or other impediments. Their chairs should be adjustable in both height and armrest configuration. An uncomfortable chair and/or impediments in the physical workspace induce fatigue as operators will be constantly aware of these areas and will work to avoid them.

- If an operator must communicate verbally with others on a regular basis, the designer should try to accommodate that with proper orientation of the console or should install a communication system of some kind.

- Some space should be allocated for document storage. Operators should be able to quickly access hard-copy operating or emergency procedures or other documentation.

- Operators should be provided with some flat space for laying out drawings.

3. Control Room Design Summary

Whether a computer system is being used or not, the real computer that runs the plant is the operator. From the color scheme to the type of lights to use, operators' comfort should be considered and their input sought. It is, after all, where they are likely to spend a large percentage of their time. The key is to make it easy for them to do the right thing by arranging the environment in a way that removes obstacles to that end.

B. Termination Room

The termination room is where all the field wiring converges and is mated to the control system. This room is called many things, including the marshalling room and the spreading room. Whatever it is called, it is the room where all the computer equipment resides and where the field cables are marshaled and terminated. Figure 3-107 shows three different configurations for this area.

This "room" can, in fact, be spread across more than one physical space. For example, the computer cabinetry can be located in its own termination room or with the operator in the control room. Or it may be found in the electrical room alongside the MCCs. Wherever it is physically located, there are some basic considerations when approaching the design task.

The termination room is the electrical and I&C technicians' domain. It should be designed with them in mind. Electrical outlets for plugging in test equipment should be readily available. If possible, a desk with a network connection to facilitate online troubleshooting should be provided. Storage space for drawing sets should be considered, as should a spare parts locker.

1. Environmental Issues

a. Lighting

As in the control room, there should be enough lights to eliminate shadowing. However, whereas lighting in the control room is subdued, the

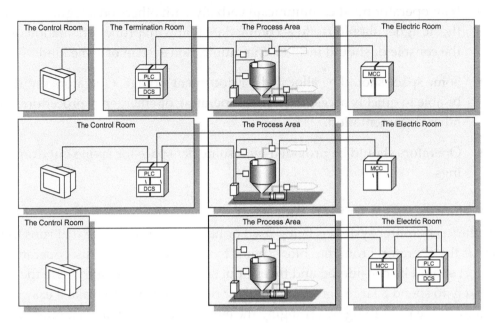

FIGURE 3-107. THREE TERMINATION ROOM CONFIGURATIONS

lighting in the termination room should be very bright. There should be no area in the room where supplemental lighting is needed. Glare is not a major issue and can be minimized by using flat paints.

b. **Heating, Ventilation, and Air Conditioning**
Temperature and humidity should be maintained to protect the equipment, as indicated in the equipment specification. This often makes the temperature of the room cooler than would be ideal for long-term occupation, though this room will probably not be constantly manned in any case. Care should be taken to ensure moisture from air conditioning system condensation does not collect in the room.

c. **Piping**
Process and water-service piping should be routed around the termination room.

d. **Computer Floors**
Computer floors facilitate making wiring adjustments easily and quickly while keeping the nest of cables from constant view. They also provide a convenient way to inject cool air into the equipment cabinets. There are a couple of issues to consider:

- Cables need to be plenum rated.

- Smoke detectors may need to be installed beneath the flooring.

2. Furniture and Equipment Arrangement

The following are some guidelines for arranging furniture and equipment in the termination room:

a. Personnel Clearances

Equipment cabinets should be arranged with electrical clearances in mind. National Electrical Code (NEC) article 110.26(A)(1)[10] applies to most I&C installations (600 volts or less). This article states minimum clearance between electrical enclosures is 0.9 m, or 3 ft, provided no voltage greater than 150 volts is present. If a voltage greater than 150 volts is present, and if there are exposed live parts on both sides of personnel working between the enclosures, then the distance should widen to 1.2 m, or 4 ft.

b. Ingress and Egress

Termination rooms should have two entrances, unless there is an unobstructed exit,[11] or unless there is twice the minimum space between cabinets.[12] It is also good practice to orient equipment doors so they are hinged to close in the direction of flight. For example, if an electrical fire starts in a cabinet that is being worked on, its door will likely be open. The design should not block the maintenance person's efforts to escape the situation.

3. Termination Room Design Summary

The E/I&C controls technician will likely spend a lot of time in the termination room performing troubleshooting tasks, upgrades, and other activities. The room should accommodate the technician as well as fulfill the environmental requirements of the equipment. It should be accessible, and should have direct access to the control room, not via a process area.

C. Process Area (Instrument Location Plan)

Instrument location plan drawings depict the location and identity of instrumentation and control equipment and provide conduit routing and cabling information. The location plan is typically spawned from an equipment arrangement or rendered from a CADD 3-D piping arrangement. Once the background is completed, showing the equipment arranged on the floor and enough steel detail to indicate location within the plant, the I&C designer layers on the instruments and shows conduit routings and contents. Junction boxes are depicted, and cable routing schedules are produced. The graphics presentation can differ from one customer to the next, depending on standards. One approach is illustrated in Figure 3-108.

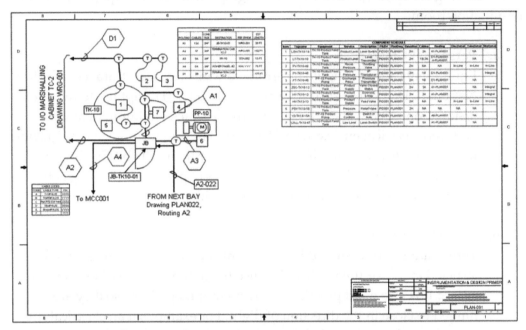

FIGURE 3-108. SAMPLE INSTRUMENT LOCATION PLAN DRAWING

This drawing is a finished plan view of the system that was described in Chapter 7. It shows our TK-10 product tank, a typical vessel with a level control system, a fixed-speed discharge pump, and a recirculation system. The pump has an HOA switch and a contact closure to provide pump status to the control system. Instrument station boxes represent the instruments. The instrument station concept lets one box represent several instruments, provided the instruments are physically grouped. In our simplified example, one box represents one instrument. However, more than one signal may be generated by that instrument.

Conduit bodies are depicted with a circled letter. The letter indicates the general type of conduit fitting as denoted by the Crouse-Hinds nomenclature. The letter T indicates a tee-type fitting that could be the Crouse-Hinds T-style body or the explosionproof (GUAT-type) fitting depending on the area classification. It is a good idea to include a legend on the drawing that matches the conduit type to the symbol to avoid confusion.

Rigid conduit then carries the cabling to its destination. A component schedule describes the instruments represented by the station boxes and details the cabling that feeds them. Items 4 and 5 each have two conduit connections, one for

AC power and one for DC signal. Ultimately, the conduit size is determined and layered onto the drawing or listed in a cable schedule.

1. Why Produce Instrument Location Plan Drawings?

Producing instrument location plans can be an expensive process. To justify them as a deliverable, the customer must be convinced they are necessary. The designer must remember the plan drawing is primarily used as a construction aid, not as a maintained document the client cares much about in the long term. An exception to this is a greenfield (all-new facility, from the ground up) project. In such a project, instrument location plan drawings provide a record of the instrumentation at startup that remains useful as a training aid or orientation tool and, as such, is easily justified to most customers.

So given the high cost of their production, why are instrument location plans even considered for retrofit work? The answer to that lies in the complexity of the project, as their cost-versus-benefit relationship can change from one job to the next. If the drawings help the constructor save more money than the engineer expends in their production, then they are justified in terms of overall project cost.

These drawings are useful in several ways. During the engineering phase of the project (Phase 2), they are used as the base documents for producing bills of materials for purchasing items, such as cable, conduit, and junction boxes. This can ultimately save materials cost since the quantity of materials purchased will be closer to the amount actually needed. But, as mentioned, their primary use is as a construction aid. Construction personnel find these drawings invaluable for planning and for execution. They greatly reduce the amount of management resources that must be allocated to conduit and cable installation and increase the chances of having a successful installation.

In keeping with the need to minimize detail, we must ask ourselves if alternatives exist. Depending on the situation, the answer is yes. So this chapter not only describes the instrument location plan drawing in detail, it also presents some viable alternatives that can be considered as cost-cutting measures.

2. Anatomy of an Instrument Location Plan

As mentioned, location plan symbols and the level of detail can differ from one customer to the next. But location plans generally consist of a graphic representation of some portion of the process floor, a component schedule, and possibly some detail sketches for clarification. These drawings are usually fairly large (22 by 34 inches; Size D, for example). In general, the graphic presentation should cover approximately 50% of the drawing area (usually the left side of the drawing). The scale of the graphical portion can vary widely since the only limitation is the instrument density and level of detail. The remaining 50% of the drawing area is generally reserved for a component schedule and various notes and details as needed.

The following are some of the key components of the instrument plan drawing (Figure 3-108):

- Major pieces of equipment, such as tanks and pumps, should be shown for orientation purposes. These should be in a light line weight and appear as background. They should be identified in some low-key manner such as with the vessel tag number or equipment name.

- Instruments should be depicted near their expected final location, either with bubbles or numbered boxes. Instrument elevation and instrument installation detail assignments should be noted either on the body of the plan or in the component schedule.

- Field junction boxes should be properly depicted and labeled.

- Conduit should be represented in some fashion. Conduit size should be shown, especially at reduction points, and conduit runs (also called *cable routings*) should be tagged.

- A component schedule must provide the detail that the simplified graphic presentation lacks. The most efficient way to generate the component schedule is from the instrument database as a report, but it can be maintained in a spreadsheet or any other suitable medium. If a database solution is sought, the instrument plan report (see Chapter 9) can be easily modified as the drawing develops and then be pulled onto the drawing just before it is finalized.

3. Design Considerations

During production of the plan drawing, the designer would be wise to consider the following issues:

- *Hazardous Area Classifications:* You must understand the ramifications of the area classification in effect for each process area. Be sensitive to these requirements when locating remote transmitters and field junction boxes. See Chapter 5 of this book, "Hazardous Area Classifications and Effect on Design," for more information.

- *Ergonomic Considerations:* Plant personnel must be able to transit the area, either on foot or while operating fork trucks or other wheeled vehicles. Junction boxes and instrument stands should be situated away from high-traffic areas to protect them from damage from fork trucks or other vehicles. Refrain from blocking aisles and ladders.

- *Maintenance Accessibility:* Be aware of any maintenance issues that might require extraordinary requirements for accessibility in the process area. For example, a vessel might need to be removed periodically for cleaning or refurbishing. Avoid mounting equipment around temporary installations, and certainly avoid using any such item for support.

- *Operational Accessibility:* Be aware of any operational issues that might require accessibility in the process area. Avoid mounting equipment near vessel manways or cleanouts.

- *Coordination with Piping Group:* Proper coordination with the piping group is essential. While it is true that the instrument location plan is diagrammatic, it does not help when the mechanical designers arrive to install a 10 inch process pipe and find a junction box installed and already wired at that location. Rerouting an occasional conduit is to be expected, particularly if it is a fast-track project with instrument mechanics installing before the pipers have finished. But there is no excuse for installing field junction boxes without coordinating with the pipers. In fact, such boxes should really appear on the equipment arrangements. Chapter 8 offers more information on this topic.

4. Drawing Production Technique

The following is one way to prepare an instrument location plan (process area plan) drawing with a minimum of extraneous graphic detail, using the instrument database as a design tool:

a. **Step One: Initialize Drawing (Generate drawing background)**
   If a project is being modeled, then a two-dimensional background can be spawned from the model. This method is very efficient since the inline piped instrumentation is usually depicted in the model, allowing these items to be perfectly and automatically spotted (placed) on the plan drawing.

   If a project is not being modeled, the best background to use is probably the equipment arrangement. In the old days, a vellum[8] copy of the drafted equipment arrangement would be generated, and then the instrument locations would be drafted over that. Today it is done in CADD. Either way, the concept is the same. The background information should appear

---

8. Vellum is a semitransparent material that could be used as an "original" drawing from which blueprints could be generated. It was popular in the "old days" because it was possible to reproduce a Mylar or linen original onto the cheaper vellum material. In this way, a mechanical arrangement could be used as a reference drawing, providing a background onto which new information could be layered—just as is done today in the CADD environment.

in a very light line weight that lets the information of interest be visually separated from it. Any extraneous information that is carried over from the equipment arrangements should be deleted, except, perhaps, for the major dimensioning between column lines.

The area to be illustrated should cover about 50% but no more than two-thirds of the drawing surface (Figure 3-109). Room must be left for the instrument schedule, a legend, a key plan, and any notes that might be necessary. The instrument schedule usually appears on the right side of the drawing, starting at the top and covering as much room as necessary down to the title block.

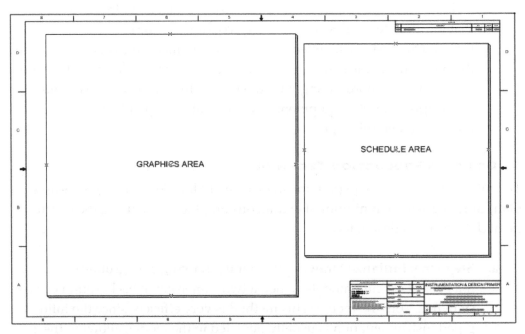

FIGURE 3-109. INITIALIZE DRAWING

Once the border is set, the equipment arrangement drawing may be imported as a reference file (if the drawing is to be a CADD drawing). Using the equipment arrangement or piping orthographic drawing as a background, the designer should delete all data except the major equipment items (Figure 3-110) while leaving enough civil detail (such as column lines) to allow plan users to get their bearings within the plant.

b. **Step Two: Spot and Classify Instruments and Instrument Groups**
This procedure assumes the drawing is produced manually as opposed to being rendered (using CADD) from a computer model. As a first pass

PART III – CHAPTER 13: PHYSICAL DRAWINGS 419

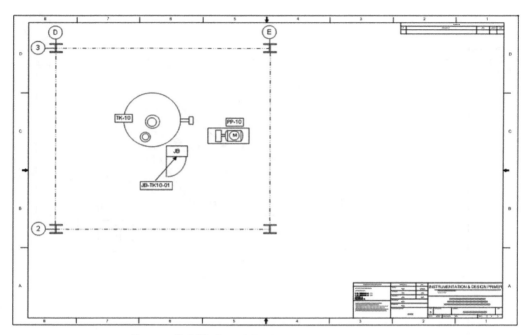

FIGURE 3-110. LOCATE MAJOR EQUIPMENT ITEMS

through the floor plans, a full-size background should be printed out. Then, as the instruments are found, their location should be penciled in as a heavy dot with the tag number noted beside it and some indication of elevation (e.g., H = high, M = medium, L = low) (Figure 3-111). Color-coded pencils could be used for this purpose as well.

Much can be done in the way of instrument locating with just a P&ID and plan drawing background based on the equipment arrangement. Usually, the P&ID provides a general indication of instrument orientation. For example, a vessel's discharge valve is shown at the bottom of the tank, and its fill valve at the top. There may be several instruments on the top that can be presented as one instrument station box. This can be placed anywhere within the vessel's outline on the plan drawing with no other corroborative evidence. The same applies to the bottom of the vessel.

This approach can be used on virtually any piece of equipment as a first-pass approximation of instrument location. Inline, piped instruments such as mass flowmeters are a different matter. If the drawing is spawned from the equipment arrangement instead of a 3-D model, then the piping drawings themselves must be consulted to find the instrument location. P&IDs are of little use for this task.

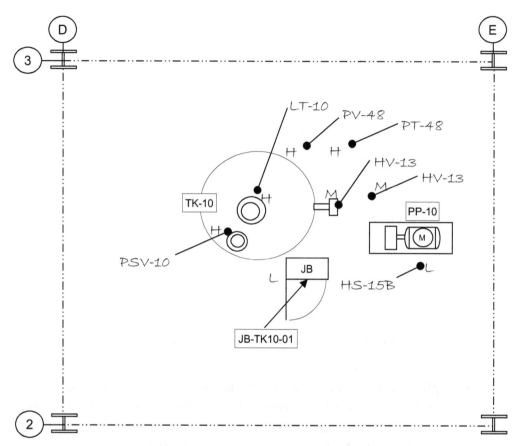

FIGURE 3-111. LOCATE INSTRUMENT ITEMS

Once all instruments are spotted, then judgments can be made as to grouping. Remember instruments can be widely separated in elevation and still appear nearby one another on the floor plan. Such instruments should probably be in different instrument groups. On the drawing, grouped instruments should be circled with some indication of their group number penciled nearby. In the simple case we are exploring here, grouping more than one instrument into one box is not necessary, so each item will be displayed as its own box (Figure 3-112).

After the instruments are grouped, the plan drawing number, the instrument group number, and the elevation code should be captured in a database or spreadsheet. In our case, we have created a new query in the instrument database called Plan Dwg Takeoff Query. We must add three new fields in the database to accommodate the new information: PlanDwg, PlanItem, and Elevation. The new query with entered data is shown in Figure 3-113.

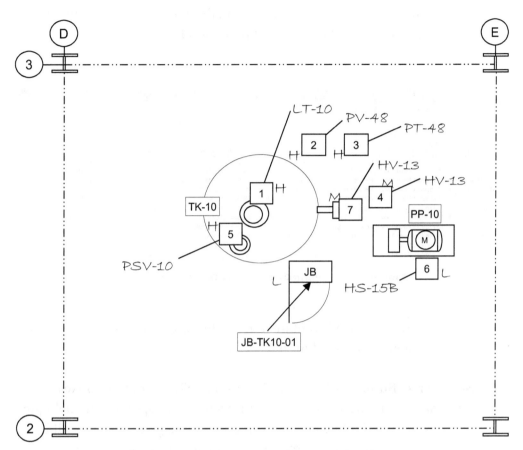

FIGURE 3-112. ADD INSTRUMENT STATIONS

FIGURE 3-113. PLANDWGTAKEOFFQUERY

The YS-TK10-15B item is not on this plan drawing but is instead fed from motor control center 001 (MCC001).

Filtering out items not found on the plan drawing and then sorting by PlanItem number yields the list in Figure 3-114.

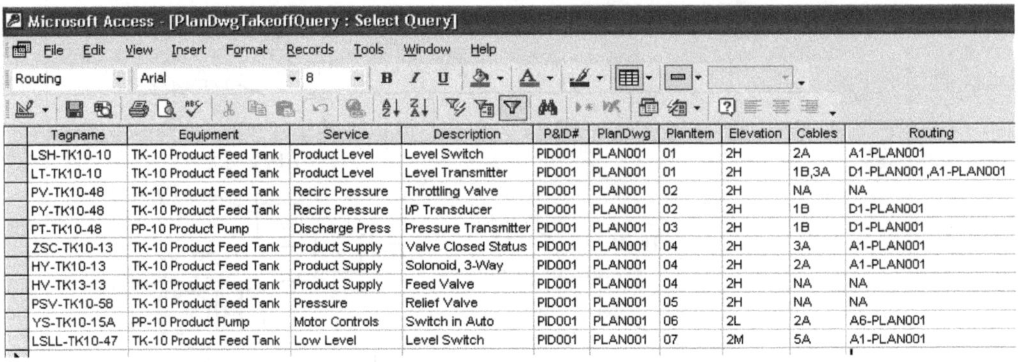

Tagname	Equipment	Service	Description	P&ID#	PlanDwg	PlanItem	Elevation	Cables	Routing
LSH-TK10-10	TK-10 Product Feed Tank	Product Level	Level Switch	PID001	PLAN001	01	2H	2A	A1-PLAN001
LT-TK10-10	TK-10 Product Feed Tank	Product Level	Level Transmitter	PID001	PLAN001	01	2H	1B,3A	D1-PLAN001,A1-PLAN001
PV-TK10-48	TK-10 Product Feed Tank	Recirc Pressure	Throttling Valve	PID001	PLAN001	02	2H	NA	NA
PY-TK10-48	TK-10 Product Feed Tank	Recirc Pressure	I/P Transducer	PID001	PLAN001	02	2H	1B	D1-PLAN001
PT-TK10-48	PP-10 Product Pump	Discharge Press	Pressure Transmitter	PID001	PLAN001	03	2H	1B	D1-PLAN001
ZSC-TK10-13	TK-10 Product Feed Tank	Product Supply	Valve Closed Status	PID001	PLAN001	04	2H	3A	A1-PLAN001
HY-TK10-13	TK-10 Product Feed Tank	Product Supply	Solonoid, 3-Way	PID001	PLAN001	04	2H	2A	A1-PLAN001
HV-TK13-13	TK-10 Product Feed Tank	Product Supply	Feed Valve	PID001	PLAN001	04	2H	NA	NA
PSV-TK10-58	TK-10 Product Feed Tank	Pressure	Relief Valve	PID001	PLAN001	05	2H	NA	NA
YS-TK10-15A	PP-10 Product Pump	Motor Controls	Switch in Auto	PID001	PLAN001	06	2L	2A	A6-PLAN001
LSLL-TK10-47	TK-10 Product Feed Tank	Low Level	Level Switch	PID001	PLAN001	07	2M	5A	A1-PLAN001

FIGURE 3-114. PLANDWGTAKEOFFQUERY, FILTERED

Notice there are indeed combined items. In the case of Item 04, there are three line items combined in that station.

c. **Step Three: Build Cable Schedule and Size Field Junction Boxes**
The drawing should be scanned to find a tentative spot for any field junction boxes that might be needed. Generally it is a good idea to split instrumentation fed from AC sources from those fed from DC sources. Discriminating between analog and digital circuits may be done on occasion but is not as important as differentiating between AC and DC circuits and among high- voltage (>300 V) and low-voltage circuits (< 300 V).

The designer should tabulate the cables that will probably be landing in each box in order to derive a wire count. On new construction, the designer should plan on at least 50% more terminals than needed. A spreadsheet program or some other tool should be used to lay out the terminals (see Chapter 15 for guidelines and methodology). An enclosure vendor catalog might be consulted to determine the standard backplane sizes to accommodate the layout (unless custom cabinets are to be constructed), and the backplane cross-referenced to an appropriate enclosure. The outside dimensions of the enclosure should be noted on the plan drawing.

d. **Step Four: Spot Field Junction Boxes**
This is a step best done in cooperation with the mechanical engineering group or whomever is responsible for the equipment arrangements. When

an ideal location has been determined based on instrument station groupings, and the anticipated size of the box is known, the junction box can be tentatively spotted on the instrument plan. A judgment can be made based on expected traffic patterns, the availability of support structure, and NEC requirements. Once spotted, the box locations should be cleared with anyone else who might use that space, such as the mechanical and structural groups. Maintenance access and traffic flow concerns must be considered before finalizing the location of the boxes.

When the location has been finalized, it is highly recommended that the junction boxes be plotted as permanent fixtures on the equipment arrangements. Otherwise, conflicts can arise over that space. Pipers will run pipes through it, for example. Remember the space to be allocated does not just include the box, but must take into account door swing, conduit access areas, and NEC requirements regarding the personnel access envelope.[13]

e. **Step Five: Route Conduit**

If you picture a conduit run in three dimensions, it typically exits a junction box, rises to the overhead, travels along the overhead supported by structural beams or some other support structure, and then begins to branch (Figure 3-115). As the branches approach instrument stations or instrument station clusters, the conduit drops along some vertical support to the instrument station. When viewed from above, such conduit runs appear similar to the branches of a tree. Therefore the term *trunk* is used to describe the part of the conduit run that is common to all the branches. *Branches* are common to groups of instrument stations, and drops (not to be confused with I/O drops) refer to the vertical sections of conduit that feed the instrument stations themselves.

Conduit location and routing paths are depicted as an approximation, so the time spent routing the conduit should be kept to a minimum. The designer should try to keep in mind the relative elevations of the instruments as instrument stations are gathered up into conduit trunks. Two adjacent stations separated by 15 feet in elevation are unlikely to be in the same conduit unless there is some kind of vertical support, like a tank or a vertical support member, between them. Also, AC circuits and DC circuits should be kept in separate conduit trunks. They will likely be sourced from different locations. Doing this protects the DC circuits from 60 Hz radio-frequency-induced electrical noise. The conduit arrangement for this project is shown in Figure 3-116.

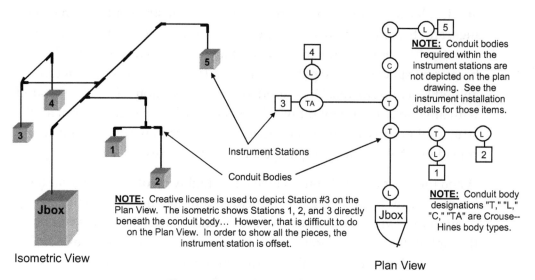

FIGURE 3-115. 3D TO 2D AND BACK

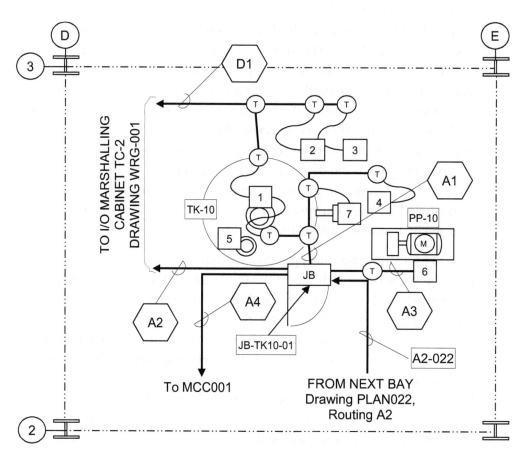

FIGURE 3-116. ADD CONDUIT DETAIL

f. **Step Six: Calculate Conduit Fill**
    To make a plan drawing or conduit schedule useful to construction, it must provide conduit fill capacities that allow successful pulling to occur. Clearly, a conduit sized just slightly larger than the cables it contains would make pulling difficult, if not impossible. Likewise, certain cable combinations are more difficult to pull than others. For more information on conduit sizing, see Part II, Chapter 5, Section A.

g. **Adjust Conduit Routings for Legal Fill**
    When the fill calculations are completed for each conduit trunk, the routing assignments already established will probably need to be adjusted to accommodate overfilled (or underfilled) conduits. Overfilled conduits should be split and the affected fill calculations repeated.

h. **Assign Conduit Routing Numbers**
    After all calculations have been made, all conduit fills are confirmed to be legal (within code), and conduit routings have been finalized, then conduit routing numbers may be assigned. A good practice is to use an alphanumeric scheme that quickly indicates the kind of circuits inside the conduit (Figure 3-117). For example, an AC (Class 2) routing code might begin with an A, and a DC (Class 1) routing might begin with a D followed by some sequential number to arrive at a unique tag number.[9] Label leaders should point to the largest section of conduit, and the conduit size at that point should be noted.

    This scheme requires that conduit routing numbers be unique to the plan drawing from which they originate. Numbers are therefore assigned at the field end. The routing number should consist of some unique designation (e.g., A21) followed by enough of the drawing number for the drawing to be identified from the tag.

    If the conduit routing never leaves the drawing, a hexagon is used as a shorthand technique, and just the routing number need be shown. If, however, the conduit originates on drawing X and then leaves that drawing and reappears on drawing Y, then drawing Y shows the conduit tag as a rectangle with the conduit number and whatever portion of the drawing number that is necessary to point to the right document.

    For example, a field junction box may have several conduits running to it from field devices. Each of those conduits will have a number unique to the plan drawing that shows the most remote instrument serviced by that

---

9. See NEC1999 article 725-15 for more information on circuit classes.

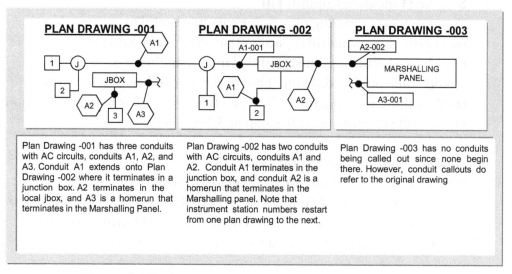

FIGURE 3-117. RECOMMENDED CONDUIT TAGGING CONVENTION

conduit. Homerun cables that run from that field junction box to the marshalling cabinet will be numbered at the field junction box drawing. Cables running from the marshalling cabinet to the control room will be numbered from the marshalling cabinet drawing.

Figure 3-118 shows a typical floor arrangement with conduit routing numbers added.

i. **Generate Component Schedule**
The instrument location plan component schedule is a table that provides detail about each instrument station. This table usually appears in the upper right corner of the drawing. A row of information is provided for each element represented by the instrument station boxes. As previously noted, a single instrument station box can represent multiple instruments. Therefore, instrument station numbers might be repeated several times in the table. Each instrument has its own record in the schedule.

Beyond cross-referencing the instrument station numbers to their associated instruments, the component schedule crosses (cross-references) each instrument to its cable code. A cable code cross-reference chart (Figure 3-119) should be devised early in the project that allows shorthand codes to be linked to the list of approved cables. This chart should appear on each plan drawing that depends on the code. Type A cable, in our example, is a single conductor, 14 gauge wire, while type B indicates a twisted pair, 16 gauge cable with an overall shield. (Please note these type classifications are arbitrary. Actual type descriptions will probably differ.)

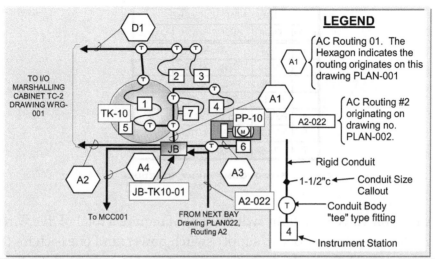

CABLE CODES		
CODE	CABLE TYPE	P/N
A	1/C#14,US	XXX
B	TWPR#16,OS	YYY
C	3/C#14,US	ZZZ
D	16/C#16, US	AAAA
E	16TWPR#16,IS	BBBB

CONDUIT SCHEDULE		
CONDUIT	CONTENTS	SIZE
A1-001	13A	3/4"
A2-001	1D	1"
A3-001	5A	3/4"
A4-001	2C	3/4"
D1-001	3B	1"

COMPONENT SCHEDULE			
STA#	TAG#	DESCRIPTION	CABLE
1	LT-10	LEVEL XMTR	1B,3A
2	PV-48	CTRL VALVE	1B
3	PT-48	PRESS XMTR	1B
4	HV-13	FILL VALVE	5A
5	PSV-58	RELIEF VALVE	NA
6	YS-15A	HOA IN AUTO	4A
7	LSLL47	LEVEL SWITCH	4A

FIGURE 3-118. INSTRUMENT ARRANGEMENT WITH SUPPORT DATA

CABLE CODES		
CODE	CABLE TYPE	P/N
A	1/C#14,US	XXX
B	TWPR#16,OS	YYY
C	3/C#14,US	ZZZ
D	16/C#16, US	AAAA
E	16TWPR#16,IS	BBBB

FIGURE 3-119. CABLE CODE CROSS-REFERENCE CHART

A table allows these shorthand codes to be used to simplify the process of determining the cable requirements of each instrument. For example, in the component schedule shown in Figure 3-120, instrument station 5 has a level transmitter with three type A and one type B cable. Also, a level switch on that same transmitter requires an additional four type A cables. In our example, a type A cable is actually a single conductor wire, but the principle holds true nonetheless.

COMPONENT SCHEDULE			
STA#	TAG#	DESCRIPTION	CABLE
1	LT-10	LEVEL XMTR	1B,3A
2	PV-48	CTRL VALVE	1B
3	PT-48	PRESS XMTR	1B
4	HV-13	FILL VALVE	5A
5	PSV-58	RELIEF VALVE	NA
6	YS-15A	HOA IN AUTO	4A
7	LSLL47	LEVEL SWITCH	4A

FIGURE 3-120. COMPONENT SCHEDULE

An on/off valve with limit switches might require a total of five wires: two for the solenoid, one to supply switch power, and one each for the open/closed limit switches. In this example, cable code 5A describes the wiring needs of the valve, denoting five 14 gauge, single conductor wires. An ultrasonic level transmitter needs one shielded cable for DC analog (a type B cable in our example) and three single conductor wires for AC power. Thus, the cable code for this instrument is 1B, 3A.

The component schedule usually appears on the drawing but can be delivered as a document in its own right. More than just listing cables, the cable schedule also provides instrument station cross-references to installation details, various nonphysical drawings, such as P&IDs and instrument elementaries, and other information that may be deemed important. The component schedule should, at minimum, provide a cross reference between the instrument station numbers that appear on the body of the drawing and the instruments represented by those station numbers along with their cabling requirements.

This type of information is best handled in a database or spreadsheet and then pulled onto the drawing just prior to its being finalized and issued. Using the database to manage this information provides the most flexibility during Phase 2, and makes the information more broadly available. Figure 3-121 is the finished component schedule for this project. It was taken directly from the database being developed for this project.

This component schedule will be placed on the plan drawing. It has the following features:

- Cable types and routing numbers are provided for each item.

- Multiple items are listed per instrument station, and each item is within close physical proximity of the others.

COMPONENT SCHEDULE

Item	Tagname	Equipment	Service	Description	P&ID#	PlanDwg	Elevation	Cables	Routing	ElecDetail	TubeDetail	MntDetail
1	LSH-TK10-10	TK-10 Product Feed Tank	Product Level	Level Switch	PID001	PLAN001	2H	2A	A1-PLAN001		NA	
1	LT-TK10-10	TK-10 Product Feed Tank	Product Level	Level Transmitter	PID001	PLAN001	2H	1B,3A	D1-PLAN001 A-PLAN001		NA	
2	PV-TK10-48	TK-10 Product Feed Tank	Recirc Pressure	Throttling Valve	PID001	PLAN001	2H	NA	NA	In-Line	In-Line	In-Line
2	PY-TK10-48	TK-10 Product Feed Tank	Recirc Pressure	I/P Transducer	PID001	PLAN001	2H	1B	D1-PLAN001			Integral
3	PT-TK10-48	PP-10 Product Pump	Discharge Press	Pressure Transmitter	PID001	PLAN001	2H	1B	D1-PLAN001		NA	
4	ZSC-TK10-13	TK-10 Product Feed Tank	Product Supply	Valve Closed Status	PID001	PLAN001	2H	3A	A1-PLAN001		NA	Integral
4	HY-TK10-13	TK-10 Product Feed Tank	Product Supply	Solonoid, 3-Way	PID001	PLAN001	2H	2A	A1-PLAN001			Integral
4	HV-TK13-13	TK-10 Product Feed Tank	Product Supply	Feed Valve	PID001	PLAN001	2H	NA	NA	In-Line	In-Line	In-Line
5	PSV-TK10-58	TK-10 Product Feed Tank	Pressure	Relief Valve	PID001	PLAN001	2H	NA	NA	NA	NA	
6	YS-TK10-15A	PP-10 Product Pump	Motor Controls	Switch in Auto	PID001	PLAN001	2L	2A	A6-PLAN001		NA	
7	LSLL-TK10-47	TK-10 Product Feed Tank	Low Level	Level Switch	PID001	PLAN001	2M	5A	A1-PLAN001		NA	

FIGURE 3-121. PLAN001 COMPONENT SCHEDULE

- Installation details will be listed when they are assigned. As is evident, not all items need all types of installation detail. There are columns for electrical cable and conduit, instrument air tubing and fittings, and mounting. The details listed here will greatly help the installers collect material and install the right configuration.

5. MATERIAL TAKEOFF

The utility of the plan drawing for the engineering firm is as an aid in building a bill of materials. Under-purchasing or over-purchasing material beyond a small allowable error margin is problematic for the construction group and the financial managers. In the worst case, construction group could be idle waiting for materials.

Therefore, the material takeoff chart is a critical aspect of the designer's job. This section describes a relatively painless method of taking off (determining the quantities of) the cable and the conduit materials with a minimum of wasted effort.

Figure 3-122 depicts a typical conduit routing. The cable requirements are known, and conduits have already been sized. Now it is time to tabulate the material quantities. Note the two types of cables: type A and type B. The notes at the station boxes indicate each station's cable needs and average elevation. It is assumed that the two-dimensional view of the conduit shows only the conduit at the ceiling level of 17 ft (minus 2 ft to get below structural steel) = 15 ft. Drops to the instrument stations are assumed to be vertical and are therefore invisible on this view, but they must be considered.

Again, this can all be done in Microsoft® Excel, except, of course, for the plan drawing itself, which is included here for reference. The designer must first label each leg of the conduit that appears on the plan drawing, and then, if in Excel, enter the basic information in the takeoff table.

The designer should scale each leg, estimating the up and down movements of the conduit as well. For example, leg A appears to be only 13 feet long accord-

430   SUCCESSFUL INSTRUMENTATION AND CONTROL SYSTEMS DESIGN

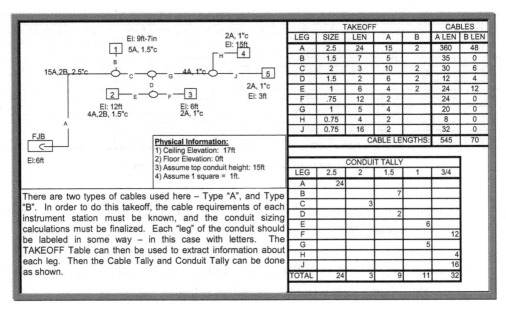

FIGURE 3-122. CABLE AND CONDUIT TAKEOFF APPROACH

ing to the scaled grid (1 block = 1 square foot), but the designer has listed 24 feet. In that leg, there are 15 type A cables and two type B cables. The length of each type is calculated as follows: 15 × 24 = 360 feet of type A and 2 × 24 = 48 feet of type B. Each leg of each conduit should be analyzed individually. In the end, the cable table provides an automatically calculated cable quantity. In a similar vein, the conduit tally table allows a quick calculation of the amount of conduit that is likely to be needed.

This method might take a little practice, but it yields an accurate tally with a minimum of fuss. It also provides a means of making quick adjustments as the design develops. If a vessel moves to another floor, for example, or if a pump gets deleted from the project scope, takeoff charts give the designer the ability to respond quickly to the change. The end result is a more accurate bill of materials with fewer hours expended. Using this method on our application yields the information given in Figure 3-123.

This information shows we need to purchase at least 790 feet of type A cable (single conductor, #14, unshielded), 490 feet of type B cable (twisted pair #16, shielded), and 200 feet of type D Cable (16 conductor #16, unshielded).

Now we need to size the conduit for each leg using the conduit sizing calculator described earlier. Leg A has three type B cables. The calculator results for leg A are shown in Figure 3-124.

A ¾-inch conduit will suffice for 40% fill, but note the caution message that is displayed. There are three cables, so an additional step is necessary.

TAKEOFF						CABLES		
LEG	SIZE	LEN	A	B	D	A LEN	B LEN	D LEN
A	1"	150		3		0	450	0
B	3/4"	10		1		0	10	0
C	3/4"	3		2		0	6	0
D	3/4"	6		1		0	6	0
E	3/4"	8		1		0	8	0
F	3/4"	150			1	0	0	150
G	3/4"	75	6			450	0	0
H	3/4"	50			1	0	0	50
J	3/4"	10	5			50	0	0
K	3/4"	12	15			180	0	0
L	3/4"	8	5			40	0	0
M	3/4"	3	10			30	0	0
N	3/4"	8	5			40	0	0
O	3/4"	8	5			40	0	0
CABLE LENGTHS:						830	480	200

CONDUIT TALLY								
LEG	2.5	2	1.5	1.25	1	3/4		TOT
A								0
B								0
C								0
D								0
E								0
F								0
G								0
H								0
J								0
F								0
G								0
H								0
J								0
TOTAL	0	0	0		0	0		0

Physical Information:
1) Ceiling Elevation: 17ft
2) Floor Elevation: 0ft
3) Assume top conduit height: 15ft
4) Assume 1 square = 1ft.

CABLE CODES		
CODE	CABLE TYPE	P/N
A	1/C#14,US	XXXX
B	TWPR#16,OS	YYYY
C	Plat.RTD Ext Wire	ZZZZ
D	16c#16,US	SSSS
E	8twpr#16,OS	VVVV
F		TTTT

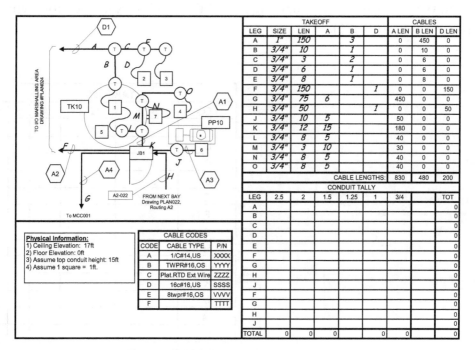

Figure 3-123. Cable Takeoff Method

	CONDUIT SIZING CALCULATOR										
	CONDUIT CONTENTS		CABLE SPECIFICATION DATA								
	CABLE QTY	Calc'd Area	Cable	Awg	Shld?	Manuf.	Rated Volts	Part No.	Svc	COST/FT	O.D.
A	-	0.000	1-cond	14	NA	var	600	?	Cond.	$0.01	0.133
B	3	0.197	1-twpr	18	IS	Belden	600	1120A	Tray		0.289
D	-	0.000	16-cond	16	US	Anixter	600	2A-1616	Tray	$0.98	0.602
	3	0.197	<<< CAUTION - CHECK AREA RATIOS PER NEC, CH9, TABLE 1, FPN #2... (Problem exists if Ratio is between 2.8 and 3.2...)								
				Complex Fill for Rigid Metal Conduit (Contents Mixed)							
				SIZE	1/2"	3/4"	1"	1-1/4"	1-1/2"	2"	3"
				AREA	0.314	0.549	0.887	1.526	2.071	3.408	7.499
				FILL	62.7%	35.8%	22.2%	12.9%	9.5%	5.8%	2.6%
				CALCULATED RATIO :	1.60	2.79	4.51	7.75	10.52	17.32	38.11

Figure 3-124. Conduit Sizing Calculator Results

In the bottom message window, a calculated ratio appears that compares the conduit ID with the cable area. The inner area of the ¾-inch conduit is 0.549 in^2 and the outer diameter of the cable is 0.197 in^2:

$$0.549/0.197 = 2.7868.$$

This falls just under the 2.8 binding limitation (described in Chapter 5, Part II). For safety's sake, it might be wise to upsize the conduit to eliminate this possible problem. We can size all the remaining legs of the conduit system by following this process (Figure 3-125).

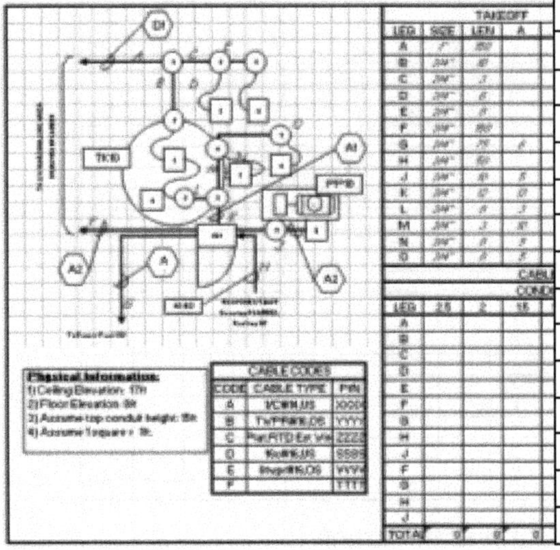

	TAKEOFF				
LEG	SIZE	LEN	A	B	D
A	1"	150		3	
B	3/4"	10		1	
C	3/4"	3		2	
D	3/4"	6		1	
E	3/4"	8		1	
F	3/4"	150			1
G	3/4"	75	6		
H	3/4"	50			1
J	3/4"	10	5		
K	3/4"	12	13		
L	3/4"	8	3		
M	3/4"	3	10		
N	3/4"	8	5		
O	3/4"	8	5		

FIGURE 3-125. CABLE TAKEOFF BY LEG

CAUTION: The Excel-based calculator depicted in Figures 2-25 and 3-124 is provided in the CD* that accompanies the book. Cable data shown in the table could be dated and is provided here for exercise purposes only. Actual values should be gathered from cable and conduit manufacturers and plugged into the calculator. Fill calculations are based on Rigid Metal Conduit (RMC) values listed in the NEC.

After all the conduit sections have been properly sized, the conduit takeoff can be done by simply moving the leg length (LEN) value into the conduit tally table under the proper conduit size heading. For example, leg A is a 1 inch conduit with a 150 foot length. So move 150 feet into the "1 inch" column of the conduit tally table for leg A, as shown in Figure 3-126.

It is as simple as that. When finished, the result is a pretty good estimate for the amount of conduit needed. In this case, we will need 150 feet of 1 inch and 312 feet of ¾ inch conduit.

This leaves us needing fittings and miscellaneous hardware, but that will be picked up later in the bill of materials task. All the hard data generated should be stored until needed for that task.

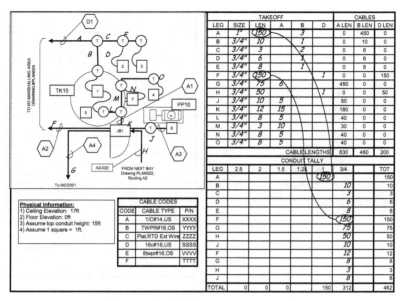

FIGURE 3-126. CONDUIT TAKEOFF

D. INSTRUMENT INSTALLATION DETAILS

This chapter introduced the instrument process area plan drawing and showed how that drawing can be used to compile quantities of materials, such as cable and conduit. However, the plan drawing views the problem from the standpoint of the process area "common" material only. It gets the signal, or power, to the instrument station. But the instrument station is itself a composite. The symbol used on the plan drawing is a simple box. The box may represent any number of instruments, with each possibly requiring different materials and/or a different configuration. How do we plug this gap? We build a set of instrument installation details.

An instrument installation detail satisfies two needs:

- Material takeoff

- Guidance for installers

Installation details provide a means for quantifying the material required at an instrument. This is done by creating generic sketches of process situations, giving each sketch a unique number, and then linking the proper sketch to the instrument by entering the detail number in the instrument database. An installation detail is shown in Figure 3-127.

Installation details are typically done on regular notebook-size paper (Size A, 8 ½ × 11). The detail in Figure 3-127 depicts the electrical hookup of a certain type of instrument. A bill of materials is provided. If the project calls for installing 50

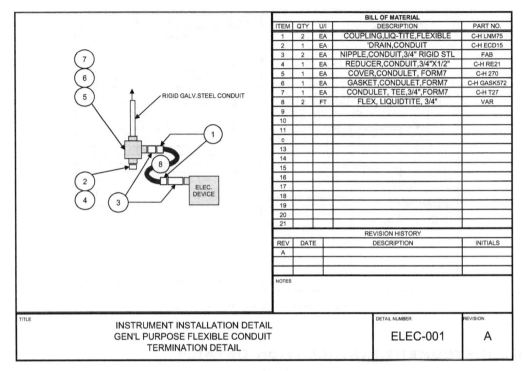

Figure 3-127. Instrument conduit installation detail

such instruments, then each item on the bill of materials is multiplied by 50. For example, we would need 100 feet of Liquidtite® flex conduit to meet that need.

There are three basic types of standard installation detail:

- Electrical to provide detail on conduit termination.

- Mechanical to provide detail on instrument air-tube fittings or mechanical linkages, etc.

- Mounting to provide detail on vessel trim or pipe-stand mounting.

In addition to the standard details, other details may find their way into the project. Special installation details are sometimes needed for those occasional design "opportunities" that may arise. Those details may become quite involved, being done on larger drawings. We will assume our little project will have none of those.

Let's briefly look at each type of detail.

1. Electrical Installation Details

We have discussed an electrical installation detail above, showing the electrical hookup of an instrument. This detail shows the conduit drop arriving in ¾-inch rigid conduit. Up to that point, the material required is calculated from the plan drawing. The detail lists the components from the tee fitting to the device (Figure 3-128). In this case, it includes a tee fitting (Item 7).

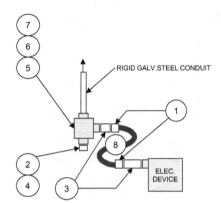

FIGURE 3-128. INSTRUMENT ELECTRICAL INSTALLATION DETAIL

A tee is used at the lowest points of each drop. Each tee needs a gasket and a cover. The tee is oriented with the base port on the horizontal plane. Flexible conduit ("flex") is mated to this port by the use of a short conduit nipple, which makes the transition from fitting to conduit coupling. The flex is then mated to the end device with a second nipple. The bottom port of the tee has a drain that prevents the buildup of condensate, which sometimes collects and eats away the conduit from the inside or finds its way into the electronic enclosure.

This represents the simplest application. In a hazardous area, conduit seals may be required along with explosionproof fittings. Part II, Chapter 5.C of this book explained some of the issues to consider for hazardous areas.

2. TUBING DETAILS

Any pneumatic instrument needs instrument air. Figure 3-129 is an installation detail for an on/off valve with an integral solenoid. It has a filter/regulator to set the ideal operating pressure for the valve and to clear the air of debris. Sometimes moisture is a concern, and a trap also needs to be installed. This particular valve is a fail-closed valve. The air is applied below a flexible diaphragm (the actuator), which is pushed closed by a spring. When air pressure overcomes the spring, the valve opens.

The material listed here is quite generic; when a detail is used on a real project, the material needs to be researched and the component schedule updated. Figure 3-130 is a second detail showing instrument air being provided to a throttling valve. In this case, there is a transducer instead of a solenoid, and the spring is on

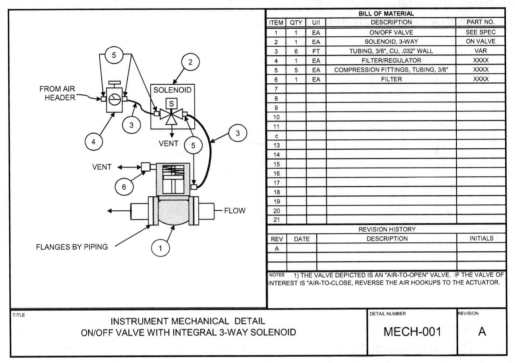

Figure 3-129. Instrument mechanical hookup detail

the other side of the actuator, making it a fail-open valve. Otherwise, the two are fairly similar.

Now let's look at a mounting detail.

3. Mounting Details

Mounting details show the mounting hardware and the configuration of the mount. In Figure 3-131, the mount is on a flange on the top of a vessel. The material listed on this detail is called vessel trim and includes nuts, bolts, gaskets, and blind flanges to be drilled and tapped to fit the instrument. Other types of mounting details revolve around the commonly used 2 inch instrument pipe stand, which can be adapted to virtually any circumstance.

4. Related Database Activities

Of course, now that we have these fine sketches, we need to do something with them. Each instrument needs to be evaluated as to the type of detail(s) that are appropriate for it. As details are matched to the instruments, they need to be logged in the database, as shown in Figure 3-132.

Notice YS-15A gets a Mount-002, which is a 2 inch pipe-stand mount. Also, note a few of the other terms in the mount column. The term *integral* means the item is a subassembly; in other words, it is part of something else and has nothing

Part III – Chapter 13: Physical Drawings

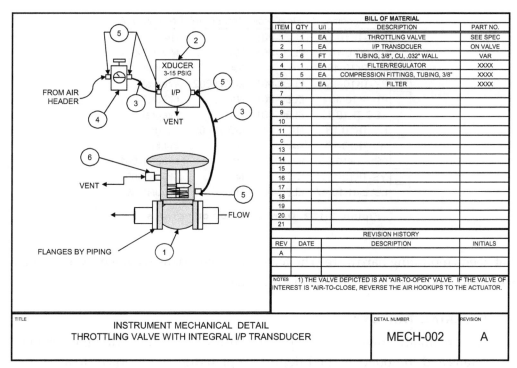

Figure 3-130. Instrument mechanical detail with throttling valve

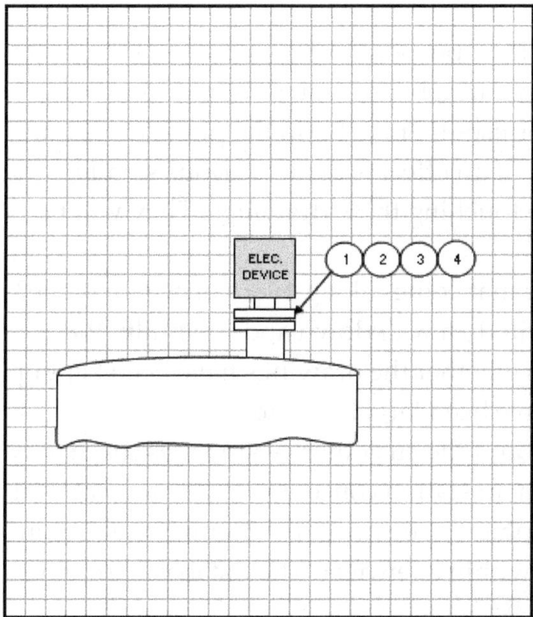

Figure 3-131. Instrument mounting detail

Tagname	ElecDetail	MechDetail	MountDetail
LSH-TK10-10	ELEC-001	NA	NA
LT-TK10-10	ELEC-001	NA	MOUNT-001
PV-TK10-48	NA	NA	In-Line
PY-TK10-48	ELEC-001	MECH-002	Integral
PT-TK10-48	ELEC-001	NA	MOUNT-001
ZSC-TK10-13	ELEC-001	NA	Integral
HY-TK10-13	NA	MECH-001	Integral
HV-TK13-13	NA	NA	In-Line
PSV-TK10-58	NA	NA	MOUNT-001
YS-TK10-15A	NA	NA	MOUNT-002
LSLL-TK10-47	ELEC-001	NA	MOUNT-001

FIGURE 3-132. DATABASE LOG OF DETAILS

that needs mounting hardware. The term *inline* means (as we have seen) the item is in a pipeline. The piping group mounts that item.

5. MATERIAL TAKEOFF

Well, we are finally down to it. The physical drawing set, including the installation details and component schedules on the plan drawing, leads us to a list of materials to purchase. This topic is picked up in Chapter 16, Procurement.

E. SUMMARY

The physical drawing set is primarily for the purpose of construction. The design team uses these drawings initially to generate a bill of materials and to validate the design concept. Then, the constructor uses them to build the plant. Once built, these drawings are rarely updated. Even if an effort is made, various small in-house projects are done without anyone touching these drawings. They will be out of date before you know it!

What about the major capital project that comes along every once in a while? Well, the A&E firm that wins that bid will do a walkdown and "as-built" the drawings. They will, quite likely, do this whether the drawings have been kept up or not. So the maintenance effort will have been largely wasted.

So from the designer's perspective—and the customer's—it makes sense to minimize the level of detail. The instrument station concept presents a great deal of information for a relatively small graphics investment. Positioning the station box can be diagrammatical, accurate to within a few feet. The database provides the detail lacking on the diagram itself.

Regarding the database, it is important that designers be fully capable of capturing the data in the database as they progress through the design. This is not something that should be left as a clerical function at the end because ground will have to be retraced and things always seem to get lost in translation. Having the database active in the designer's workspace at all times makes it convenient to log the information as it is discovered. By the time the drawings are marked up, the database will be done.

To summarize the products generated under this topic, we created a process area instrument plan drawing, which details the cable and conduit and the approximate locations of each instrument. We also generated some installation details. These documents are used to build a bill of materials, which is covered in Chapter 16, Procurement.

PART III – CHAPTER 14: INSTRUMENT AND CONTROL WIRING

Chapter 5 (Design Practice) described basic wiring techniques. Topics such as signal sinking versus sourcing, wire numbering techniques, and other topics pertinent to the art of wiring a control system were discussed. The NFPA 79 standard was discussed as well. This chapter is about the practical aspects of generating a set of I&C wiring drawings suitable for construction and plant maintenance.

What constitutes the I&C wiring set of deliverables? Figure 3-133 describes the relationships among the different types of drawings.

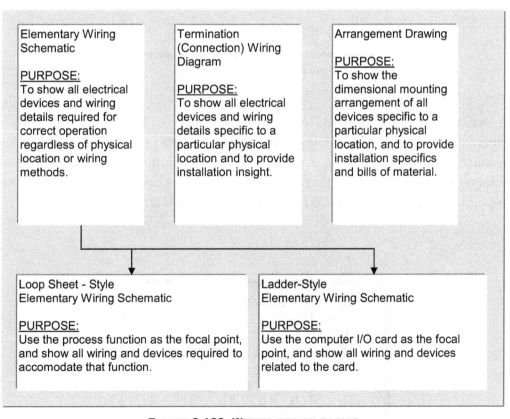

FIGURE 3-133. WIRING DESIGN BASICS

Why three different types of drawings? The answer: Each type has its purpose. Three are needed to cover three different perspectives.

First, the panel fabricator needs guidance in building the cabinetry. The fabrication function may be performed by a vendor or the constructor. While this may not be wiring per se, it is so closely tied to the wiring task as to be inseparable. The design of the enclosure must accommodate the operability requirements (Figure 3-134).

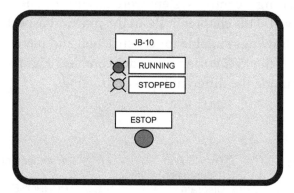

FIGURE 3-134. FABRICATION

Next, the constructor needs to be provided for. A set of drawings should be generated that presents the wiring information from the perspective of the wiring installer. An installer working in a junction box, for example, only needs point-to-point wiring information for that box, not for the whole circuit. This kind of point-to-point wiring drawing is called a *connection diagram* (Figure 3-135).

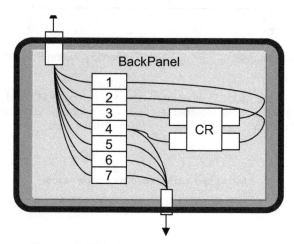

FIGURE 3-135. WIRING INTERCONNECTIONS

The maintenance technician must also be considered; his needs are discussed in Section A below). While the connection diagram is a good tool if a problem has been isolated to a particular location, it is woefully inadequate for troubleshooting the entire system. It is impossible, in most cases, to see an entire electrical circuit on one page. The big picture must be pieced together after the signal has been tracked through drawing after drawing.

The elementary wiring diagram, whether in the form of a loop sheet, ladder diagram, or wiring schematic, solves this problem. The elementary shows the entire circuit from power source to load, usually on a single sheet (Figure 3-136).

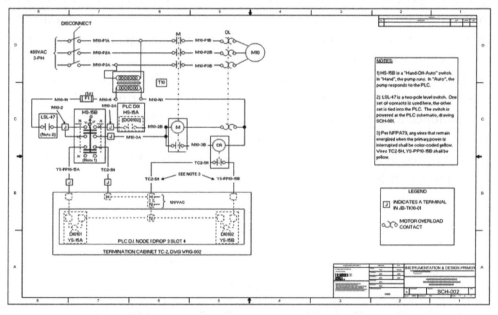

FIGURE 3-136. ELEMENTARY WIRING DIAGRAM

As intimated in the previous paragraph, there are several styles of elementary (also called schematic) wiring diagram, including instrument ladder diagrams and loop sheets, with several variations of each. As with P&IDs, the primary intent of an elementary is to convey functional information, at the expense of the physical if necessary. Loop sheets show the physical, to a degree. They depict the entire circuit on one drawing, but the components are organized by physical location.

Because the loop sheet and the instrument ladder diagram are frequently used in tandem—the loop sheet depicting the analog circuits and the ladder diagram the digital circuits—this chapter presents both. And since both depend on the wiring diagram to provide enough physical wiring detail for construction, a wiring diagram is presented along with a fabrication layout.

A. Instrument Elementary (Ladder) Diagram

Instrument elementary diagrams form the foundation of the electrical documentation set. These drawings show the complete circuit at a glance, allowing the viewer to see the entire path of electron flow. Unlike the wiring (connection) diagram, which provides a mechanical "connect the dots" approach to the physical act of wiring, the elementary diagram does not focus on the physical location of the equipment. Instead, it focuses on the electrical functionality of the system.

The instrument elementary *(ladder) diagram* focuses on one aspect of the control system. The PLC or DCS I/O module is often the focal point. In some cases, a piece of equipment or system being controlled is the focal point. It is generally a large drawing (22 by 34 inches; Size D) that shows all the wiring associated with each circuit. Historically, this drawing has been the method of choice for diagramming all electrical wiring, whether for motors, lights, garbage disposals, or instrument and control circuits.

In Figure 3-137, a typical instrument elementary is shown. This drawing sample has field devices (switches) pulling in relays that are in turn feeding the signals to annunciators and other control circuits. Lettered boxes indicate physical locations and circuit locations. Feed-through terminals are depicted as lettered boxes embedded in the wiring. Device locators are lettered boxes placed near devices. The letters indicate their location.

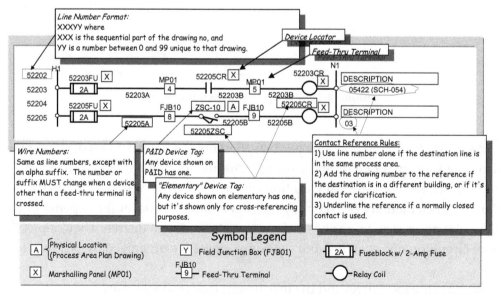

FIGURE 3-137. TYPICAL INSTRUMENT ELEMENTARY CONTENT

Line numbers are located in a column to the left of the graphic section of the drawing. These numbers are used as vectors. Each number is unique. In some cases—and this is probably the best scheme—a portion of the drawing number forms the first part of the rung number. Then a two-digit suffix is added to make the number unique, to give each drawing a unique 100-number block. These numbers are used to generate wire numbers and equipment designators called *electrical and instrumentation (E&I) numbers*. E&I numbers provide a shorthand method for navigating through the drawing set. This is not to be confused with P&I numbers, which are assigned on P&IDs. Each device may have either, neither, or both numbers assigned to it. Not all electrical devices show up on the P&ID, and an E&I number is required only if it is referenced by another drawing.

Wire numbers, at least in the scheme presented above, are assigned based on their proximity to rung numbers. Each wire segment has a number, though the number is shown only if it changes. Figure 3-137 shows wire numbers at every wire segment for clarification purposes. Notice that by convention, the wire number does not change as it crosses a feed-through terminal. It only changes if it crosses a device, such as a fuse, switch, or relay. If the wire number needs to be changed, the rung portion does not necessarily change. If the wire remains on the same horizontal plane, then the wire number is only modified by incrementing a letter suffix. If, however, the wire moves vertically and connects to a circuit on a different rung, the wire number changes to the new rung number the next time it crosses a device.

This scheme has some major advantages. Probably the foremost is the elimination of the drawing list as a primary document that must always be available to the maintenance personnel. In this scheme, finding a wire number automatically leads the technician to the proper drawing.

Relay contact references are assigned based on where they appear. Ideally, the contact and its coil are linked, as shown in Figure 3-138. In that case, there is no need to provide reference to contact locations because they are linked to the coil. But this is a rare luxury. Usually the contacts are embedded in logic that appears elsewhere, perhaps on different drawings entirely. In those cases, a method of cross-referencing needs to be employed.

In Figure 3-139, a four-pole relay coil is shown with references to the locations of its contacts. These references appear as the rung number where each contact is shown. The rung reference 10327 is underlined to indicate that a normally closed contact is used at that location.

At rung 10327, the contact for this relay is shown (Figure 3-140). Wire 10327B leads to relay 52203CR, contact block 4, and common terminal C4. The contacts are closed to pass the signal only if the relay coil is off. Otherwise, the contacts are open, thus blocking the signal. The signal leaves the relay on normally closed terminal NC4, and the wire number changes to 10327C.

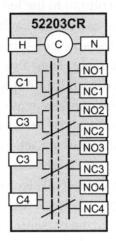

Figure 3-138. 4-pole relay coil with contacts

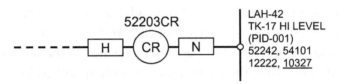

Figure 3-139. Four-pole relay coil with cross-references to its contacts

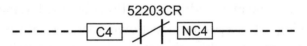

Figure 3-140. Four-pole relay contacts with cross-reference to its coil

A ladder elementary is usually used to show PLC or DCS digital circuit wiring to devices such as on/off valves, motors, and solenoids. Analog circuits may be shown in this format as well, though some prefer the loop-sheet format for those elements.

Rung numbering is optional, depending on the scheme employed. Some prefer using the instrument tag as a part of the number rather than a drawing number sequence or even some independent sequence tied to no methodology. While deriving rung numbers from the drawing number is probably best overall and would work great for a new facility, existing systems rarely employ that scheme. As a result, we will use a derivative of the instrument tag for our numbering process.

In our project example, the ladder diagrams are used to convey wiring information regarding power distribution and discrete (on/off) control circuits. We will use loop sheets for the analog circuits.

The following is a discussion of the primary elementary formats:

1. MOTOR ELEMENTARIES

Since the motor elementary (Figure 3-141) is a product of the electrical design team, we will not spend a great deal of time on it. However, the controls team does play a significant role in its production.

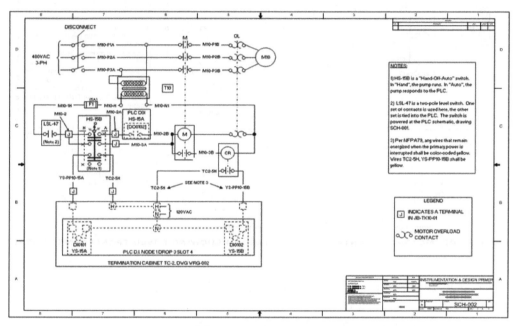

FIGURE 3-141. MOTOR ELEMENTARY WIRING DIAGRAM

This motor control circuit is self-powered in that the control logic power source is the motor winding voltage. A 4:1 stepdown transformer sits across two of the three phases and steps the voltage down to single phase 120 VAC. If the disconnect is thrown at the motor bucket,[10]* power is killed to the entire bucket, except where our controls are tied in. We will discuss this in more depth shortly.

The output of the transformer is fused (Figure 3-142), after which the interlock chain is usually powered up. In this case, the only hardwired interlock is the TK-10 low-level level switch LSLL-47. An HOA switch is next. If the operator rotates the switch to the Hand position, the pump starts and runs until the interlock is lost. If the switch is in the Off position, the pump stops no matter what other conditions are in place. If it is in Auto, the control system has control of the pump motor through HS-15A.

---

10. A "motor bucket" is a compartment in the MCC that contains the controls hardware for the motor.

448   SUCCESSFUL INSTRUMENTATION AND CONTROL SYSTEMS DESIGN

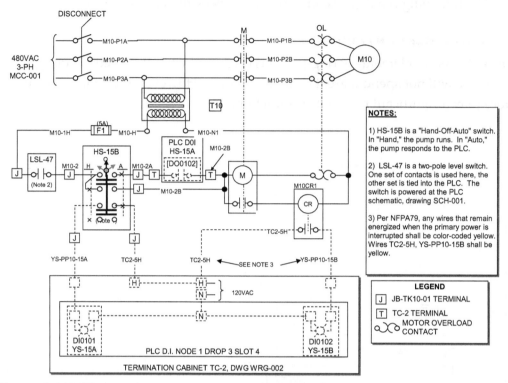

FIGURE 3-142. MOTOR ELEMENTARY WIRING DIAGRAM SHOWING FUSED TRANSFORMER OUTPUT

Two sets of contacts send status information back to the control system. The first, YS-15A, provides HOA switch status, "true" if in Auto, back to the PLC via digital input point DI0101. The other input, HS-15B, provides a run indication through the motor contactor's (M-coil) auxiliary contacts. *Note:* The "linkage" line is a mechanical link to all components that are touched by it.

Now, back to the disconnect discussion. It is desirable, though not a requirement, that the motor disconnect switch be able to kill all the power in the bucket. This is a safety consideration that minimizes the hazard to electricians, who may assume all power is off just because the motor disconnect has been pulled. This creates a design problem because of the PLC hardware we have selected. In order to accommodate this feature, we have selected a digital output module with output contacts that are electrically isolated from any PLC power, so they can be embedded directly into the control logic of the motor [assuming the inrush (starting) current of the motor coil is not too high]. The digital input modules, however, are another thing entirely.

The digital input modules are 120 VAC sinking, which means each I/O point on the modules shares a common neutral that is internal to the module (see the discussion in Chapter 5.D). For our purposes, we will assume we have no other module configurations available. This prevents us from embedding the digital inputs directly into the motor circuitry. We now have two choices:

1. Extend motor control power to the TC2 termination cabinet and install interposing (isolation) relays there to convert the signal power from the motor circuit to the PLC circuit.

2. Extend PLC power to the motor bucket and do the interfacing there.

In this case, we have elected to mix the power in the motor bucket. By doing this, we only need one relay since the HOA switch already has a spare set of contacts that we can use, and it is not physically in the motor bucket anyway. Note the bottom set of HOA contacts. These are completely isolated from the motor control circuit, yet when the operator rotates the switch to the Auto position, the contacts close (as indicated by the x on the mechanical linkage line).

The other signal we need comes right from the bucket. It is the "motor is running" status signal, which is generated by the motor starter relay's *auxiliary* contacts. These are called auxiliary (aux) because the primary contacts on this starter are very heavy ones, rated for 480 VAC duty, whereas the smaller aux contacts are rated for 120 volt service. If we could have had two sets of aux contacts, we would be all right. But, we only have one, and that one is needed as the seal-in contacts for the motor starter itself. That leaves us with the need to interpose a second relay between the motor power distribution and the PLC power distribution systems.

This interposing relay is labeled CR and is fed by wire number M10-3A. The name of this relay, when referred to elsewhere, is relay M10CR1. Since we are mixing power in this bucket, the externally powered wires (TC2-5H and YS-PP10-15B) need to be yellow in color to provide cautionary information to the electrician. This information is provided in Note 3 on the drawing.

2. AC Power Distribution Schematic

Power distribution needs to come first, if possible, to prevent rework on the other drawings later. In the power distribution ladder diagram shown in Figure 3-143, power is fed from a three-phase 480 VAC source. The three hot and one neutral wires are terminated at instrument power panel IPP-6. Notice one of the shortcomings of the ladder diagram: The neutral wire is not shown as landing in the IPP. That is understood. It is a good idea to show a terminal somewhere on the drawing to represent that neutral terminal. We will do that on the main drawing later.

In any case, power is fed through 10 amp breaker CB1 to the field devices via termination cabinet TC-2, terminal strip TS-1, and fuses F1 through F8. PLC rack power is fed through CB2 and fuse F9.

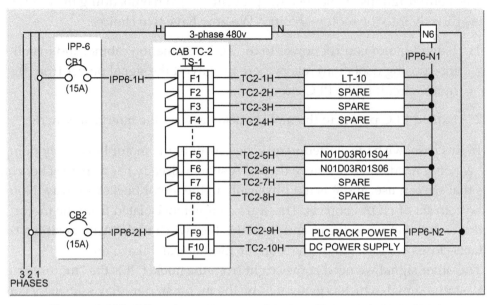

FIGURE 3-143. AC POWER DISTRIBUTION ELEMENTARY WIRING DIAGRAM

Notice the wire numbers, in our ideal world, take you back to the termination cabinet and fuse. This is a great aid when troubleshooting the system and should be employed if possible. Also, where the neutrals are concerned, we are violating our guideline that says we do not change wire numbers across feed-through terminals. This is a good place for an exception to our rule. It is easy to lose track of the amount of current flowing through our neutrals (returns). Many times the designer pays good attention to distributing power properly through the hot side, using fuses and/or breakers to manage the current. Then, he gathers multiple circuits up on a single return wire, ignoring the fact that all the currents for each circuit sum up when gathered onto a single return. If the return wire is common to several "hots," as ours is in this example, the return wire will need to be sized accordingly. In our case, we decided to use a single return wire from our I/O modules and a second return wire from our power supplies, giving each a unique number to help us manage them. Unfortunately, the numbering scheme is sometimes set by the customer and is not negotiable. In such cases, the designer must be highly organized in approach. It is easy to lose the handle on your power distribution system, so every available tool should be used.

Which brings up a good question: What is a good way to keep up with power-panel loading? The chart in Figure 3-144 is a very good method. It provides a reference to help locate the load and an estimate in watts for each breaker. This helps in the effort to balance the load across the three phases of the power source.

This information mimics the power distribution elementary somewhat, but goes a little further in helping manage the loading. This chart should be given a drawing number and maintained. Another good thing about this chart is that it can be kept in the power panel breaker box.

INSTR. POWER PANEL IPP-6						
DESCRIPTION	CKT No.	WATTS/PHASE			CKT No.	DESCRIPTION
		A	B	C		
TK-10 FLD DEVICES	1	60			9	
TC-02 PLC EQUIP.	2		180		10	
TK-11 FLD DEVICES	3			100	11	AT-66 ANALYZER
REACTOR FLD DEV.	4	120			12	
	5				13	
	6				14	
	7				15	
	8				16	
TOTAL WATTS		180	180	100	TOTAL WATTS	
NOTE: When assigning circuits, balance the phase load if possible.						

FIGURE 3-144. AC POWER PANEL LOADING CHART

3. DC POWER DISTRIBUTION SCHEMATIC

DC power is distributed in much the same way as AC power, except that the DC power is usually generated from a local DC power supply (Figure 3-145). This power supply output is then distributed (in our example) via a set of fuses in terminal strip 2 (TS-2).

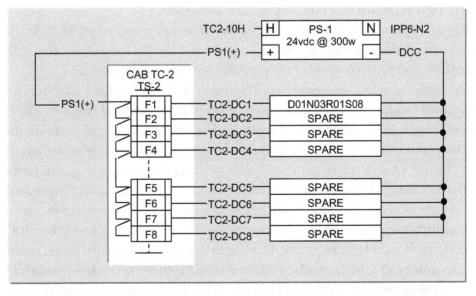

FIGURE 3-145. DC POWER DISTRIBUTION ELEMENTARY WIRING DIAGRAM

Again, note the wire numbers have meaning, taking the field technician back to the originating cabinet and fuse. The power supply feeds circuits through the various I/O modules, except for the analog output module, which sources its own current from backplane power. That is typical for analog output modules.

4. PLC Ladder Diagram (Elementary)

The PLC ladder elementary provides overall wiring information from the perspective of a specific equipment item. For a PLC ladder elementary, a PLC module is shown with all the wiring associated with it, including I/O point wiring. For a motor elementary, the wiring is from the perspective of the motor circuit, with all associated wiring shown. This approach does a good job of conveying information in a format that is relatively useful for construction, at least from the perspective of the equipment item, and that is very useful and efficient for maintenance. It is also probably the best format for efficient instrumentation design.

The limitation of this method is its inability to depict all the control wiring for a particular system on one drawing, which is *most* desirable for maintenance. For example, this method may show six transmitters, all from different tanks, as opposed to one transmitter and two valves, and seven other instruments, all on the same tank. An on/off valve, for example, is likely to have its solenoid shown on one drawing and its limit switches on another. It is incumbent upon the designer to provide good cross-referencing in such cases.

In Figure 3-146, the PLC is the focal point of the drawing. As has been mentioned, most ladder elementaries are large (at least 22 × 34 inches; Size D) and so can show a lot of information. A good rule of thumb is to assume 32 I/O points per size D drawing. That leaves room for all the wiring detail that might be needed, plus room for labeling, comments, and notes.

The conceptual ladder elementary shown in the figure is a hybrid style that has developed since the advent of the loop sheet. Like the loop sheet, this style of elementary is organized around the physical location of the device.

The traditional ladder elementary (Figure 3-147) was just that, a ladder. It had two vertical "supports" (power feeds) connected by horizontal rungs of logic. Current flowed from left to right. Depending on the level of detail desired, the depiction could show no terminals, or all of them, along with a reference vector to the enclosure. One of the major drawbacks to the earlier ladder diagram format was the fact that it did not represent wiring in a very realistic way. Single conductor wiring was at one time the norm but is no longer. Most instruments now use two-conductor shielded cables. The traditional ladder diagram had difficulty depicting such a cable as an entity. Having to wire using paired wires is more restrictive and can affect how the wiring should be shown. The hybrid sketch of Figure 3-146 shows the improved version with power distribution moved inside

PART III – CHAPTER 14: INSTRUMENT AND CONTROL WIRING   453

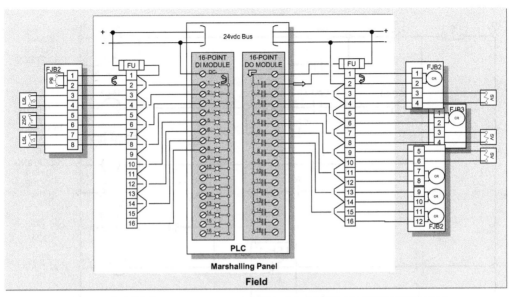

FIGURE 3-146. INSTRUMENT ELEMENTARY WIRING DIAGRAM CONCEPT

the field device, distributed in the marshalling panel, and returned through the PLC—just as it would really happen.

So for our little task, where do we start? The answer to that, once again, is with the database. In our Access database (see page 523 for instructions on obtaining this database), we previously entered the information needed for this task. Rather than create a new query for this task, we will work directly from the instrument and I/O list table. We can hide a few fields to reduce the clutter.

Open the table, and click on Format, then Unhide Columns. The Unhide selection opens a window that lets you hide or unhide (Figure 3-148).

We cannot eliminate much because the elementary is the repository for the entire load information. After our table is prepared, we may begin.

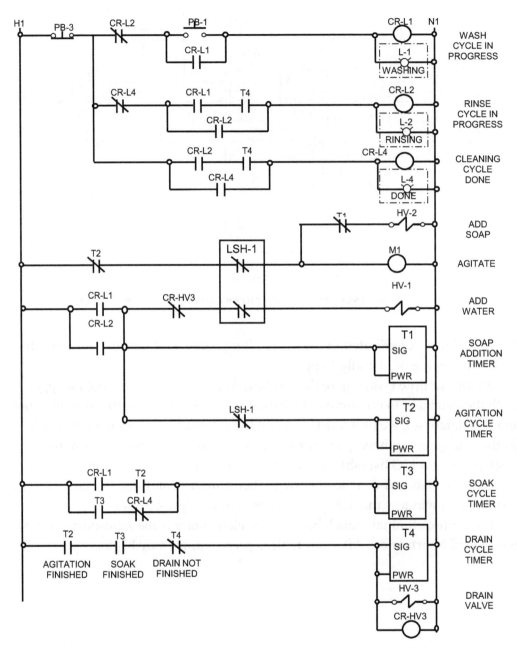

Figure 3-147. Traditional ladder elementary—Washing machine application

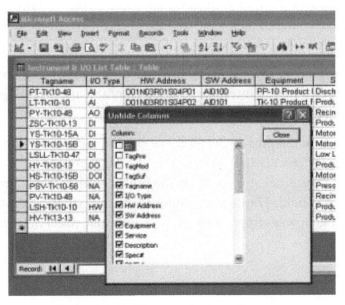

FIGURE 3-148. "UNHIDE COLUMNS" WINDOW

a. **Discrete (Digital) Inputs**

Select an I/O type of digital input (DI) by selecting a DI data cell and then picking the Filter by Selection button on the main menu (Figure 3-149). Doing this hides any data records other than I/O type DI.

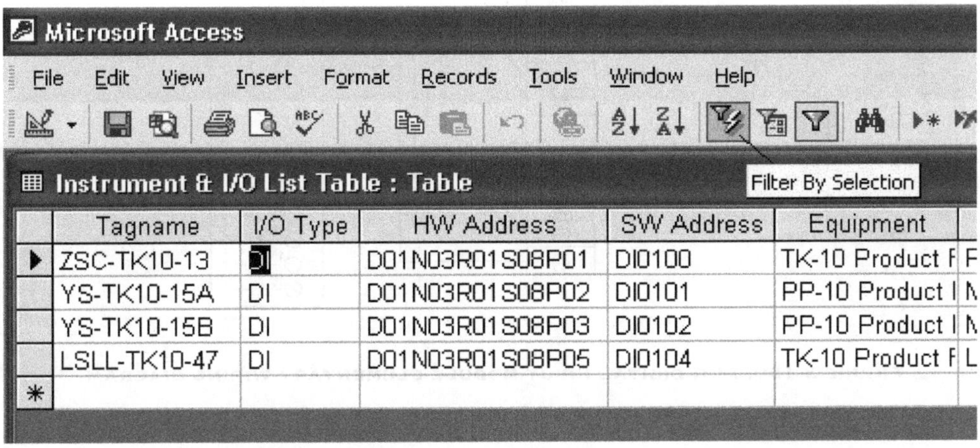

FIGURE 3-149. INSTRUMENT AND I/O LIST TABLE, FILTER BY SELECTION

This brings us to the first wiring task: the digital inputs. We must first wire up the digital input module and get it powered up and ready for the I/O. **Beware**! The PLC rack power supply should not be used to power I/O, even with 24-VDC I/O such as we have here. This is to prevent a problem in the field from taking down the entire PLC. It is best to have *signal* power separate from *backplane* power.

According to the I/O list, there are four digital inputs. The module we have selected has an internally bussed common and so is a *sinking* module (see "Sinking and Sourcing," Chapter 5.D.1a).

Starting with the power distribution, we find our AC power source at IPP-6, circuit 1 (Figure 3-150). Wires 1H and N1 supply power to our I/O. Fuse F5 on that terminal strip protects the power supply from shorts that may occur in the field. The current rating of the fuse should be low enough to protect the power supply and, at the same time, prevent heating up the wires.

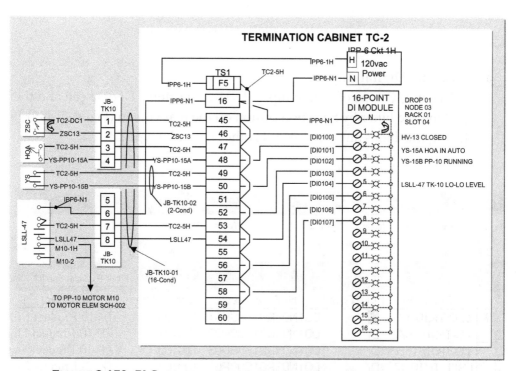

FIGURE 3-150. PLC DIGITAL INPUT MODULE ELEMENTARY WIRING DIAGRAM

The wire number changes across the fuse to TC2-5H to provide a link back to the fuse F5 in TC2 should power be interrupted to the inputs for this module. This hot wire is jumpered down the terminal strip to distribute power to each I/O point.

The first eight I/O points are prewired from the module to terminal strip TS-1, beginning at terminal 46. The wires are tagged with the software address rather than the actual wire number here. The tag is in brackets to indicate that it is a tag, not a wire number. This is a significant distinction because we want the field signal wiring to reflect the field device and the power wiring to reflect the power source. Those are the actual tag numbers. But we want to pre-wire between the I/O module and the terminal strip, with many of those points not yet assigned to a field device. So the software address, being a unique number and good information to have, is a good way to tag that wiring.

As we have seen, the rule for wire numbering is to change the number every time the circuit crosses a device. A device is a relay, fuse, solenoid, transmitter, or other electrical element—anything except a feed-through terminal or butt splice. Feed-through terminals and butt splices do not change the properties of the signal and should not interrupt it, and so are transparent to the numbering process (the exception being neutral wires, as discussed earlier). Hence our use of brackets to designate a tag as opposed to a number.

Wiring between termination cabinet TC-2 and field junction box JB-TK10 is via one 16-conductor cable, only part of which is shown on this drawing. The cable number takes its identity from the junction box number. All cables are identified from the field end. This facilitates identification in the termination (marshalling) panel.

Notice LSLL-47. This instrument is 120 VAC powered. The signal power has been jumpered at the device onto the H terminal to power up the unit. A neutral wire IPP6-N1 is included in the homerun cable to provide a return for the switch from TC-2, circuit 2.

Notice also the reference to the I/O module's place in the network at the upper right of the module. This information is sometimes neglected, to the ultimate dismay of the maintenance department.

b. **Digital Outputs, Isolated**
In the I/O list table, clear the previous filter by pressing the apply filter icon. This brings back the entire list. Pick one of the digital output, isolated (DOI) records and filter by selection again. The DOI records should appear as shown in Figure 3-151.

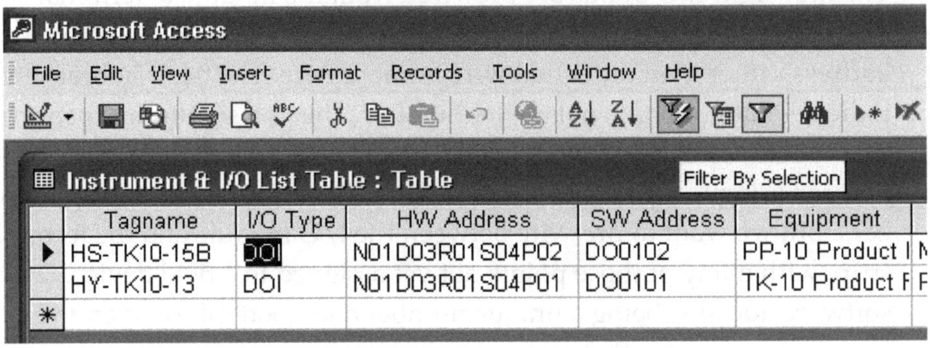

FIGURE 3-151. FILTERED ON DOI (DIGITAL OUTPUT, ISOLATED)

This gets us to the second wiring task: the digital outputs. As before, we must first wire up the digital output module and get it powered up and ready for the I/O.

Figure 3-152 shows an isolated digital output module. Each I/O point is electrically isolated from the other, with no internal jumpering. It is tempting to jumper the hot side at the module, but that would keep users from embedding the digital outputs into a circuit directly. So the module was prewired to a terminal strip, where wiring modifications are easier. Also, provisions were made for each I/O point to have its own fuse.

This module is linked to the AC terminal strip, though, technically, the signal type does not matter. Each I/O point is a "dry" (i.e., unpowered) contact. Note that the terminal and fuse numbers pick up where the discrete input module leaves off.

B. LOOP SHEET (REF: ISA-5.4-1991)[14]

The loop sheet provides probably the best compromise between the needs of engineering, construction, and maintenance (Figure 3-153). While not all instrumentation on a particular equipment item is depicted, at least one aspect of it is shown in detail. For example, if a maintenance technician troubleshoots a problem with the tank's level controls, odds are he will be able to pull one loop sheet for the task. The designer must handle a lot of paper, with fewer instruments shown per sheet, but much of that can be cloned. For example, all level control loops are graphically similar, so only the text needs to be modified. And if intuitive numbering schemes are employed, even modifying text can become a minor issue. The constructor must still rely heavily on physical drawings, but the loop sheet is very useful.

Like the ladder diagram, the loop sheet provides information about the complete circuit. Power flow is easy to follow from the power source to the load and

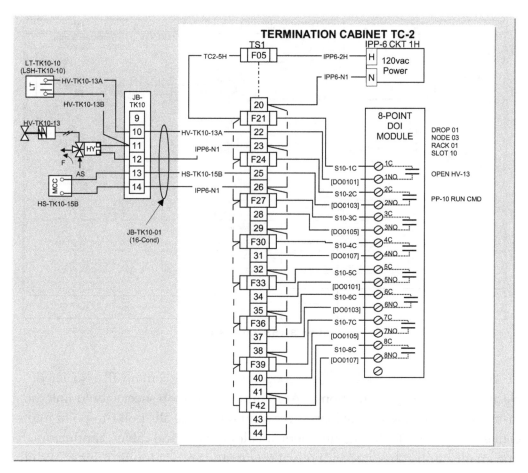

FIGURE 3-152. DIGITAL OUTPUT (ISOLATED) PLC OUTPUT MODULE ELEMENTARY WIRING DIAGRAM

back. While discrete (on-off) information can be presented very well on the loop sheet, it is optimized for analog equipment. The loop sheet usually has information regarding the process connections and sometimes provides information usually found only on a specification sheet, such as manufacturer and model number and scaling and calibration information.

The loop sheet is generally an 11- by 17-inch (Size B) drawing that shows all the wiring necessary for a particular process control loop. The term *control loop* is generally associated with a sensor, a controller, and an output device of some kind. But loop sheets also depict wiring required for individual instruments, such as pressure transmitters, that might not be linked to an output device. The loop sheet organizes wired devices by their physical location. The order is generally left to right from the field to the controller or bottom to top from the controller to the field.

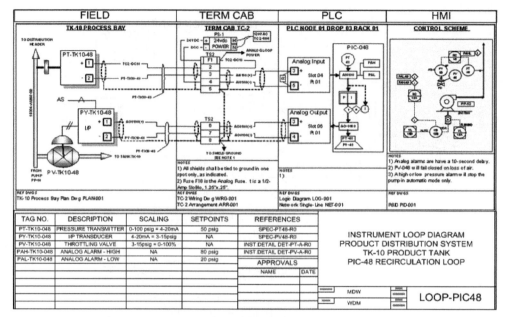

FIGURE 3-153. LOOP SHEET

According to ISA-5.4-1991[14], the loop sheet should "contain all associated electrical and piping connections." At minimum, the loop sheet should link the instrument to the P&ID. The loop sheet should provide all "point-to-point interconnections with identifying numbers or color of electrical cables, conductors, pneumatic multi-tubes, and individual pneumatic and hydraulic tubing." While there is more in that specification, the key point is that the loop sheet differs from the elementary in the level of mechanical detail that should be present.

Once again, the first thing to do is to consult the database. This time, we need both the analog input and the analog output. This calls for an advanced filter/sort (Figure 3-154).

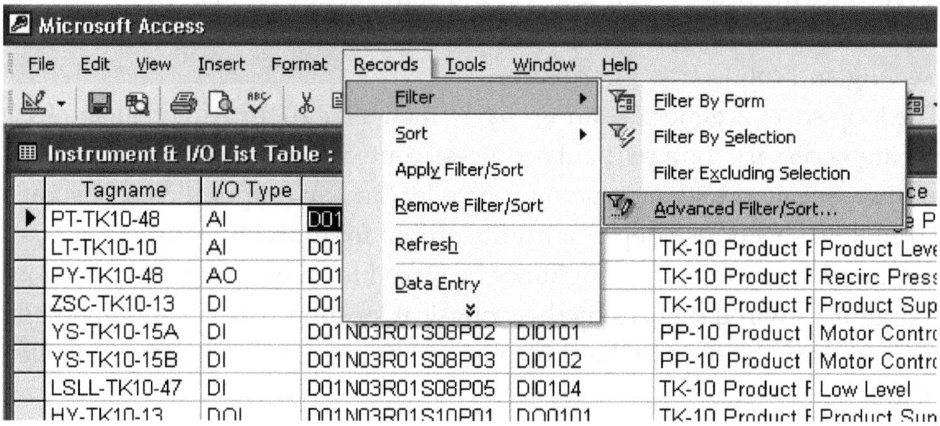

FIGURE 3-154. ADVANCED FILTER/SORT

The advanced filter/sort opens a design view in which the user may pick the fields to use (Figure 3-155). We have selected two fields: SW Address and HW Address. We filter out anything that does not start with an A and then sort by HW Address.

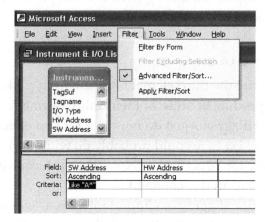

FIGURE 3-155. ADVANCED FILTER/SORT, FIELD SELECTION

After we apply the filter/sort, the listing appears as shown in Figure 3-156.

Tagname	I/O Type	HW Address	SW Address	Equipment	Service	Description	Spec#	P&ID#
PT-TK10-48	AI	N01D03R01S08P01	AI0100	PP-10 Product	Discharge Pres	Pressure Transr		PID001
LT-TK10-10	AI	N01D03R01S08P02	AI0101	TK-10 Product F	Product Level	Level Transmitte		PID001
PY-TK10-48	AO	N01D03R01S10P01	AO0100	TK-10 Product F	Recirc Pressure	I/P Transducer	NA	PID001

FIGURE 3-156. RESULTS OF ADVANCED FILTER/SORT

Given the information already captured in the database, it is possible to begin work on the loop sheet (Figure 3-153). Loop sheets are organized by physical location. As mentioned above, the information usually progresses from the field device on the left to the control system and logic information on the right.

The loop diagram can be a beautiful drawing if the designer has the time to put into it. This example (Figure 3-153) shows how one can be done right, with all the bells and whistles. Economics often dictates a more basic product. But even if that is the case, the loop sheet is still a useful tool in the hands of the maintenance technician.

A drawback to the loop sheet is the way the process wiring gets "chopped up" into a bunch of small packages. The loop sheet increases the number of documents that must be handled and is not useful to the installer unless only one loop is being installed. Mass termination (where an electrician terminates hundreds of wires in a single panel) is definitely difficult if the only tool available is the instrument loop sheet. For all these reasons, cross-referencing into the other drawing sets is important.

An alternative to the loop sheet is the ladder diagram already discussed. An example of a ladder diagram for analog circuits is not provided here, but through a review of discrete ladder diagrams, an analog version can easily be envisioned.

C. Connection Diagrams

Connection diagrams are the third element of the wiring package. They are a virtually indispensable product because of their utility during construction. Ladder elementaries and loop diagrams just do not cut it for cost effective mass-termination tasks. Our little project here could probably do without connection diagrams, but they are a necessity for larger tasks. (Note that the terms *connection* and *termination* are interchangeable in this context.)

Why is the connection diagram so important? It is a "connect the dots" diagram that guides the wiring installer through the construction process. If connection diagrams are not provided by the design engineer, then the construction foreman invariably has to create them from scratch in the field. To minimize cost, the construction foreman will probably assign the task to the most junior personnel, who are the least capable of seeing the big picture or correctly interpreting drawings.

Given the importance of connection diagrams, why not use them exclusively and forget about the other two wiring products? Some sites do use them exclusively. But the connection diagrams do not provide a reasonable maintenance package. It is very difficult to tell, for example, if there is a complete path for current flow. In those cases, the maintenance technician may end up sketching the circuit while tracking the wiring through the drawing set—not a good place to be with the plant down at 2:00 a.m.!

Why is it better to create them in the engineering office as opposed to in the field with construction personnel? Because the engineering team is better able to see the big picture. They will conduct a design check and can include this product in the check to make sure the connection diagrams are properly integrated into the rest of the design package. The larger the project, the bigger the benefit.

So how do we begin? We must first have a complete set of elementary diagrams, or nearly so, and/or a set of logic diagrams. Instruments should be spotted on the process area plans to get a feel for the physical orientation of the instruments. Also, enclosure sizes should be tentatively chosen to align with the number of terminations that will likely be available. The enclosure configuration is not concluded until this wiring design task is complete, however.

We will start with the field junction box. In our case, ferreting out the information from the drawing set is simple—only AC circuits will be fed through this box.

First of all, we know that we have a 16-conductor homerun cable that connects the FJB with the termination cabinet. This cable is a good place to begin.

1. JUNCTION BOX JB-TK10-1: INITIAL LAYOUT

Probably the best place to lay out a connection diagram is in Microsoft® Excel. Open a workbook and click on the upper left row/column block. (It has no row/column designation.) The entire spreadsheet will turn black. Then position the cursor over one of the column header buttons and right click. Select Width and type in 1.71. Hit return, and voilà! You have graph paper with perfectly sized boxes. Note: The actual value to type in may differ from computer to computer, just experiment until a reasonable square is achieved, and then see what the value should be by right-clicking on the column header, and selecting "width." See what that number is, and use it throughout.

Somewhere in the middle of the page, select a group of three boxes by clicking inside one cell and then dragging across two more. Then, select Merge and Center from the menu (Figure 3-157). The three cells are now one.

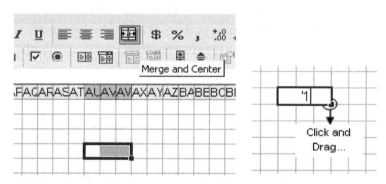

FIGURE 3-157. CREATING A CONNECTION DIAGRAM

Key in an apostrophe and a 1 ('1) in the cell that was just merged, and then grab the handle and drag down until there are 16 merged cells, numbered 1 through 16. Your terminal strip was just created. While the terminal strip is selected, use the All Borders button to draw lines around the terminals (Figure 3-158).

To the left of the top terminal, skip one column of cells and again enter '1. Then drag down again as before. This creates the conductor numbers for the 16-conductor homerun cable.

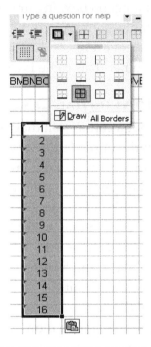

FIGURE 3-158. TERMINAL STRIP CREATION

Now we need to consult the instrument elementary diagrams generated earlier (Figure 3-159). Starting with the digital input module, it appears that we will require seven wires from TS-1 in the termination cabinet to terminals 1 through 7 in the junction box.

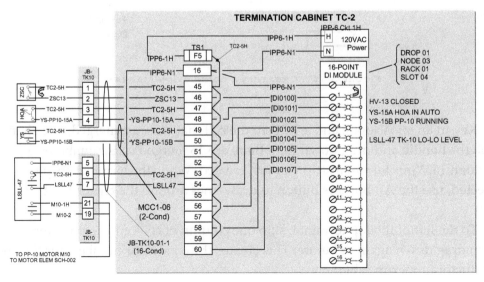

FIGURE 3-159. INSTRUMENT ELEMENTARY DIAGRAM, DIGITAL INPUT MODULE

Note that level switch LSLL-47 requires logic power. The hot terminal can be powered from the common side of the alarm relay, using TC2-5H. But there is no neutral. So we must add a neutral wire (IPP6-N1) to the wire bundle going from TC-2 to JB-TK10-01 terminal 5. Likewise, we should consult the elementary diagram for the digital output module (Figure 3-160). It appears that terminals 8 through 10 will be taken with these circuits. Now we can begin configuring the wiring diagram using the spreadsheet. We will do it by the numbers (Figure 3-161):

1. Configure the terminal strip as described above. Set up about 26 terminals.

2. Designate the left side of the terminal strip as the field side and the right side the *homerun* side. List the wire numbers shown on the elementary drawings on the proper side of the terminals. Then, on the right side of the terminal strip (column J), terminate the 16 conductor homerun cable that will be routed to termination cabinet TC-2.

3. In column K, use the following format to create a routing code: XX-Y-ZZ, where XX is the routing number, Y is the cable number within the route, and ZZ is the wire number (or color) within the cable. In our case, we will use conductor numbers rather than colors. Try to keep the conductors together, but we will soon see that terminal 9 is a field terminal and will need to be skipped.

4. Next, check the analog circuits for any that may require power other than loop power (i.e., circuits containing four-wire devices). One such exists, though we did not do a loop sheet for it. Level transmitter LT-10 requires 120 VAC power. Looking at the power distribution elementary (Figure 3-143), we see that LT-10 is fed from fuse TS1-F1, so we must add TS2-1H and IPP6-N1 to our terminal strip as well. This power is being fed from TC-2 and so will need to arrive via the homerun cable. Use conductors 15 and 16 at terminals 16 and 17.

5. In column C, list the field device ID number (i.e., instrument tag). Use the Merge Cells icon to group the wires to the device (see cells C5 and C6, for example).

6. In column D, list the terminal number of each device. If the actual terminal number is not known, list the function (e.g., H for hot).

7. In column E, enter the field cable routing code. This code is of the same format described for the homerun cable.

8. Paint the entire sheet white by clicking the blank Row/Column button at the upper left corner of the spreadsheet. Then, use the paint bucket icon to paint the spreadsheet.

9. Use the inside border icon to put lines around the text. Dress up the header as desired, and voilà!

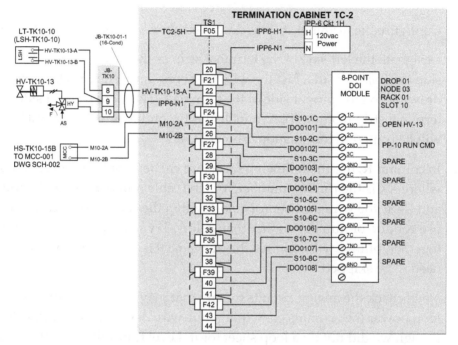

FIGURE 3-160. INSTRUMENT ELEMENTARY DIAGRAM, DIGITAL OUTPUT MODULE

A connection diagram (wiring schedule; termination chart) has now been produced for the junction box that may be used throughout the life of the project (Figure 3-162). At the end of the project, this connection diagram may be used to quickly generate a wiring diagram.

As additional cables are added, new columns may be inserted. For example, cable 1 of routing A2 is the homerun cable. If a second cable is needed for that route, then a new column may be added to the right labeled A2-2. New routings may be added in the same way.

Our labors have resulted in a spreadsheet showing the wire numbers, the terminal numbers, and the routing codes. Why, you might ask, are there so many of the same wire in the homerun cable? There are four IPP6-N1 wires and five TC2-5H wires. Each of these is a current-carrying conductor. As we have discussed, if we were to combine all four N1 wires, that wire would carry the sum of the four currents. So rather than having to keep track of the additive currents, and to

	A	B	C	D	E	F	G	H	I	J	K	L	M	N
1														
2														
3			(5)	(6)	(7)	(2)		(1)		(2)	(3)			
4			DEVICE	TERM	ROUTE	WIRE#		TS-1		WIRE#	ROUTE			
5			ZSC-13	C	A1-01	TC2-5H		1		TC2-5H	A2-1-01			
6				NO	A1-02	ZSC13		2		ZSC13	A2-1-02			
7			YS-PP10-15A	C	A3-01	TC2-5H		3		TC2-5H	A2-1-03			
8				NO	A3-02	YS-PP10-15A		4		YS-PP10-15A	A2-1-04			
9				N	A1-03	IPP6-N1		5		IPP6-N1	A2-1-05			
10			LSLL-47	H	A1-04	TC2-5H		6		TC2-5H	A2-1-06			
11				NO1	A1-05	LSLL47		7		LSLL47	A2-1-07			
12			LSH-TK10-10	C	A1-06	HV-TK10-13-A		8		HV-TK10-13-A	A2-1-08			
13			HV-TK10-13	H	A1-07	HV-TK10-13-B		9						
14				N	A1-08	IPP6-N1		10		IPP6-N1	A2-1-09			
15					A1-09			11			A2-1-10			
16								12			A2-1-11			
17								13			A2-1-12			
18			YS-PP10-15A	A1	A3-03	M10-2A		14		M10-2A	A2-1-13			
19								15			A2-1-14			
20			LT-TK10-10	H	A1-09	TC2-5H		16		TC2-1H	A2-1-15			
21				N	A1-10	IPP6-N1		17		IPP6-N1	A2-1-16			
22								18						
23								19						
24								20						
25								21						
26								22						
27								23						
28								24						
29								25						
30								26						

FIGURE 3-161. TERMINATION DRAWING SETUP

prevent "piling on" as the maintenance folks add circuits, it is best to just go ahead and run them separately.

Now we must layer on the remaining routing information. We know from our earlier work on the instrument location plan drawing that several conduits exit this junction box. Revisiting the plan drawing shows us five conduits entering or exiting the junction box (Figure 3-163).

We have completed the wiring to our field devices. However, we aren't finished yet! There are still two undefined conduits hitting our box. Routing A2-022 connects the TK10 process bay to the adjacent bay. Drawing -022 provides information as to the contents (not included here). Routing A4 is going to the MCC.

We know from looking at the motor elementary that several wires need to be routed to the HOA switch mounted at the motor. We have accounted for the two HOA wires that go to TC-2. However, there is still some wiring to pass to the MCC.

Revisiting the motor elementary, a few things become evident (Figure 3-163). First, we aren't done with LSLL-47! We have captured the signal that feeds the PLC, but not the signal for the hardwired output. Wire M10-1H arrives from the MCC and must be passed to the level switch via routing A1. Wire M10-2 comes

JB-TK10-01 WIRING SCHEDULE

DEVICE	FIELD SIDE			TS-1	HOMERUN SIDE	
	TERM	ROUTE	WIRE#		WIRE#	ROUTE
ZSC-13	C	A1-01	TC2-5H	1	TC2-5H	A2-1-01
	NO	A1-02	ZSC13	2	ZSC13	A2-1-02
YS-PP10-15A	C	A3-01	TC2-5H	3	TC2-5H	A2-1-03
	NO	A3-02	YS-PP10-15A	4	YS-PP10-15A	A2-1-04
LSLL-47	N	A1-03	IPP6-N1	5	IPP6-N1	A2-1-05
	H	A1-04	TC2-5H	6	TC2-5H	A2-1-06
	NO1	A1-05	LSLL47	7	LSLL47	A2-1-07
LSH-TK10-10	C	A1-06	HV-TK10-13-A	8	HV-TK10-13-A	A2-1-08
HV-TK10-13	H	A1-07	HV-TK10-13-B	9		
	N	A1-08	IPP6-N1	10	IPP6-N1	A2-1-09
		A1-09		11		A2-1-10
				12		A2-1-11
				13		A2-1-12
YS-PP10-15A	A1	A3-03	M10-2A	14	M10-2A	A2-1-13
				15		A2-1-14
LT-TK10-10	H	A1-09	TC2-5H	16	TC2-1H	A2-1-15
	N	A1-10	IPP6-N1	17	IPP6-N1	A2-1-16
				18		
				19		
				20		
				21		
				22		
				23		
				24		
				25		
				26		

FIGURE 3-162. TERMINATION CHART

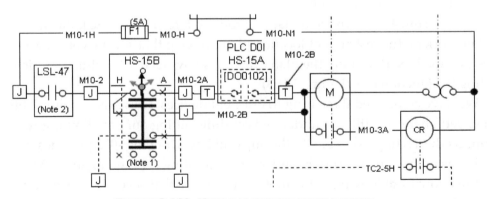

FIGURE 3-163. MOTOR ELEMENTARY FRAGMENT

back from the level switch and is passed to the HOA switch via routing A3. Two wires come back from the HOA switch. The first, M10-2A, is passed to TC-2 via routing A2. The second, M10-2B, is sent back to the MCC via routing A4. We must capture these moves in our termination chart (Figure 3-164).

| JB-TK10-01 WIRING SCHEDULE ||||||||
| FIELD SIDE |||| TS-1 | HOMERUN SIDE |||
DEVICE	TERM	ROUTE	WIRE#		WIRE#	ROUTE
ZSC-13	C	A1-01	TC2-5H	1	TC2-5H	A2-1-01
	NO	A1-02	ZSC13	2	ZSC13	A2-1-02
YS-PP10-15A	C	A3-01	TC2-5H	3	TC2-5H	A2-1-03
	NO	A3-02	YS-PP10-15A	4	YS-PP10-15A	A2-1-04
LSLL-47	N	A1-03	IPP6-N1	5	IPP6-N1	A2-1-05
	H	A1-04	TC2-5H	6	TC2-5H	A2-1-06
	NO1	A1-05	LSLL47	7	LSLL47	A2-1-07
LSH-TK10-10	C	A1-06	HV-TK10-13-A	8	HV-TK10-13-A	A2-1-08
HV-TK10-13	H	A1-07	HV-TK10-13-B	9		
	N	A1-08	IPP6-N1	10	IPP6-N1	A2-1-09
		A1-09		11		A2-1-10
				12		A2-1-11
				13		A2-1-12
YS-PP10-15A	A1	A3-03	M10-2A	14	M10-2A	A2-1-13
				15		A2-1-14
LT-TK10-10	H	A1-09	TC2-5H	16	TC2-1H	A2-1-15
	N	A1-10	IPP6-N1	17	IPP6-N1	A2-1-16
				18		
YS-PP10-15A, LSLL-47	C1,C2 C2	A3-04, A1-11	M10-2	19		
HS-PP10-15A	H1	A3-05	M10-2B	20	M10-2B	A4-1-01
LSLL-47	NO2	A1-12	M10-1H	21	M10-1H	A4-1-02
				22		A4-1-03
				23		A4-2-01
				24		A4-2-02
				25		A4-2-03
				26		

FIGURE 3-164. FINISHED TERMINATION CHART

This chart is ready to hand off to someone for creation of the junction box drawing. It is recommended that the actual drafting task be done during the latter stages of the design job, particularly if there is the possibility of some changes. It is much easier to manipulate information while it is in spreadsheet form.

After all this, generating the junction box wiring diagram shown in Figure 3-165 is relatively simple.

2. TERMINATION CABINET TC-2

The termination cabinet is a bit more complex just because there are more wires plus some equipment. A thumbnail sketch of the inner panel is shown in Figure 3-166. We know we will have two terminal strips because we have a mix of power: AC and DC. The smaller DC strip is on the right side of the panel. The PLC rack is top center.

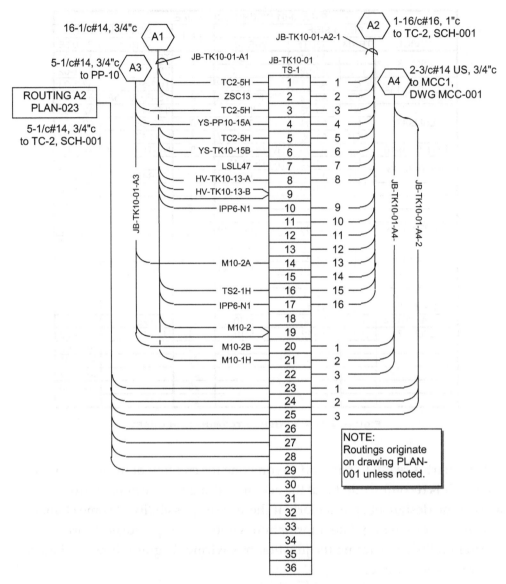

FIGURE 3-165. JUNCTION BOX WIRING DIAGRAM

Because it is simpler, we will work on the DC side first:

a. **DC Circuits (TS-2)**

We'll work on TS-2 first because it should go quickly. Again, we will start in Microsoft® Excel, but the principles are the same regardless of the spreadsheet software used.

In reviewing the DC power distribution schematic, we see that eight fuses have been detailed. We will start with the first one since it looks like it may be the only one needed.

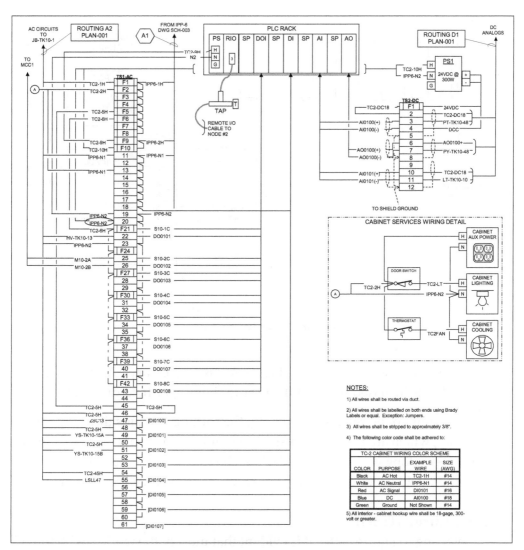

FIGURE 3-166. INNER PANEL, CABINET TC2

We have created one of the two loop sheets. The one we created covers two of the three analog instruments. We can proceed and anticipate the level transmitter's needs without the other loop sheet.

In the segment of the wiring diagram shown in Figure 3-167, we see DC terminal strip TS-2. The strip is powered from DC power supply PS1, which is, in turn, fed from fuse 10 in the AC terminal strip (note wire number TC2-10H).

Starting at the field end, conduit routing D1 arrives at the cabinet. We can tell that the routing originates on another drawing because it is a rectangle

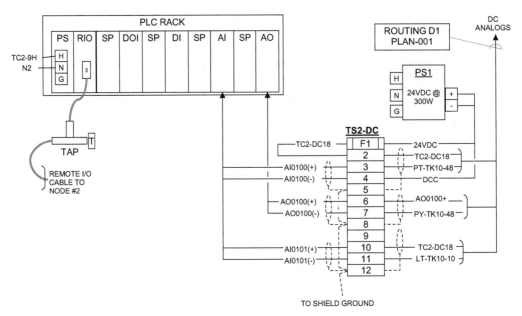

FIGURE 3-167. DC WIRING

rather than a hexagon (refer to Figure 3-118 if necessary). And the drawing number is included to direct the user to the proper field drawing.

According to the reference drawing, the routing contains three type B cables, which are twisted-pair shielded cables. These cables are terminated on TS-2, along with their shields. The shields are gathered and terminated at the shield ground bus bar. The signals are then forwarded to the PLC.

b. **AC Circuits (TS-1)**

Referring to Figure 3-166, it is evident that most of the work is with the AC circuits, which are to the left on TS-1. All of the discrete (on/off) circuits are driven by AC power. Also, notice that there are quite a few empty terminals. It is always good practice to prewire the connections between the control system hardware and the marshalling terminal strip(s). The connectors on the I/O modules usually present a wiring challenge, requiring more manual dexterity than most electricians can display when effecting a quick repair on a hot system. It is much easier to work with the wiring after it is broken out on a terminal strip.

The wiring appears to be quite involved, but it is really a "connect the dots" proposition if the installer has the proper documentation. Most installers would rather spend their time exhibiting a good termination technique than deciphering a set of elementary drawings.

So let's spend a minute looking at the situation from the installer's point of view. For the installer, a good termination drawing exhibits the following characteristics:

- The drawing should be focused. It should present just the needed information and no more.

- The components should be oriented properly. Ideally, the termination drawing should show the components in roughly the correct position and in exactly the right orientation from the wiring perspective. It is not as important to exactly match the arrangement drawing as it is for the wires to be shown in the right orientation. For example, if a wire is shown hitting the left side of the terminal, make sure it is possible to terminate to that side. We'll talk more about that in a moment.

- The drawing should be well cross-referenced, indicating conduit routings and wiring destinations. Also, there should be an index to the set of elementaries, loop sheets, or other such schematics in case they are needed for clarification.

There are also a few things to stay away from:

- Don't put too many wires on one terminal. Know how many wires can be terminated on one side of the terminal or fuse block. Usually, the terminal manufacturer can provide a breakdown of how many wires per gauge can be terminated. There is also a min and max for wire gauge. Be sure you aren't asking the installer to do the impossible. And remember, a jumper costs two terminals.

- Don't wrap the wiring around the terminal strip to change sides if it can be avoided. That makes it harder to see and harder to troubleshoot after the duct cover is in place. Provide a feed-through terminal to get a signal from one side of the strip to the other. In the section of TC-1 shown in Figure 3-168, TC2-5H is powering a string of jumpers. But the jumpers needed to be on the right side of the strip. So we used a single terminal to feed the wire through rather than loop it around.

Remember, the installer has physical space limitations and is unable to just wrap wires around anywhere. Such problems need to be anticipated.

In reality, the wiring problem presented here is really rather simple. Let's zoom in a bit and discuss some of the particulars.

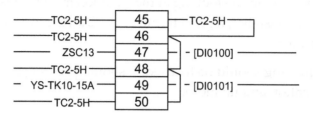

FIGURE 3-168. WIRING DIAGRAM SECTION OF TC-1

c. **Incoming Power**

Power is fed to the cabinet from instrument power panel IPP-6, part of which is shown in Figure 3-143.

This figure depicts cabinet power originating at IPP-6. There are two circuits allocated. Circuit 1 is reserved for instrument power. Of course, only LT-10 is powered directly from the TC-1 fuses at fuse F1. IPP-6, circuit 2, is for the TC-2 control system power, which is distributed to the PLC on fuse F9 and to the DC power supply on fuse F10. Notice also that the neutrals change numbers, which is a violation of the requirement to change numbers only across devices, not feed-through terminals. As we have seen, this is an exception to the rule. It assists in managing the return current flowing on each individual wire and also in isolating problems. It is difficult to find a specific wire on a bus bar when they are all numbered the same. **Caution:** This is an exception that will need to be approved by the customer beforehand and well documented if it is applied.

Now, looking at the TS-1 AC terminal strip shown in Figure 3-169, notice how the wiring reflects the Figure 3-168 power distribution schematic. Cabling is extended from IPP-6 to TC-2 via routing A1. The A1 routing comes from TC-2 cabinet drawing WRG-001, so the full number is A1-WRG001.

Wire numbers in this scheme reflect the circuit number, so IPP6-1H originates at circuit 1 of IPP-6. It distributes to the first eight fuses. Notice that fuses F1, F5, F6, F9, and F10 are used, just as depicted on the schematic, and that the fuses are jumpered as shown. The difference is that the wiring on the connection diagram is shown accurately in terms of position, as opposed to accurately in terms of function.

Notice also that the load for fuse F1 is not in the cabinet. The wire leaving fuse F1 (TC2-1H) joins a multiconductor cable in routing A3-PLAN001, which is the instrument location plan drawing. A detail of this plan

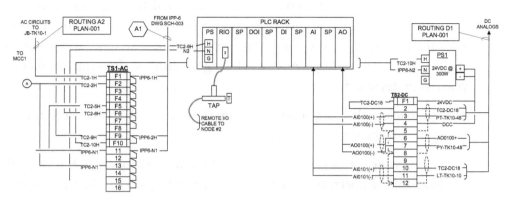

FIGURE 3-169. AC POWER DISTRIBUTION

drawing is shown in Figure 3-170. Notice the conduit exiting the bottom of LT-10. That is the power feed that originates at the junction box.

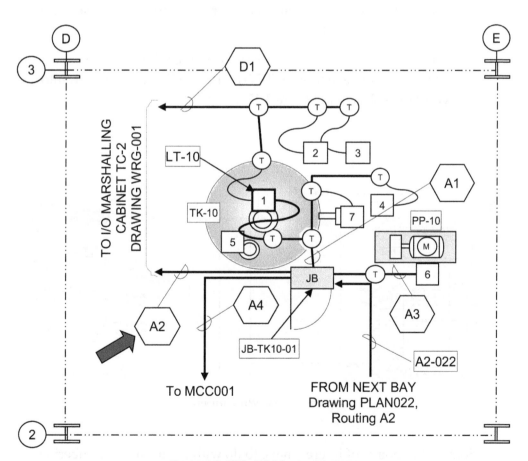

FIGURE 3-170. LT-10 POWER FEED

Coming out of the junction box is routing A3, which is our routing of interest. Notice the difference in routing symbols: the first, on the connection diagram, spells out the entire routing callout, whereas the routing on the originating drawing is merely a hexagon with the drawing number portion understood.

Let's go back to the connection diagram (Figure 3-169). Notice a couple of things regarding drafting practice:

- All crossing wires (or at least most of them) are broken on the vertical. None of the horizontal wires are broken. It is good drafting practice to pick one style and stick with it.

- There are no "node dots." An electrical node is a point at which several wires are attached, each with the same voltage. In a connection drawing, using node dots to indicate joined wires is deceptive and causes the designer to make assumptions about the number of terminals to include. This type of drawing should reflect the exact physical situation facing the installer.

- Wires may be joined into a wire run provided they are broken out on both ends. Notice that wires TC2-10H and TC2-N2 are joined into a wire run as shown in Figure 3-171.

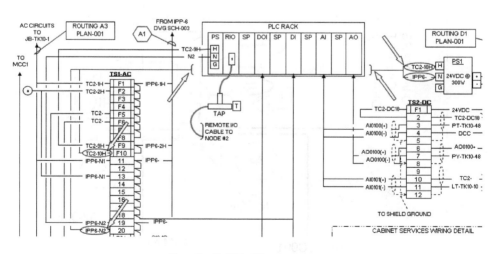

FIGURE 3-171. WIRE RUNS=

Some other points of interest have to do with the terminal numbering scheme. The diagram in Figure 3-172 is split to provide two sections of the drawing, top versus middle. Notice that the fuses are numbered in sequence with the terminals rather than given a unique fuse number.

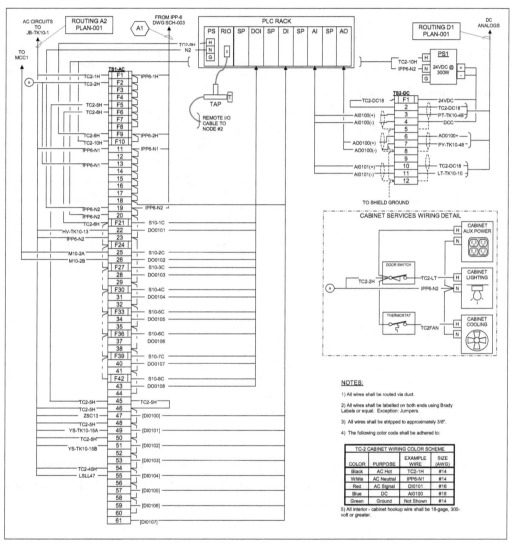

FIGURE 3-172. FUSE/TERMINAL NUMBERING SEQUENCE

Why numbers and not colors? Colors are okay except that some folks are color-blind. Also, it is entirely possible that the cable that the design assumes will be used won't be. Sometimes, unless a recognized color standard, such as NFPA79, is being applied, a different cable of equal properties is purchased instead, thus making the color scheme invalid.

NFPA79 is a standard that applies to many aspects of this task. Chapter 15, paragraphs 2 and 3 of the 1997 edition of NFPA79 deal with wire colors as listed in the chart shown in Figure 3-173.

NFPA79 WIRE COLOR GUIDELINES					
IEC & EUROPE		NFPA		PURPOSE	AUTHOR'S COMMENTS
BASE	STRIPE	BASE	STRIPE		
Green-and-Yellow	none	Green	Yellow	Ground	Stripe is optional
		Black	none	Ungrounded line, load, and control at line voltage	
		Red	none	Ungrounded AC control conductors at less than line voltage	
		Blue	none	Ungrounded DC Control conductors	Used for Intrinsically Safe Circuits, etc.
Orange	none	Yellow	none	Ungrounded Control circuit conductors that may remain energized when the main disconnecting means [for the equipment] is in the OFF position...	Example: Motor-Run status auxiliary contacts that have computer power running thru them. These will remain energized after motor main is opened.
		White or Natural Grey	none	Grounded circuit conductor	
Light Blue	none	White	Blue	Grounded (current-carrying) DC Circuit conductors	
		White	Yellow	Grounded (current-carrying) AC Circuit conductors that remain energized when the disconnecting means is off.	

FIGURE 3-173. NFPA WIRE COLOR SCHEME

Before the termination drawing gets published, the designer should make a second pass to indicate the colors of hookup wire or simply place a chart similar to that presented in Figure 3-174 on the drawing.

TC-2 CABINET WIRING COLOR SCHEME		
COLOR	PURPOSE	EXAMPLE WIRE
Black	AC-Hot	IPP6-1H, TC2-1H
White	AC-Neut	IPP6-N1
Red	AC Signal	DO0101
Blue	DC	AI0101(-)

FIGURE 3-174. TC-2 WIRING COLOR SCHEME

D. Wiring Summary

The wiring task is intricate. If it is done right, maintenance technicians should be able to track a signal in any direction with just one bit of information. If the system presented here is used, the wire numbers will direct them to the power source, the marshalling cabinet, or the instrument (if it is a signal wire). Once at the instrument, the power source is easy to figure out because it is listed on the wire numbers.

This same ability to track information should hold true throughout the drawing package. Properly done, each wire leaving a drawing has its destination drawing referenced, plus its routing number. (By the way, the routing number is the same as the cable number if there is only one cable in a routing. If more than one cable is in the routing, a suffix should be added to the routing number to differentiate the cables.) A cable schedule is produced later in this book (see Figure 3-192 "Wire and Cable Calculation Table").

Elementary wiring diagrams, loop sheets, and other such schematics are drawings that provide functional detail for a circuit. If necessary, the complete circuit should be readily available to the user at the expense of any physical orientation information.

Connection diagrams provide physical orientation. They are physical representations of the wiring produced primarily for installers and so should be optimized for their needs. Generating connection drawings is an absorbing but interesting task that requires a mix of practical aptitude and artistic flair. The connection diagram, coupled with the mechanical arrangement, tells the complete story about the wiring in an enclosure or piece of equipment.

We have created several drawings in this section. They are reproduced below. Figure 3-175 is the resulting TC-2 PLC cabinet connection diagram WRG-001.

The field junction box will end up with a mechanical arrangement and a bill of materials, which will be produced in the next chapter. In the meantime, junction box diagram WRG-002 shows its current state (Figure 3-176).

We have also partial motor elementary wiring diagram SCH-002 (Figure 3-177), which will be completed by the electrical group.

We now have power distribution information SCH-003 (Figure 3-178).

And we have generated three instrument wiring schematics. The first is the loop sheet for pressure control loop PIC-48, drawing LOOP-PIC48, shown in Figure 3-179.

The loop sheet is, in reality, an 11- by 17-inch drawing. The second schematic drawing is the ladder diagram for the PLC discrete modules in slots 04 and 06 of our rack (Figure 3-180).

Now we must make sure we have captured everything in the database. First, let's look at the document control table (Figure 3-181).

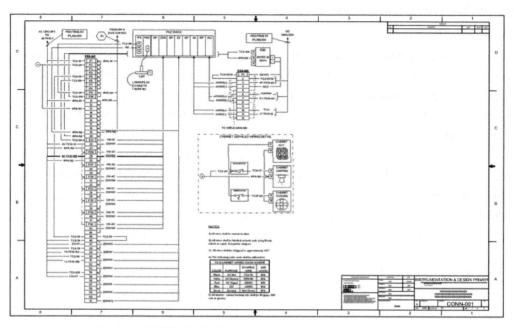

FIGURE 3-175. TC-2 PLC CABINET CONNECTION DIAGRAM

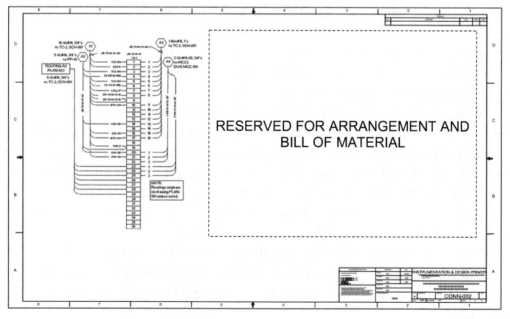

FIGURE 3-176. PARTIAL JUNCTION BOX DIAGRAM

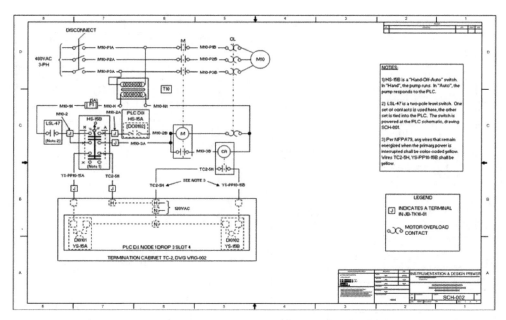

FIGURE 3-177. PARTIAL MOTOR ELEMENTARY WIRING DIAGRAM

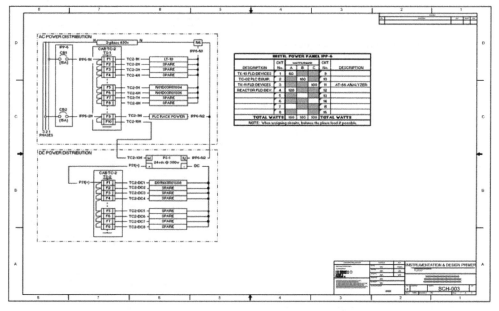

FIGURE 3-178. POWER DISTRIBUTION INFORMATION

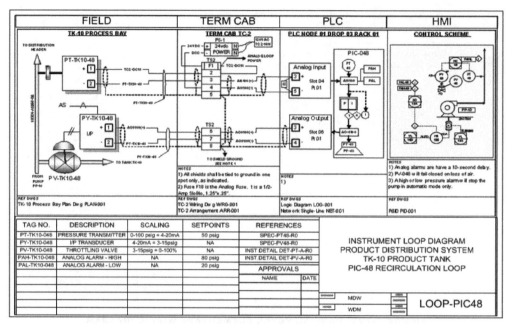

Figure 3-179. Pressure control loop PIC-48 loop sheet

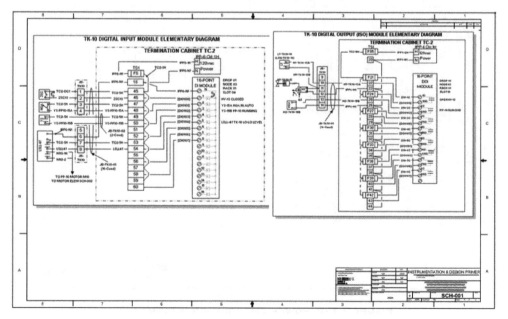

Figure 3-180. Ladder diagram for discrete modules

PART III – CHAPTER 14: INSTRUMENT AND CONTROL WIRING 483

ID	Document#	Rev#	Title Line1	Title Line2	Title Line3	Title Line4	Short Title
1	PID-001	0	Piping & Instrumentationn Diagram	Product Distribution System	TK-10 Product Tank		TK-10 P&ID
2	WRG-001	0	Connection Diagram	Product Distribution System	Node 01, Drop 03	Termination Cab TC-2	TC-2 Wiring
3	WRG-002	0	Connection Diagram	Product Distribution System	TK-10 Product Tank	Junction Box JB-TK10-1	JB-TK10-1 AC Wiring
4	ARR-001	0	Cabinet Arrangement	Product Distribution System	Node 01, Drop 03	Termination Cab TC-2	TC-2 Arrangement
5	ARR-002	0	Cabinet Arrangement	Product Distribution System	TK-10 Product Tank	Junction Box JB-TK10-1	JB-TK10-1 Arrangement
7	NET-001	0	Network Single-Line	Product Distribution System	Node 01		Network Sgl Line
8	LOG-001	0	Logic Diagram	Product Distribution System	TK-10 Product Tank	Junction Box JB-TK10-1	JB-TK10-1 Arrangement
9	PLAN-001	0	Process Area Instrument Plan	Product Distribution System	TK-10 Product Tank	Process Bay	TK10 Process Bay Plan
10	LOOP-PIC48	0	Instrument Loop Diagram	Product Distribution System	TK-10 Product Tank	Recirc Pressure Loop PIC-048	PIC-048 Loop Sheet
11	SCH-001	0	Instrument Elementary Diagram	Product Distribution System	Remote I/O Drop 03, Slots 04 &	120vac Discrete PLC Circuits	PLC I/O Schematic
12	SCH-002	0	Motor Control Diagram	Product Distribution System	PP-10 Product Pump		PP-10 Motor Schematic
13	SCH-003	0	120vac Power Distribution Schema	Product Distribution System	Instrument Power Panel IPP-6		IPP-6 Power Panel Schem

FIGURE 3-181. DOCUMENT CONTROL TABLE

It looks like all the drawings are there. Now, how about the instrument list? Figure 3-182 illustrates what we will see if we bring both the document control table and the instrument and I/O list table into our field of view.

Tagname	I/O Type	HW Address	SW Address	Schematic#	LoopSheet#	MarshCab#	MarshCabArg#	MarshC
PT-TK10-48	AI	N01D03R01S08P01	AI0100					
LT-TK10-10	AI	N01D03R01S08P02	AI0101					
PY-TK10-48	AO	N01D03R01S10P01	AO0100					
ZSC-TK10-13	DI	N01D03R01S06P01	DI0100					
YS-TK10-15A	DI	N01D03R01S06P02	DI0101					
YS-TK10-15B	DI	N01D03R01S06P03	DI0102					
LSLL-TK10-47	DI	N01D03R01S06P05	DI0104					
HY-TK10-13	DO1	N01D03R01S04P01	DO0101					
HS-TK10-15B	DO1	N01D03R01S04P02	DO0102					
LSH-TK10-10	HW	NA	NA					
PSV-TK10-58	NA	NA	NA					
PV-TK10-48	NA	NA	NA					
HV-TK13-13	NA	NA	NA					

Record: 7 of 13

ID	Document#	Rev#	Title Line1	Title Line2	Title Line3	Title Line4	
1	PID-001	0	Piping & Instrumentationn Diagram	Product Distribution System	TK-10 Product Tank		TK-10
2	WRG-001	0	Connection Diagram	Product Distribution System	Node 01, Drop 03	Termination Cab TC-2	TC-2 V
3	WRG-002	0	Connection Diagram	Product Distribution System	TK-10 Product Tank	Junction Box JB-TK10-1	JB-TK
4	ARR-001	0	Cabinet Arrangement	Product Distribution System	Node 01, Drop 03	Termination Cab TC-2	TC-2 A
5	ARR-002	0	Cabinet Arrangement	Product Distribution System	TK-10 Product Tank	Junction Box JB-TK10-1	JB-TK
7	NET-001	0	Network Single-Line	Product Distribution System	Node 01		Netwo
8	LOG-001	0	Logic Diagram	Product Distribution System	TK-10 Product Tank	Junction Box JB-TK10-1	JB-TK
9	PLAN-001	0	Process Area Instrument Plan	Product Distribution System	TK-10 Product Tank	Process Bay	TK10 F
10	LOOP-PIC48	0	Instrument Loop Diagram	Product Distribution System	TK-10 Product Tank	Recirc Pressure Loop PIC-048	PIC-04
11	SCH-001	0	Instrument Elementary Diagram	Product Distribution System	Remote I/O Drop 03, Slots 04 &	120vac Discrete PLC Circuits	PLC I/C
12	SCH-002	0	Motor Control Diagram	Product Distribution System	PP-10 Product Pump		PP-10
13	SCH-003	0	120vac Power Distribution Schema	Product Distribution System	Instrument Power Panel IPP-6		IPP-6 F

FIGURE 3-182. DOCUMENT CONTROL TABLE AND INSTRUMENT AND I/O LIST TABLE

In this case, we have frozen the first four columns of information on the instrument list. This allows the other fields to slide beneath, letting us keep the key data visible.

The database that results after entering all the data we can capture is shown in Figure 3-183. Notice that no fields are empty, except for the MCC-related ones. An empty field means there is work to do. NA in a field means that the item was evaluated and deemed to be of no consequence for that aspect.

Tagname	I/O Ty	HW Address	SW Address	P&ID#	Schematic	LoopSheet#	MarshJB	MarshCab	MarshCab	FieldJBox#	FieldJBoxA	FieldJBoxWrg
PT-TK10-48	AI	N01D03R01S08P01	AI0100	PID001	NA	LOOP-PIC48	TC-2	ARR-001	WRG-001	NA	NA	NA
LT-TK10-10	AI	N01D03R01S08P02	AI0101	PID001	NA	LOOP-LT-10	TC-2	ARR-001	WRG-001	NA	NA	NA
PY-TK10-48	AO	N01D03R01S10P01	AO0100	PID001	NA	LOOP-PIC48	TC-2	ARR-001	WRG-001	NA	NA	NA
ZSC-TK10-13	DI	N01D03R01S06P01	DI0100	PID001	SCH-003	NA	TC-2	ARR-001	WRG-001	FJB-TK10-01	ARR-002	WRG-002
YS-TK10-15A	DI	N01D03R01S06P02	DI0101	PID001	SCH-003	NA	TC-2	ARR-001	WRG-001	FJB-TK10-01	ARR-002	WRG-002
YS-TK10-15B	DI	N01D03R01S06P03	DI0102	PID001	SCH-003	NA	TC-2	ARR-001	WRG-001	MCC		
LSLL-TK10-47	DI	N01D03R01S06P05	DI0104	PID001	SCH-003	NA	TC-2	ARR-001	WRG-001	FJB-TK10-01	ARR-002	WRG-002
HY-TK10-13	DOI	N01D03R01S04P01	DO0101	PID001	SCH-003	NA	TC-2	ARR-001	WRG-001	FJB-TK10-01	ARR-002	WRG-002
HS-TK10-15B	DOI	N01D03R01S04P02	DO0102	PID001	SCH-003	NA	TC-2	ARR-001	WRG-001	MCC		
LSH-TK10-10	HW	NA	NA	PID001	NA	NA	NA		NA	FJB-TK10-01	ARR-002	WRG-002
PSV-TK10-58	NA	NA	NA	PID001	NA	NA	NA		NA	NA	NA	NA
PV-TK10-48	NA	NA	NA	PID001	NA	NA	NA		NA	NA	NA	NA
HV-TK13-13	NA	NA	NA	PID001	NA	NA	NA		NA	NA	NA	NA

FIGURE 3-183. INSTRUMENT AND I/O LIST TABLE

Now that the wiring task is complete, it's time for the panel arrangement (also called cabinet arrangement) task.

Part III – Chapter 15: Panel Arrangements

Panel arrangement drawings (panel arrangements, cabinet arrangements) depict equipment that is mounted on back panels inside electrical enclosures, along with any front-panel penetrations. These arrangements accomplish several things:

- They convey construction detail about the contents and placement of front-panel equipment, such as control switches and lights.

- They convey construction detail about the contents and placement of back-panel equipment.

- They show the placement and contents of any engraved labels.

- They provide a bill of materials for each item in the panel that must be purchased.

- They allow the enclosure (i.e., the can, the box) to be sized with confidence.

The anatomy of the drawing is very simple. It should show a graphic of the panel, usually on the left side, and a bill of materials, usually on the right. It is strictly a physical drawing showing the relative locations of the items in the panel. Graphic detail should be minimized, and no wiring detail need be included. A not–to-scale drawing is usually sufficient, provided the designer is sure that any potential conflicts have been identified and scrutinized and that any front-panel penetrations have been sufficiently detailed. However, in today's design environment, it is usually just as easy to draw to scale as not. Most component manufacturers have CADD renderings of their components that can be obtained or even downloaded off the internet. If drawn on a 1:1 scale, all these elements can be used, then the resulting drawing can be scaled to fit the border. Technically, the drawing is still "not-to-scale," since a physical scale can't be used to determine size as compared to the scale on the drawing border, but the drawing contents will be proportionally accurate.

A. Procedure

One quick initial design method for creating these drawings uses Microsoft® Excel as a design aid. The following is a procedure for turning your Excel spreadsheet software into a design tool:

1. Turn the spreadsheet into a sheet of graph paper as described in Section C.1.

2. While the entire spreadsheet is selected, set the font size to 8 pt, and format the cells for "center-center" text.

3. Get rough size dimensions of the equipment to be mounted in the cabinet, and decide on a per-square scale for your sketch. Usually 1 inch per square is good enough, which will be our choice here.

4. If the drawing toolbar is not visible, select View on the top menu, then select Toolbars, and activate the drawing toolbar.

5. Pick cell D4 as the upper corner of the panel and type an apostrophe plus a one ('1) and hit return. You will see a 1 displayed, but the leading apostrophe tells Excel to treat the cell as a text data type. Click on the cell, grab the handle on the lower right of the cell, and pull to the right. A scale of horizontal inches appears.

6. Pick cell C5, type in '1, grab the handle, and pull down. This results in a scale of vertical inches.

7. The upper left corner of cell D5 is now the upper left corner of your panel. Draw the panel contents to scale.

8. On the drawing menu, select Rectangle. With the ALT key depressed, draw a rectangle. Here are some things to note about drawing a rectangle using Excel:
    a. The grid is the upper left corner of the selected cell. The first pick is done with the ALT key down, which puts the element on the grid (the ALT key can be released prior to the second pick). In this manner a device can be properly depicted that is of intermediate size.
    b. Note: If an element is drawn with the Shift key depressed, perfect circles and squares result.
    c. Note: Most elements from the drawing menu can contain text. Just draw the element and start typing.

9. Draw all the major components, including any wiring ducts that may be desired. Draw them as near to scale as possible, but they can be general in

shape. A group of terminal blocks that are each 5 inches long by 1.5 inches wide can be drawn as one rectangle, for example. Be sure to leave adequate access room between devices. For example, leaving a minimum of 1.5 inches between a terminal block and an adjacent wiring duct is good practice.

10. Arrange the components within the scaled area. Consult the enclosure manufacturer for standard sizes of backplanes, and select the size that would comfortably contain your equipment. Cross-reference the backplane size to the enclosure size.

11. Use sheet 2 of the Excel workbook to generate your bill of materials. This can be rough or very polished. Excel provides probably the best format for developing a bill of materials due to its ease in editing. Once the bill of materials is finished, it can be pulled onto a CADD drawing with no problem.

12. Email your workbook to your drafter to formalize it into a CADD drawing.

B. Junction Box JB-TK10-01 Arrangement Drawing ARR-002

To make use of the procedure outlined above, some research must be done. The big question for this small junction box is: How many terminals are needed? Also, what kind of terminal blocks should be used, and what are their dimensions?

The terminal block we have selected is shown in Figure 3-184. It requires a DIN-rail mount and one endpiece to cover the last block in a group. (Note that "DIN" stands for Deutches Institut fir Normung, the German national standards organization. A DIN Rail is a specially designed rail that accepts clip-on components.) The terminal block is 0.33 inch high and 1.5 inches wide. It accepts a wire range of #22 to #8 AWG (American Wire Gauge). Another accessory we will pick up is standoffs to raise the DIN-rail 2 inches to make it easier to get to the wires.

Figure 3-184. Terminal block

That's about all that is going into our panel except for some wiring duct, which is really optional for this task. So let's design the box!

First of all, after the graph grid is done, we need to decide on a scale. Since the terminals are approximately three to the inch, let's use three blocks per inch. We need to select a size that is slightly smaller than the blocks. It is best to be conservative and end up with a little extra space.

Let's find out how tall our box needs to be. Consulting our wiring diagram for this box, we see that we need a minimum of 24 terminals: $24 \times 0.33 = 8$. So this number of terminals takes up only 8 inches of vertical space. That, then, is our minimum starting dimension.

Now how about side to side? Our terminal blocks are 1.5 inches wide. We need a minimum of 1.5 inches on either side of the terminal blocks, preferably 2 inches. So using 2 inches gives us an inner minimum width of 5.5 inches. Plus, we have decided to use wire duct to support the wires and dress up the box a bit. We will use 1½ inch duct to surround the terminals, so that adds 3 inches to both dimensions.

To summarize, we have just designed a box that requires an inner panel that is at least 11 inches tall and 8.5 inches wide. Let's move to our Excel sheet for the rest of it.

1. **SET UP A SCALE**

    - At the Excel sheet, pick cell C2. Enter the value '1 and hit return.

    - Select B2, B3, and B4, and press the Merge and Center icon. This merges those three cells, which represents 1 inch. Now, pull those across to the right, and the number of inches appears across the top of the page (Figure 3-185).

    - Do the same thing again, except vertically, starting at cell B3. If desired, those cells may be formatted to display the numeric values in the center of the cell by right clicking the mouse and selecting Format Cells and Alignment.

2. **DESIGN THE PANEL**

    - Pick Text Box from the drawing menu. Hold down the ALT key to activate the grid, place the text box, and size it to cover a width of 4½ cells and a height of 24 cells. While the box is highlighted, configure it for vertical text and key in "24 terminals." The space allocated should be 8 inches by 1.5 inches (Figure 3-186).

FIGURE 3-185. SETTING UP A SCALE

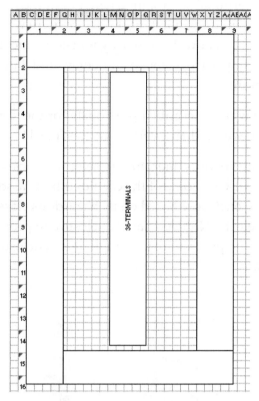

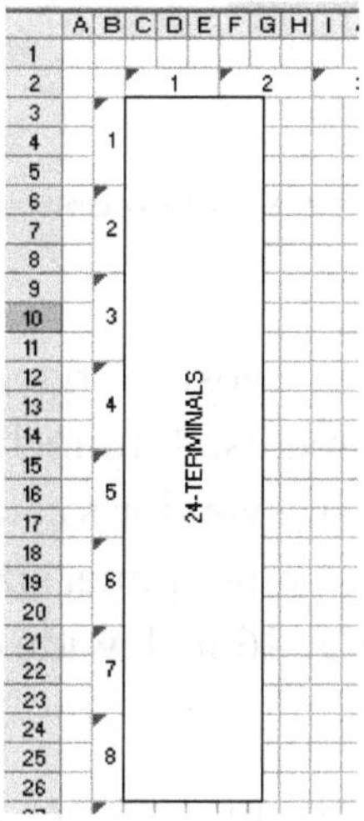

FIGURE 3-186. INITIAL LAYOUT

- Always provide spare capacity. In fact, most specifications require 50% spare capacity on a new installation. That lengthens our terminal strip by 4 inches. So we must adjust our strip length from 8 inches to 12 inches.

- Now place the terminal strip somewhere out of the way, and let's draw our wire way (wire duct). First, we know the wire way surrounds the terminal strip, and we know the wire way will be 1.5 inches wide. We need to add approximately 0.5 inch above and below and 2 inches side to side around the terminals.

- Draw the wire way as described previously for the terminals. After doing so, we find that we need 15.5 vertical inches, not 11.

- Now consult the vendors' literature to select an enclosure. (Note that Chapter 5.C dealt with some of the considerations associated with specifying an electrical enclosure based on its hazardous classification.) In our case, we will select a NEMA-4-rated enclosure to protect against water washdowns in the process area. A single-door enclosure will do nicely (Figure 3-187).

FIGURE 3-187. SINGLE-DOOR ENCLOSURE

Now that we have selected an enclosure style, we must find an inner panel to accommodate our wiring requirements. A 16-inch high by 9-inch wide panel will work. After consulting the enclosure vendor's catalog, we discover that there is no 16- by 9-inch panel. The nearest inner panel size is a 17- by 13-inch panel, which brings us to a 20-inch high by 16-inch wide enclosure. Thus, selection of the inner panel leads us to the enclosure size.

- Return to Excel, sketch in the inner panel, and manipulate the wire duct and terminal strips to take the most advantage of the extra space. In this case, there will be a 4-inch area to the right of the termination zone that can be used for relays or even additional terminals should the need arise.

3. Generate a Bill of Materials

For each item represented on the diagram, place at least one bubble and leader (Figure 3-188). To make a bubble, pick Oval from the drawing menu and hold down the Shift key while drawing to make a perfect circle. Then, while the bubble is selected, key in the desired bubble number.

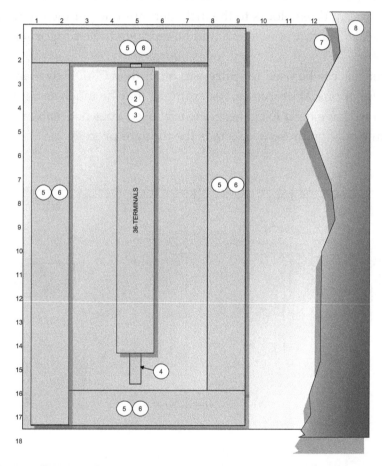

			COMPONENT SCHEDULE		
ITEM	QTY	U/I	DESCRIPTION	MANUFACTURER	PART NO.
1	36	EA	TERMINALS	VARIOUS	SSSS
2	1	EA	BARRIERS	VARIOUS	SSSS
3	2	EA	ANCHORS	VARIOUS	SSSS
4	3	FT	RAI., DIN	VARIOUS	SSSS
5	8	FT	DUCT, 1.5 X 2"	VARIOUS	SSSS
6	8	FT	COVER, DUCT, 1.5"	VARIOUS	SSSS
7	1	EA	PANEL, 17X13	VARIOUS	SSSS
8	1	EA	ENCLOSURE, 20X16X08	VARIOUS	SSSS

FIGURE 3-188. JUNCTION BOX WITH BILL OF MATERIALS

C. Summary

Our finished panel arrangement, as shown in Figure 3-189, is now ready to turn over to the drafting department, where it will be added to the drawing that already details the interconnections.

The shorthand technique we used generated a decent sketch of a small junction box. This same technique works just as well on a larger scale and may be easily adapted to generating a TC-2 sketch. The biggest difference is the scaling that is set up. Instead of three blocks to the inch, perhaps one per inch would work, or some other scale. All components should be redrawn to be proportional to that scale.

Such sketches can be used for purposes other than CADD drawing development. Concept approval processes, for example, may be much easier with their use. Though we have used Excel here, the tool used does not really matter. The important information presented here is the thought process.

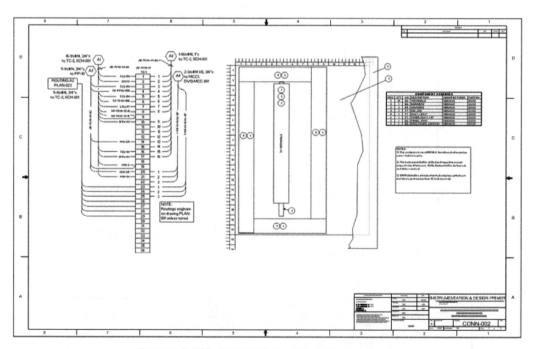

FIGURE 3-189. FINISHED PANEL ARRANGEMENT

Part III – Chapter 16: Procurement

The procurement of materials (as separate from instruments and equipment) cuts to the chase. It is pretty much the reason for most of the other engineering tasks. A design package is built from the conceptual to the actual, and procurement is about as *actual* as it gets. A frequent lament from the construction partner is, "How can I build something if I don't have the parts?" The design team needs to put lots of thought into the procurement process early in the design phase.

Who does the actual purchasing? It doesn't matter. The design team must perform virtually the same functions regardless. Of course, one organization may like the information organized differently from another, but that is the problem of the cost engineering group. The cost group fulfills the bill of material, finding sources for the material, and sometimes building bid specifications and managing the bid process in order to insure the prices paid are competitive. Usually, unless the design engineer is running a small operation, the cost group handles those details.

Primarily two types of materials are specified by the I&C design team:

- Electrical materials, including cable, conduit, enclosures, and termination hardware.

- Mechanical materials, including pneumatic tubing, flanges, gaskets, and mounting brackets.

These types of materials are called bulk materials.

Purchasing and handling bulk materials, while not necessarily a sexy topic, is one of those basic tasks that needs to be done right to achieve success. Bulk materials consist of materials that are considered to be commodities[11], such as those mentioned above and nuts, bolts, pull boxes, junction boxes, tubing, tube fittings, and vessel trim. Instruments, engineered items that require long lead times for their manufacture, or equipment items that require individual specifications do not fall into the bulk materials category.

Constructors spend a tremendous amount of time handling this type of material. While some bulk materials, such as conduit, can be piled up and used as

---

11. In this context, commodities are common items that are mass produced and can be purchased directly from the vendor or manufacturer without a detailed specification.

needed, other items are specific to their purpose and must be handled differently. Once they are received on site, which must be well before they are needed, such items must be tagged and stored in a rational way that allows them to be located at the right time by the right people. Look at virtually any construction site, and you will note constructors walking around with clipboards in a frustrated search for their materials. Proper materials handling, which includes planning, can alleviate this frustration.

Proper materials purchasing and handling is a team effort that begins with the design team, extends through the purchasing group and the vendors, and ends with the constructor. If all elements work properly, the amount of time wasted on the construction site can be minimized.

This chapter focuses on the product of the design team.

A. Typical Purchasing Cycle

When a design is nearly complete, the design team begins the takeoff, which means someone sits down with the drawing set and begins building a bill of materials for purchase.

As discussed in Chapter 1, Part I, a bill of materials (B/M, or BOM) is a document that lists all the materials required for construction. For manageability, this list is generally tied to a specific equipment item, such as a control panel, or to a particular location on the process floor. It is assembled without regard to manufacturer or type of item and includes all the items needed. Ideally, the bill of materials also lists as references any associated non-commodity items (e.g., instruments) and points to the specification sheet or purchase order to which those items are tied. Ideally, a constructor should be able to use the bill of materials to gather all the materials necessary for a specific task.

On larger projects, it is critical that each BOM be given a unique designation and that a bill of materials index be maintained. This index should include the BOM number, any WBS number that might apply, the intended use of the materials, and any drawings used in the material takeoff activity.

The design team generates a set of bills of materials that are specific to a drawing or set of drawings. It is important to organize these lists with the constructor in view. For example, it makes little sense to generate a bill of materials containing both conduit items and pneumatic tubing items. These should be split into separate listings because the construction team rarely has the same group do both tasks. Some suggestions for bill of materials categories are made later in this chapter.

For the remainder of this discussion, refer to Figure 3-190.

Once a bill of materials, or a set of them, is produced, it is given to the purchasing department. Bills of materials are organized by purpose. Purchasing

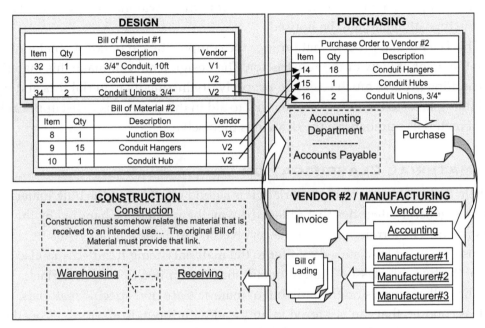

FIGURE 3-190. TYPICAL PROCUREMENT CYCLE

parses the materials, reorganizing them according to vendor. A purchase order (PO) is then produced for each vendor.[12] When this occurs, any link between the particular item to be purchased and its intended use is lost. So it is a good idea to include a column on the bill of materials called Purchase Order. When a PO is created, the designer should be given an opportunity to check it for accuracy. At that time, it is a good idea for the designer to log the PO number of each item on the bill of materials. Creating this link between the original bill of materials and the PO greatly assists the constructor when it is time to collect materials.

Vendors often represent several manufacturers. Newark Electronics, for example, publishes a catalog that contains thousands of items manufactured by perhaps a hundred different entities. Newark gives each item a unique Newark catalog number by which any item can be identified within their system. So while a single PO to Newark Electronics probably results in a single invoice back to the purchasing department, it could also result in multiple shipments from many different manufacturers.

The common thread through all of this is the PO number. Each of these various shipments will arrive at the construction site with a bill of lading. The bill of lading references the original PO number. It is the construction team's responsibility to properly receive these materials and to log them in against the original PO. One way to do this is to assign a warehouse bin area to each PO. The con-

---

12. Frequently, a purchase order is written directly to the manufacturer, bypassing the vendor. What is presented in this discussion is a worst-case scenario from the standpoint of material tracking.

structor's approach must accommodate the amount of warehouse space available and the type of material to be housed.

This is a normal process that typifies the complexities in handling materials from the point of view of the constructor. The design team can make this materials handling process much easier by properly organizing its material lists and by backfilling links to the POs. Another way to aid in this process is to list applicable bill of materials numbers on the drawings themselves.

B. Material Classification

To help the constructor track materials, it is a good idea to split the bulk materials into two general categories that reflect the way they should be handled by the constructor. The terminology that has evolved here can be a little confusing, because the word *bulk* is used to refer to the materials being handled and also to one of the two sub-classes (bulk or detail) that are applied to bulk materials.

The bulk materials on the bulk bill of materials are non-specific materials, such as conduit, that can be stored in bulk and withdrawn from stores as needed. Conversely, the detail bill of materials is for materials that have specific uses as shown on detail drawings.

These drawings typically incorporate component schedules that list specific materials that are depicted in the body of the drawing. Materials shown on instrument installation details, for example, would be listed as a detail bill of materials, as would materials appearing in a component schedule on a control panel arrangement drawing.

Both types of bills of material are discussed in the sections that follow.

C. Bulk Bill of Materials

A bulk bill of materials covers materials that can be purchased in bulk. That is, materials that do not need to be linked to a specific equipment specification or purpose. Cable, conduit, conduit fittings, cable tray, nuts and bolts, conduit hangers, and hookup wire are examples of materials that may be purchased in bulk.

Bulk material items are frequently purchased with a single bulk bill of materials. There are several reasons for not breaking up an order for this type of material:

- Most vendors allow quantity discounts. The amounts of these discounts vary as the quantity rises, so it is desirable to make large orders.

- Storage space for bulk materials is usually limited. For example, a single large pile of 2-inch conduit will suffice for the entire project.

For a large facility, it probably is wise to find a way to subdivide the material listings by location of use, realizing that the purchasing department can combine these listings to minimize the number of POs that must be written (Figure 3-190). To this end, a conversation with the constructor is beneficial. If the constructor plans to have a different construction team for each floor, for example, that would be a rational way to break the bills of materials.

So where do we start? Actually, we have a pretty good beginning already. We have generated a bill of materials for junction box JB-TK10-01 and for instrument location plan drawing PLAN-001. That should be enough material for us to play with.

The bulk electrical materials takeoff worksheet is shown in Figure 3-191.

If you have read Chapter 3, you are already familiar with this worksheet. As we have seen, it has several tables for calculating the quantities of various types of material:

- Wire and cable
- Terminations, junction boxes, and cabinets
- Instrument air piping
- Conduit and fittings (general purpose)
- Conduit and fittings (in this example, Class I, Division 2)

We will go through each of these.

First, let's review the plan drawing takeoff sketch we made for plan drawing PLAN-001. On that sketch, we detailed the cable and conduit lengths. We have checked with the plant, and the area is non-hazardous, so we can use the general-purpose conduit table (Figure 3-126).

Creating the takeoff is simply a matter of fetching the totals and then rounding up to the next standard unit of issue. For example, it is sometimes more cost effective to buy a 1000 foot spool of cable than it is to buy a 750 foot spool. Such decisions are best made by the cost engineering department, or by the procurement officer after talking with the vendor. The Bill of Material should only display what is actually needed, plus a percentage for wastage. Usually a 20-50% factor should be added for wastage. For example, if seven fuseblocks are needed, then the bill of material should call out seven blocks. Even if the designer knows the fuseblocks come in boxes of fifty, only the seven should be called out.

Let's concentrate on the cables (Note: the Type designations shown are purely arbitrary, but are reflected on the Cable Type chart on Figure 3-126):

- *Type A:* According to the takeoff chart, we need 830 feet of type A wire, which is single-conductor #14. This type of wire is for power, so it will be

BULK MATERIAL TAKEOFF WORKSHEET

Customer:	ADDISON PLASTICS	Charge No:	332211
Project:	FEED WATER SYS	Design Team	I&C#2
CP No.	999444	Revision:	A
WBS No.	001	Date:	2-15-02
Location:	BOONIESVILLE, TX	By:	ZZZ

COST SUMMARY

CABLE	$727.72
CABINET	$2,280.15
CONDUIT	$453.78
TUBING / DETAILS	
INSTR. AIR PIPING	$0.00
GRAND TOTAL:	**$3,461.65**

CONSTRUCTION LABOR DATA

- 3696 Wiring Terminations
- 66 Tubing Terminations
- 101 Conduit Terminations

Are there hazardous areas? If so, what percentage of the equipment will be in hazardous locations? **0%**

WIRE & CABLE CALCULATION TABLE

QTY.	U/M	UNIT PRICE	TOTAL PRICE	DESCRIPTION	STOCK NUMBER	SOURCE
	FT	$0.39	$0.00	3/C #18 AWG UNSHLD CABLE, 300V		
480	FT	$0.35	$168.00	2/C #16 AWG SHLD CABLE, 300V		
	FT	$0.30	$0.00	2/C #16 AWG UNSH CABLE, 300V		
277	FT	$0.12	$33.24	1/c #14, 600V, Hookup Wire Black		
277	FT	$0.12	$33.24	1/c #14, 600V, Hookup Wire White		
277	FT	$0.12	$33.24	1/c #14, 600V, Hookup Wire Green		
500	FT	$0.12	$60.00	1/c #16, 600V, Hookup Wire Red		
500	FT	$0.12	$60.00	1/c #16, 600V, Hookup Wire Blue		
200	FT	$1.25	$250.00	16twpr #18, Indiv Shld for Analog Ckt Homeruns		
	FT	$1.25	$0.00	4-triad, Indiv Shld for RTD Ckt Homeruns		
	FT	$1.00	$0.00	2/C Thermocouple Extension Wire		
150	FT	$0.60	$90.00	Network Communications Cable		
	SUBTOTAL		$728			

TERMINATIONS, JUNCTION BOXES & CABINETS

QTY.	U/M	UNIT PRICE	TOTAL PRICE	DESCRIPTION	STOCK NUMBER	SOURCE
19	EA	$5.00	$95.00	Fuse blocks		
89	EA	$1.35	$120.15	Terminals		
1	EA	$400	$400.00	Digital Junction box, 20" x 16" w/ Panel		
	EA	$400	$0.00	Analog Junction Box		
1	EA	$1,200	$1,200.00	PLC Cabinet, 49x72x18 w/ Panel		
1	EA	$300	$300.00	24VDC Power Supply		
1	EA	$100	$100.00	Fan Kit w/ Thermostat		
1	EA	$50	$50.00	Light Kit w/ Door Switch		
1	EA	$15	$15.00	4-Gang Outlet w/ Hardware		
	SUBTOTAL		$2,280			

I&C PROJECT ESTIMATE - INSTRUMENT AIR PIPING

QTY.	U/M	UNIT PRICE	TOTAL PRICE	DESCRIPTION	STOCK NUMBER	SOURCE
	FT	$2.20	$0.00	1/2" Pipe, Stl, Galv, Sch 40		
	EA	$0.40	$0.00	1/2" NPT, TEE, Iron, Galv, 150#		
	EA	$0.50	$0.00	1/2" NPT, Plug, Iron, Galv, 150#		
	EA	$0.58	$0.00	Hanger, Pipe, Adj.Clevis, 1/2", Grinnel Fig.260		
	FT	$0.15	$0.00	Rod, con't. Thd, Stl, Zinc-Coated		
	EA	$0.35	$0.00	1/2" NPT, Elbow, 90 Deg, Iron, Galv, 150#		
	EA	$0.02	$0.00	Nut, heavy Hex, Stl		
	EA	$0.50	$0.00	Clamp, Adj. Beam, 3/8"		
	EA	$0.50	$0.00	Bushing, Pipe, Iron, Galv, Scr, 150#, 1" x 1/2"		
	SUBTOTAL		$0			

CONDUIT & FITTINGS - GENERAL PURPOSE

QTY.	U/M	UNIT PRICE	TOTAL PRICE	DESCRIPTION	STOCK NUMBER	SOURCE
400	FT	$0.45	$180.00	3/4" ,RIGID STL CONDUIT		
2	EA	$1.25	$2.50	3/4" ,OUTLET BODY, TYPE C27		
14	EA	$1.25	$17.50	3/4" ,OUTLET BODY, TYPE T27		
2	EA	$1.25	$2.50	3/4" ,OUTLET BODY, TYPE L27		
2	EA	$1.25	$2.50	3/4" ,UNION,		
17	EA	$1.00	$17.00	COVERS & GASKETS		
12	EA	$0.97	$11.64	CONDUIT CLAMP, TYPE RC-3/4		
200	FT	$0.63	$126.00	1" ,RIGID STL CONDUIT		
	EA	$2.50	$0.00	1" ,OUTLET BODY, TYPE C37		
1	EA	$2.50	$2.50	1" ,OUTLET BODY, TYPE T37		
	EA	$2.50	$0.00	1" ,OUTLET BODY, TYPE L37		
2	FT	$2.50	$5.00	1" ,UNION,		
1	EA	$2.50	$2.50	COVERS & GASKETS		
8	EA	$2.50	$20.00	CONDUIT CLAMP, TYPE RC-1		
2	EA	$2.07	$4.14	REDUCER, 1" X 3/4"		
	FT	$0.93	$0.00	11/2" ,RIGID STL CONDUIT		
	EA	$5.00	$0.00	11/2" ,OUTLET BODY TYPE GUAC59		
	EA	$5.00	$0.00	11/2" ,OUTLET BODY TYPE GUAT59		
	EA	$5.00	$0.00	11/2" ,OUTLET BODY TYPE GUAN59		
	EA	$5.00	$0.00	11/2" ,UNION,EXPLOSION PROOF ,UNF505		
	EA	$5.00	$0.00	COVERS & GASKETS		
	EA	$5.00	$0.00	CONDUIT CLAMP, TYPE RC-1-1/2		
30	FT	$2.00	$60.00	FLEX CONDUIT		
	ft	$12.00	$0.00	CABLE TRAY		

FIGURE 3-191. BULK MATERIALS TAKEOFF WORKSHEET

QTY.	U/M	UNIT PRICE	TOTAL PRICE	DESCRIPTION	STOCK NUMBER	SOURCE
	EA	$5.00	$0.00	11/2",UNION,EXPLOSION PROOF ,UNF505		
	EA	$5.00	$0.00	COVERS & GASKETS		
	EA	$5.00	$0.00	CONDUIT CLAMP, TYPE RC-1-1/2		
30	FT	$2.00	$60.00	FLEX CONDUIT		
	ft	$12.00	$0.00	CABLE TRAY		
	SUBTOTAL		$454			

	CONDUIT & FITTINGS - CLASS I DIV II
QTY.	U/M	UNIT PRICE	TOTAL PRICE	DESCRIPTION	STOCK NUMBER	SOURCE
	FT	$0.45	$0.00	3/4" ,RIGID STL CONDUIT		
	EA	$11.60	$0.00	3/4" ,OUTLET BODY, TYPE GUAC26		
	EA	$12.60	$0.00	3/4" ,OUTLET BODY, TYPE GUAT26		
	EA	$11.60	$0.00	3/4" ,OUTLET BODY, TYPE GUAN26		
	EA	$4.25	$0.00	3/4" ,UNION,EXPLOSION PROOF UNF205		
	EA	$8.00	$0.00	CONDUIT SEAL TYPE EYS21 3/4"		
	EA	$0.97	$0.00	CONDUIT CLAMP, TYPE RC-3/4		
	FT	$0.63	$0.00	1" ,RIGID STL CONDUIT		
	EA	$14.76	$0.00	1" ,OUTLET BODY, TYPE GUAC36		
	EA	$15.52	$0.00	1" ,OUTLET BODY, TYPE GUAT37		
	EA	$14.76	$0.00	1" ,OUTLET BODY, TYPE GUAN36		
	EA	$7.62	$0.00	1" ,UNION,EXPLOSION PROOF UNF305		
	EA	$1.07	$0.00	CONDIUT CLAMP, TYPE RC-1		
	EA	$10.00	$0.00	CONDUIT SEAL TYPE EYS21 1"		
	FT	$0.93	$0.00	11/2",RIGID STL CONDUIT		
	EA	$49.55	$0.00	11/2",OUTLET BODY TYPE GUAC59		
	EA	$54.37	$0.00	11/2",OUTLET BODY TYPE GUAT59		
	EA	$49.55	$0.00	11/2",OUTLET BODY TYPE GUAN59		
	EA	$12.60	$0.00	11/2",UNION,EXPLOSION PROOF ,UNF505		
	EA	$11.60	$0.00	CONDIUT CLAMP, TYPE RC-1-1/2		
	EA	$28.00	$0.00	CONDUIT SEAL TYPE EYS21 11/2"		
	EA	$61.80	$0.00	FLEX CONDUIT, ECLK115		
	ft	$12.00	$0.00	CABLE TRAY		
	SUBTOTAL		$0			

FIGURE 3-191 (CONTINUED). BULK MATERIALS TAKEOFF WORKSHEET

evenly distributed between white, black, and green – about 270 feet of each. All of this wire will be routed via conduit between the junction box and the field devices.

- *Type B:* The takeoff chart calls for 480 feet of twisted-pair #16 shielded wire.

- *Type D:* The chart shows that this type is multiconductor, 16-pair cable, with 200 feet listed.

Let's not forget about our termination cabinet, which contains quite a bit of hookup wire. In addition to the wire used in the process area, we will order a 500-foot spool of each color for the cabinet work. Turning to our wire and cable calculation table, shown in Figure 3-192, we find we will have to spend $728 on wire and cable for this project.

Next, let's look at the cabinetry. Consulting the cabinet drawing for TC-2, we find a total of 19 fuse blocks and 53 terminals. In addition, there are 36 terminals in the junction box for a total of 89. So our next chart is beginning to take shape, as shown in Figure 3-193.

Next is the instrument air piping. If we have any air users, we need to pipe some instrument air into the area. In this case, we find an instrument air header in

WIRE & CABLE CALCULATION TABLE

QTY.	U/M	UNIT PRICE	TOTAL PRICE	DESCRIPTION	STOCK
	FT	$0.39	$0.00	3/C #18 AWG UNSHLD CABLE, 300V	
480	FT	$0.35	$168.00	2/C #16 AWG SHLD CABLE, 300V	
	FT	$0.30	$0.00	2/C #16 AWG UNSH CABLE, 300V	
277	FT	$0.12	$33.24	1/c #14, 600V, Hookup Wire Black	
277	FT	$0.12	$33.24	1/c #14, 600V, Hookup Wire White	
277	FT	$0.12	$33.24	1/c #14, 600V, Hookup Wire Green	
500	FT	$0.12	$60.00	1/c #16, 600V, Hookup Wire Red	
500	FT	$0.12	$60.00	1/c #16, 600V, Hookup Wire Blue	
200	FT	$1.25	$250.00	16twpr #18, Indiv Shld for Analog Ckt Homeruns	
	FT	$1.25	$0.00	4-triad, Indiv Shld for RTD Ckt Homeruns	
	FT	$1.00	$0.00	2/C Thermocouple Extension Wire	
150	FT	$0.60	$90.00	Network Communications Cable	
	SUBTOTAL		$728		

FIGURE 3-192. WIRE AND CABLE CALCULATION TABLE

TERMINATIONS, JUNCTION BOXES & CABINETS

QTY.	U/M	UNIT PRICE	TOTAL PRICE	DESCRIPTION	STOCK NUM
24	EA	$5.00	$120.00	Fuse blocks	
100	EA	$1.35	$135.00	Terminals	
1	EA	$400	$400.00	Digital Junction box, 20" x 16" w/ Panel	
	EA	$400	$0.00	Analog Junction Box	
1	EA	$1,200	$1,200.00	PLC Cabinet, 49x72x18 w/ Panel	
1	EA	$300	$300.00	24VDC Power Supply	
1	EA	$100	$100.00	Fan Kit w/ Thermostat	
1	EA	$50	$50.00	Light Kit w/ Door Switch	
1	EA	$15	$15.00	4-Gang Outlet w/ Hardware	
	SUBTOTAL		$2,320		

FIGURE 3-193. TERMINATIONS AND CABINETRY

the next bay, so it is a simple matter to extend tubing to our modulating valve and block valve. No galvanized air piping is needed.

Next, we have the conduit and conduit fittings (Figure 3-194). The type numbers provided in the list are Crouse-Hinds part numbers. Referring to our plan drawing takeoff, we see that we need 150 feet of 1-inch conduit and 312 feet of ¾-inch. We may also obtain a fitting count by looking at the plan drawing and doing a count of fittings. Note that only nine fittings are depicted on the drawing (Figure 3-125), while a total of nineteen will actually be purchased. An additional tee must be provided at the conduit connection at each instrument. In addition, there are some vertical drops in which the fittings are superimposed on each other. So the count must be higher. Also, note that a transition is made from 1-inch conduit to ¾-inch conduit for routing D1. Reducers will need to be purchased for that case.

Part III – Chapter 16: Procurement

QTY.	U/M	UNIT PRICE	TOTAL PRICE	CONDUIT & FITTINGS – GENERAL PURPOSE DESCRIPTION	STOCK N
400	FT	$0.45	$180.00	3/4", RIGID STL CONDUIT	
2	EA	$1.25	$2.50	3/4", OUTLET BODY, TYPE C27	
14	EA	$1.25	$17.50	3/4", OUTLET BODY, TYPE T27	
2	EA	$1.25	$2.50	3/4", OUTLET BODY, TYPE L27	
2	EA	$1.25	$2.50	3/4", UNION,	
17	EA	$1.00	$17.00	COVERS & GASKETS	
12	EA	$0.97	$11.64	CONDUIT CLAMP, TYPE RC-3/4	
200	FT	$0.63	$126.00	1", RIGID STL CONDUIT	
	EA	$2.50	$0.00	1", OUTLET BODY, TYPE C37	
1	EA	$2.50	$2.50	1", OUTLET BODY, TYPE T37	
	EA	$2.50	$0.00	1", OUTLET BODY, TYPE L37	
2	FT	$2.50	$5.00	1", UNION,	
1	EA	$2.50	$2.50	COVERS & GASKETS	
8	EA	$2.50	$20.00	CONDUIT CLAMP, TYPE RC-1	
2	EA	$2.07	$4.14	REDUCER, 1" X 3/4"	
	FT	$0.93	$0.00	1 1/2", RIGID STL CONDUIT	
	EA	$5.00	$0.00	1 1/2", OUTLET BODY TYPE GUAC59	
	EA	$5.00	$0.00	1 1/2", OUTLET BODY TYPE GUAT59	
	EA	$5.00	$0.00	1 1/2", OUTLET BODY TYPE GUAN59	
	EA	$5.00	$0.00	1 1/2", UNION, EXPLOSION PROOF, UNF505	
	EA	$5.00	$0.00	COVERS & GASKETS	
	EA	$5.00	$0.00	CONDUIT CLAMP, TYPE RC-1-1/2	
30	FT	$2.00	$60.00	FLEX CONDUIT	
	ft	$12.00	$0.00	CABLE TRAY	
	SUBTOTAL		$454		

Figure 3-194. Conduit and conduit fittings

Also, some other miscellaneous parts may need to be purchased. For example, at each down leg, a drain fitting needs to be installed to provide drainage for any condensate that may collect.

D. Detail Bill of Materials

The detail bill of materials is taken from the installation details. It is built by matching each installation detail in the database against any instrument for which it is used. Having already done that in Chapter 13, Section D, we are able to create a new query called the *installation query*. Running that query results in the database listing (installation detail assignment data) in Figure 3-195.

This tabulation lets us count the number of times a particular installation detail is used. Our new tally is shown in Figure 3-196.

The next step is to tabulate all the equipment items listed on all the bills of materials. This is not as difficult as it sounds since there usually aren't that many details. However, a minor complication is that a particular item may be used on more than one detail. This causes us to maintain a lookup table that assigns each item to the proper detail. A summary table then totals each item and places it into a bill of materials format. First, we must tabulate our material. Taking the four

Microsoft Access - [Installation Query : Select Query]

Tagname	P&ID#	PlanDwg	PlanItem	Spec	MechDetail	ElecDetail	MountDetail
YS-TK10-15B	PID001	MCC001		NA			
HS-TK10-15B	PID001	MCC001		NA			
LSH-TK10-10	PID001	PLAN001	01	NA	NA	ELEC-001	NA
LT-TK10-10	PID001	PLAN001	01		NA	ELEC-001	MOUNT-001
PV-TK10-48	PID001	PLAN001	02		NA	NA	In-Line
PY-TK10-48	PID001	PLAN001	02	NA	MECH-002	ELEC-001	Integral
PT-TK10-48	PID001	PLAN001	03		NA	ELEC-001	MOUNT-001
ZSC-TK10-13	PID001	PLAN001	04	NA	NA	ELEC-001	Integral
HY-TK10-13	PID001	PLAN001	04	NA	MECH-001	NA	Integral
HV-TK13-13	PID001	PLAN001	04		NA	NA	In-Line
PSV-TK10-58	PID001	PLAN001	05		NA	NA	MOUNT-001
YS-TK10-15A	PID001	PLAN001	06	NA	NA	NA	MOUNT-002
LSLL-TK10-47	PID001	PLAN001	07		NA	ELEC-001	MOUNT-001

FIGURE 3-195. INSTALLATION DETAIL ASSIGNMENT DATA

DETAIL	QTY
MECH-001	1
MECH-002	1
ELEC-001	6
MOUNT-001	4
MOUNT-002	1

FIGURE 3-196. NEW DETAIL SHEET TALLY

installation details we have recently finished, the material tabulation is as shown in Figure 3-197.

Notice that 3/8 inch tubing and tube fittings, the filter-regulator, and other items appear on more than one detail. So the list needs to be consolidated. Also, it is handy to have the detail quantities at the top of the columns, and subtotals to the right as shown in Figure 3-198.

Now we must make some adjustments and add fields to build our bill of materials. First, in cell V5, type the following and hit return:

= R2*R5+S2*S5+T2*T5+U2*U5

The result should be a 1 next to Throttling Valve. As a test, enter a 1 in cell T5 as if Mount-001 needed a valve. The total should increment to 2. Delete the T5

PART III – CHAPTER 16: PROCUREMENT

			QUANTITIES PER DETAIL			
	DESCRIPTION	U/I	MECH-001	MECH-002	MOUNT-001	ELEC-001
5	THROTTLING VALVE	EA		1		
6	I/P TRANSDUCER	EA		1		
7	TUBING, 3/8", CU, .032"WALL	FT	6	6		
8	FILTER/REGULATOR	EA	1	1		
9	COMPRESSION FITTINGS, TUBING, 3/8"	EA	5	5		
10	FILTER	EA	1	1		
11	ON/OFF VALVE	EA	1			
12	SOLENOID, 3-WAY	EA	1			
13	BOLTS	EA			2	
14	NUTS	EA			2	
15	2" BLIND FLANGE, DRILLED & TAPPED	EA			2	
16	GASKET	EA			1	
17	COUPLING, LIQ-TITE, FLEXIBLE	EA				2
18	DRAIN, CONDUIT	EA				1
19	NIPPLE, CONDUIT, 3/4" RIGID STEEL	EA				2
20	REDUCER, CONDUIT, 3/4"X1/2"	EA				1
21	COVER, CONDULET, FORM7	EA				1
22	GASKET, CONDULET, FORM7	EA				1
23	CONDULET, TEE, 3/4", FORM7	EA				1
24	FLEX, LIQUIDTITE, 3/4"	FT				2

FIGURE 3-197. MATERIAL TABULATION BY DETAIL

V5   fx =R$2*R5+S$2*S5+T$2*T5+U$2*U5

2	Enter Detail Quantity:		1	1	4	6	12	
3			QUANTITIES PER DETAIL				TOTAL	
4	DESCRIPTION	U/I	MECH-001	MECH-002	MOUNT-001	ELEC-001	QUAN	U/I
5	THROTTLING VALVE	EA		1			1	EA
6	I/P TRANSDUCER	EA		1				EA
7	TUBING, 3/8", CU, .032"WALL	FT	6	6				FT
8	FILTER/REGULATOR	EA	1	1				EA
9	COMPRESSION FITTINGS, TUBING, 3/8"	EA	5	5				EA
10	FILTER	EA	1	1				EA
11	ON/OFF VALVE	EA	1					EA
12	SOLENOID, 3-WAY	EA	1					EA
13	BOLTS	EA			2			EA
14	NUTS	EA			2			EA
15	2" BLIND FLANGE, DRILLED & TAPPED	EA			2			EA
16	GASKET	EA			1			EA
17	COUPLING, LIQ-TITE, FLEXIBLE	EA				2		EA
18	DRAIN, CONDUIT	EA				1		EA
19	NIPPLE, CONDUIT, 3/4" RIGID STEEL	EA				2		EA
20	REDUCER, CONDUIT, 3/4"X1/2"	EA				1		EA
21	COVER, CONDULET, FORM7	EA				1		EA
22	GASKET, CONDULET, FORM7	EA				1		EA
23	CONDULET, TEE, 3/4", FORM7	EA				1		EA
24	FLEX, LIQUIDTITE, 3/4"	FT				2		FT

FIGURE 3-198. CONSOLIDATED MATERIAL WITH DETAIL QUANTITY

value. Now we have an equation that totals the number of throttling valves needed for the project, based on the number of details and the number of items needed per detail.

Now we need to propagate that calculation to the other items. To do that, Excel must be told not to automatically increment certain elements of the equation. For example, as we pull the calculation down, we want the item row to increment but the detail row to stay where it is.

To do this, we must insert the direct address operator "$" in front of any items we do not want to increment:

$$= R\$2*R5+S\$2*S5+T\$2*T5+U\$2*U5$$

Now we can pull the equation down to copy it to each row in our table of total item quantities, as shown in Figure 3-199.

	P	Q	R	S	T	U	V	W
1								
2	Enter Detail Quantity:		1	1	4	6	12	
3			QUANTITIES PER DETAIL				TOTAL	
4	DESCRIPTION	U/I	MECH-001	MECH-002	MOUNT-001	ELEC-001	QUAN	U/I
5	THROTTLING VALVE	EA		1			1	EA
6	I/P TRANSDUCER	EA		1			1	EA
7	TUBING, 3/8", CU, .032"WALL	FT	6	6			12	FT
8	FILTER/REGULATOR	EA	1	1			2	EA
9	COMPRESSION FITTINGS, TUBING, 3/8"	EA	5	5			10	EA
10	FILTER	EA	1	1			2	EA
11	ON/OFF VALVE	EA	1				1	EA
12	SOLENOID, 3-WAY	EA	1				1	EA
13	BOLTS	EA			2		8	EA
14	NUTS	EA			2		8	EA
15	2" BLIND FLANGE, DRILLED & TAPPED	EA			2		8	EA
16	GASKET	EA			1		4	EA
17	COUPLING, LIQ-TITE, FLEXIBLE	EA				2	12	EA
18	DRAIN, CONDUIT	EA				1	6	EA
19	NIPPLE, CONDUIT, 3/4" RIGID STEEL	EA				2	12	EA
20	REDUCER, CONDUIT, 3/4"X1/2"	EA				1	6	EA
21	COVER, CONDULET, FORM7	EA				1	6	EA
22	GASKET, CONDULET, FORM7	EA				1	6	EA
23	CONDULET, TEE, 3/4", FORM7	EA				1	6	EA
24	FLEX, LIQUIDTITE, 3/4"	FT				2	12	FT
25								

FIGURE 3-199. TOTAL ITEM QUANTITIES

Next, we insert descriptive fields, like part number, manufacturer, unit price, and total price, and we now have a serviceable bill of materials spreadsheet that can be continually modified as new details are generated.

We animate the total price column and then format both the unit price and total price columns as accounting data types. To format these columns, we highlight both columns and right click the mouse, then pick Format Cells, Number, and then Accounting.

The result, before adding manufacturer and part number data, is shown in Figure 3-200.

So for the first time, we have a general idea of how much our installation cost (neglecting labor) will be. Instrument cost, including materials, is $5,097.10. Now we can clean this up a bit and add some header information about the customer and the project, and we end up with the bill of materials shown in Figure 3-201.

PART III – CHAPTER 16: PROCUREMENT

FIGURE 3-200. PART NUMBER AND PRICE

ENGINEERING BILL OF MATERIAL

CUSTOMER:	ADDISON PLASTICS	PREPARED BY:	JESSE JAMES
PROJECT:	TK-10 PRODUCT TANK	CHECKED BY:	BILLY HOLIDAY
PROJECT ID:	TK10-0001-1A	BOM #:	TK10-INST-001 REV A
PROJ. ENGR:	B. MASTERSON	DATE:	5-31-02
REFERENCES:	PID-001, PLAN-001		
COMMENTS:	This list of material lacks mounting materials for several instruments. Will follow up.		

QTY	U/I	DESCRIPTION	MANUFACTURER	PART NO.	UNIT PRICE	TOTAL PRICE
1	EA	THROTTLING VALVE			$ 2,500.00	$ 2,500.00
1	EA	I/P TRANSDUCER			$ 500.00	$ 500.00
12	FT	TUBING, 3/8", CU, .032"WALL			$ 0.50	$ 6.00
2	EA	FILTER/REGULATOR			$ 25.00	$ 50.00
10	EA	COMPRESSION FITTINGS, TUBING, 3/8"			$ 0.25	$ 2.50
2	EA	FILTER			$ 1.50	$ 3.00
1	EA	ON/OFF VALVE			$ 1,500.00	$ 1,500.00
1	EA	SOLENOID, 3-WAY			$ 150.00	$ 150.00
8	EA	BOLTS			$ 0.10	$ 0.80
8	EA	NUTS			$ 0.10	$ 0.80
8	EA	2" BLIND FLANGE, DRILLED & TAPPED			$ 10.00	$ 80.00
4	EA	GASKET			$ 0.25	$ 1.00
12	EA	COUPLING, LIQ-TITE, FLEXIBLE			$ 5.00	$ 60.00
6	EA	DRAIN, CONDUIT			$ 5.00	$ 30.00
12	EA	NIPPLE, CONDUIT, 3/4" RIGID STEEL			$ 5.00	$ 60.00
6	EA	REDUCER, CONDUIT, 3/4"X1/2"			$ 5.00	$ 30.00
6	EA	COVER, CONDULET, FORM7			$ 5.00	$ 30.00
6	EA	GASKET, CONDULET, FORM7			$ 5.00	$ 30.00
6	EA	CONDULET, TEE, 3/4", FORM7			$ 10.00	$ 60.00
12	FT	FLEX, LIQUIDTITE, 3/4"			$ 0.25	$ 3.00
					Total Price:	$ 5,097.10

FIGURE 3-201. FINAL BILL OF MATERIALS WORKSHEET

If we want to, we can sort the data by description (Figure 3-202). But we must be careful to sort the entire row, not just the area we're concerned about. Don't forget the links we have established.

Select just the data rows, not the header. On the Sort popup, pick No Header Row, and select the column the description data is in. Execute the sort, and

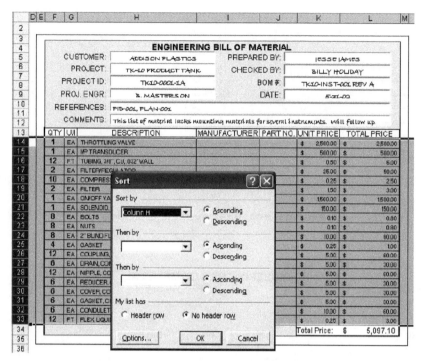

FIGURE 3-202. SORT BY DESCRIPTION

configure the print area to include the material list plus the border. The printed engineering bill of materials worksheet is shown in Figure 3-203.

E. Procurement Summary

So we have generated a set of materials lists that covers the entire spectrum of the I&C project. Admittedly, this material list is somewhat incomplete. We do not have part numbers, and the cost figures are fictional. However, the thought process conveyed here, along with basic spreadsheet tools, will help get you started.

The primary point is the ease with which the information can be extracted, provided good techniques are used when designing the drawings. The key to building an accurate and cost effective materials list is the prep work that happens before the takeoff activity even begins.

PART III – CHAPTER 16: PROCUREMENT

ENGINEERING BILL OF MATERIAL

CUSTOMER: ADDISON PLASTICS	PREPARED BY:	JESSE JAMES
PROJECT: TK-10 PRODUCT TANK	CHECKED BY:	BILLY HOLIDAY
PROJECT ID: TK10-0001-1A	BOM #:	TK10-INST-001 REV A
PROJ. ENGR: B. MASTERSON	DATE:	5-31-02
REFERENCES: PID-001, PLAN-001		
COMMENTS: This list of material lacks mounting materials for several instruments. Will follow up.		

QTY	U/I	DESCRIPTION	MANUFACTURER	PART NO.	UNIT PRICE	TOTAL PRICE
8	EA	2" BLIND FLANGE, DRILLED & TAPPED			$ 10.00	$ 80.00
8	EA	BOLTS			$ 0.10	$ 0.80
10	EA	COMPRESSION FITTINGS, TUBING, 3/8"			$ 0.25	$ 2.50
6	EA	CONDULET, TEE, 3/4", FORM7			$ 10.00	$ 60.00
12	EA	COUPLING, LIQ-TITE, FLEXIBLE			$ 5.00	$ 60.00
6	EA	COVER, CONDULET, FORM7			$ 5.00	$ 30.00
6	EA	DRAIN, CONDUIT			$ 5.00	$ 30.00
2	EA	FILTER			$ 1.50	$ 3.00
2	EA	FILTER/REGULATOR			$ 25.00	$ 50.00
12	FT	FLEX, LIQUIDTITE, 3/4"			$ 0.25	$ 3.00
4	EA	GASKET			$ 0.25	$ 1.00
6	EA	GASKET, CONDULET, FORM7			$ 5.00	$ 30.00
1	EA	I/P TRANSDUCER			$ 500.00	$ 500.00
12	EA	NIPPLE, CONDUIT, 3/4" RIGID STEEL			$ 5.00	$ 60.00
8	EA	NUTS			$ 0.10	$ 0.80
1	EA	ON/OFF VALVE			$ 1,500.00	$ 1,500.00
6	EA	REDUCER, CONDUIT, 3/4"X1/2"			$ 5.00	$ 30.00
1	EA	SOLENOID, 3-WAY			$ 150.00	$ 150.00
1	EA	THROTTLING VALVE			$ 2,500.00	$ 2,500.00
12	FT	TUBING, 3/8", CU, .032"WALL			$ 0.50	$ 6.00
					Total Price:	$ 5,097.10

FIGURE 3-203. ENGINEERING BILL OF MATERIALS

Part III – Chapter 17: Quality Control— The Integrated Design Check

Having progressed to the end of the design phase, we must now validate the work before sending it to the field to be built. Is this step really necessary? Can't we just "fire it off" and let Construction manage it from here? "After all, they can redline the corrections, and we'll fix 'em!," you might say.

All too often, that is exactly what happens. The cost of this approach, while low in terms of the project budget, is actually very high, as mistakes caught too late erode the customer's confidence and jeopardize the project. In many cases there are monetary penalties if the quality is poor or if the product is late.

As has become evident throughout the design process, there is a vast number of decisions to be made and minute details to attend to. If the design process presented in this book is followed, most of the major design flaws will be discovered during the design phase. That is because information that is presented in multiple places will originate from a common data record. Navigation through the document package will be enhanced, making problems more evident.

Nonetheless, a final design check should be integrated into the project from the beginning. If the design events are properly logged as they are executed, then the design check at the end is much easier, and much time will be saved.

What is the goal of a design check? It would be easy to say, "zero defects in the finished product," but that will not occur. Time is too short to allow the necessary ultra-detailed care and attention, not to mention the necessary repairs and revisions. However, that goal can be approximated, and as a result of the effort, the design team can rest assured that no major problems will stop the construction effort.

There are a few specific issues to consider in terms of establishing design-check criteria. Each document needs to be checked and then either highlighted as okay or redlined for repair. Two different approaches need to be considered: administrative and technical.

A. Administrative Content – Individual Checks

Nontechnical experts should be designated to page through the document set to check for specific appearance aspects:

- Is the title block correct in terms of drawing number, revision level, and title?

- Is the proper project data in place (e.g., project site, process area, project number, issue date, and so on)?

- Are the proper initials in place (e.g., designed by, checked by, and so on)?

- Were the proper CADD practices used in terms of valid levels, colors, and text fonts and sizes? This requires someone paging through the drawings at a workstation.

- Are notes legible, complete, and grammatically correct (and thereby understandable)?

B. Technical Content – Squad Check

Checking technical content on a document package is difficult to do piecemeal. The document set is interrelated among drawings and among document types. The goal is to answer the following questions:

- Do the instrument specifications reflect the intent of the P&ID, and will they meet the scope requirements?

- Do the elementaries reflect the intent of the instrument specs and the P&ID?

- Do the wiring diagrams reflect the intent of the elementaries?

- Do the instrument location plans conflict with the finished piping drawings?

- Does every bill of materials reflect the needs presented on the process plans and cabinet arrangements?

- Are all these properly reflected in the instrument database?

- Have the interfaces to other disciplines been properly addressed?

The best way to tackle this set of interrelated tasks is to do them simultaneously, using a team (squad). The ideal setup is to have one person checking one or two documents, with between three and eight personnel on the squad check team (depending upon the size of the project). These folks should shut themselves into a "war room" with a large table that can accommodate all the document sets laid out flat upon it. They should then step through the entire project, using the project database and the P&IDs as the key controlling documents. At the end, the database should be completely highlighted, indicating it has all been checked, as should the entire set of documents. Sound time consuming? It is. But so is the alternative method of using a single designated checker. The time saved using a single checker is likely to be minimal, and the simultaneous method is much more comprehensive and effective.

Of course, as the result of this check, some adjustments will need to be made in the document package. These should be made on the fly by the check team. As documents are completed by the check team, meaning they are either fully highlighted and/or marked up, they may be released to the CADD group for correction. Every revised drawing should be attached to its original markup to maintain the document trail.

When the design check is complete, the markups should be retained and stored. They should be retained for at least one year after plant startup.

C. SQUAD-CHECK ROSTER

Who should be on this squad-check team? For the most part, the design team itself. For one thing, it is the most practical approach. These individuals are already assigned to the project. They are knowledgeable about it and are motivated to see it through and to deliver a quality product.

It is highly recommended that the members not check their own work. Usually, a design team breaks down along vertical lines, one person or group doing wiring diagrams, another the floor plans. During the check, each should check the other's work. Rotating personnel away from their own work does a couple of good things. First, if a mistake was made once, it will probably be missed if the same person looks at it again. Another benefit is the cross-training aspect. After a few projects, team members should be able to float between tasks with little loss in productivity.

One other addition to the squad should be an experienced outsider. This should be a senior individual who is not a full-fledged member of the design team but who is still knowledgeable about the project and about the design process as a whole. And most important, this person should be knowledgeable about the customer and the customer's requirements.

D. DESIGN-CHECK SUMMARY

The design check is a function of quality management. It is often undervalued and underutilized. Usually, design teams rely on a single individual to check the project, which can be inefficient because this person probably does not have access to all reference drawings and may not be expert in some areas being checked. Or the design is allowed to proceed to construction unchecked, where the construction team is, in effect, the checking agent, with mistakes leading to delays, frustration, and increased cost.

A third option exists, whereby the design check is considered and allowed for throughout the life of the project. All the designers prepare for the check by logging their progress in the database, by keeping a personal log of potential problems to watch out for, and by communicating with each other. At the end of the project, the project database can be used as a checking tool that streamlines the squad-check process and maximizes the effectiveness of the time spent in its execution.

Part III – Chapter 18: Phase 3—Deployment

In this chapter, we will divorce ourselves from our small design task, and speak to a broader perspective in order to present a clearer picture of what is involved in a typical large-scale deployment. Phase 2 has ended, and Phase 3 has begun. In an ideal world, most of the design team would go on to other things, leaving a skeleton crew to interact with the constructor. However, on all but the smallest projects, things are not usually that simple. Most of the time, the design team issues their documents in packages that line up with a work breakdown structure or WBS. A WBS organizes the work and coordinates it so that the work flows smoothly. A WBS breaks the design task into pieces, each of which is handled like a small project. The design team issues its design product to construction in WBS packages.

Many mistakes can be avoided by close association between the construction and design teams. And if the designers are in the field and are known to be helpful, then the perception of competence will be enhanced to the benefit of all.

A. Construction

There are two kinds of construction support provided by the design team: onsite and offsite. It is a good idea to determine ahead of time the periods that are critical for onsite support and to have people available continuously through those periods. Otherwise, having someone on call offsite is usually acceptable, coupled with regularly scheduled site visits to clarify issues and facilitate the work. For more information on the construction team and its activities and needs, refer to Part I.

The construction support phase breaks down into several periods, each with a different set of needs as relates to the design team.

1. Kickoff Meeting

Each construction activity (as listed on the WBS document) should begin with a kickoff meeting. At this meeting, the Phase 3 lead for the design team should present the engineering package. This presentation should be made to the construction superintendent and each of the foremen. The presentation should cover, at minimum, the following topics:

- Bills of materials: status of materials and instruments and expected arrival dates, special handling issues, and problems and concerns.

- Design overview: scope of work and presentation of the design package.

- Offsite support infrastructure: offsite support phone numbers, pagers, faxes, vendor lists with contacts, and so on.

- Onsite support plan: onsite schedule, personnel list with phone numbers, and so on.

Upon completion of this presentation, there should be an open discussion period followed by a walkdown of the facility if appropriate. In some cases, this may be the constructor's first visit onsite, or it may be a group of prospective constructors who are preparing bids. In either case, materials, scope of work, infrastructure, schedule and so forth are important information.

2. CONSTRUCTION

Construction activities usually provide a bit of a lull for the design team. Some members of the team may be reassigned to other duties at this time, never to return to the project. Others may monitor the project, getting involved only when a problem arises. During certain predefined periods of particular risk to the project, design team representative(s) may travel to the site to observe and provide direct support. Onsite time for the design team should be maximized within budgetary constraints. The ideal situation is to have one representative onsite through the entire process who can call in additional support as needed.

The services likely to be needed during this time are, for the most part, clarification and direction services. Also, the constructor almost always needs help in locating materials, which is easier with the help of someone who has knowledge of the database.

If design flaws are discovered during this period, the design team must operate much like a fire brigade, assembling quickly and attacking the problem to minimize its effects. Many times it is the team's ability to react effectively that prevents a small problem from ballooning into a large one. It is best to attack these problems directly and decisively, if necessary, reassembling the squad-check team to perform an emergency recovery.

Construction activities during this time include installation and verification. After the installation of an item is complete, an inspector checks it out for mechanical completeness, and the electrical wiring is wrung out (thoroughly tested) to make sure connections have been properly made. The construction phase is over when the mechanical completion milestone has been achieved. Mechanical completion means that all equipment has been installed and wired per the design drawings and that all construction specifications have been met. Further, all punchlists have been addressed and outstanding items resolved, and the owner has performed an inspection and has agreed that the system is mechanically complete.

B. PRE-COMMISSIONING

Pre-commissioning starts when the constructor has achieved mechanical completion. At this point, the owner takes a small step towards ownership of the facility. Everything is generally still "owned" by the prime contractor, but the true owner will begin to get involved as systems are powered up, as maintenance procedures are engaged, and as preliminary checkout activities begin. Infrastructure subsystems, such as power, instrument air, water, and other service elements are charged, engaged, and otherwise activated. Equipment is lubricated, and packing is removed. Temporary structures are dismantled.

After the integrity is proven on the various infrastructure and services subsystems, some process element testing may begin. Sometimes called the "Bump & Stroke" phase, each end device can be manually tested to ensure its integrity both from a mechanical and electrical standpoint. Motors can be decoupled from their loads and *bumped* (power applied for a very short duration) in order to ensure proper direction of rotation. Pneumatic on/off valves can be stroked full-open, full-closed, generally by mechanically forcing the solenoid to change states. Throttling valves can be stroked using a handheld DC current source. During this period, the control system may or may not be intact, so these tests must generally be done locally at the instrument or end device.

The database is useful during pre-commissioning, as reports can be generated for each process area listing all the devices in that area, and providing a means for the field personnel to report the results of their tests. The E/I&C technical support team can be of assistance in providing the reports and helping the constructor manage the tests. The E/I&C team will also be available for technical support as the constructor and owner perform their tests.

During pre-commissioning, the CSI will be performing control system checks in order to bring the control system online. Computer system(s) will be energized, and a series of tests will be done in order to ensure system integrity. After all the control system elements have been confirmed active, performance test baselines can be obtained. Network data throughputs, screen update times, and other parameters should be checked and recorded for future use in determining system health.

C. Cold-Commissioning (Site Acceptance)

The term "cold" in this context implies simulation of actual process material. In some cases, such as in the material handling aspects, a slug of actual material might be used, but for the most part, the pipes and vessels will be empty. The owner generally takes another major step toward taking ownership of the facility at this point. While the CSI is in charge of this step in the process, the owner's operators generally begin to make the moves at the HMI. Getting them involved at this stage provides an important training opportunity and gives them a better handle on the system they will eventually have to operate on their own.

The Cold Functional Test is governed by the Site Acceptance Test Procedure, which may itself be based on the FAT discussed in Chapter 10. At this point, all the motors and valves have been bumped and stroked, and the service subsystems have been activated. The intent of this test is to test the control system to the fullest extent possible without expending raw materials. If water can be used instead of a solution, then water should be used.

Cold-commissioning requires a validated control system, so the pre-commissioning process must be complete. There are two levels of testing: Device Tests, and Subsystem Tests.

1. Device Tests

Each individual device associated with the control system must be tested all the way through to the HMI screen, and even into the Historian if one is being used. Temperatures, flows, pressures, and other analog inputs can be simulated, using handheld simulators in order to confirm that the sensors are connected to the proper control points. Discrete inputs can be simulated by forcing the switch closed, by placing jumpers across open sets of contacts, and/or by lifting wires to simulate switches opening under process conditions. All end devices (motors, valves, etc.) can be manually bumped and stroked by the remote operator from the HMI, with personnel located at the device to observe the equipment's action. Wherever possible, if the equipment can be operated dry, the motors will be coupled to their loads to observe their operation.

For example, a conveyor motor can be coupled to its conveyor, and the conveyor started and allowed to run. If there is a local run mode, that can be tested. Then the conveyor system can be placed in remote mode and can be started by the remote operator manually through the HMI. All the associated equipment to the conveyor (horns and beacons) can be observed to operate as desired. Drift switches, pull chords, and zero speed switches can be tested, and proper conveyor operation confirmed. All of this can be done "cold," with no material on the belt.

2. Subsystem Tests

After the individual devices are confirmed, then each subsystem should be checked as far as possible in a dry state. Using the conveyor example, let's assume that a coalyard having ten conveyors, several diverter gates, a surge bin, and several coal hoppers needs to be tested. The Subsystem Test will verify that the startup and shutdown sequences will correctly phase the conveyors, depending on the situation: that conveyors will automatically start and stop in the correct sequence, and that they will react properly to surge bin level, as simulated. All emergency tests will be conducted, and any Zone controls will be verified. All of this, again, can be done with no coal on the belts.

Other subsystems may be more difficult to test cold. For example, a boiler would require simulation of temperatures and pressures in order to test its automatic operation.

D. Hot-Commissioning (Startup)

The term "hot" in this context implies the use of actual process material, at temperature and pressure. The generation of production-quality product is the aim of this step.

At this point, the owner begins to take command of the system. The owner's operators will have received training from the systems integrator. The maintenance staff is getting familiar with the system. The construction team is available, as are the leading members of the design team. It is now time to light the candle, so to speak. It can make for some tense moments as systems begin to come up to temperature, process fluid (or more often, water) begins flowing in the pipes, and motors begin spooling up.

The owner's operations department will generally create a performance test procedure and the E/I&C and CSI groups become bystanders and consultants. If problems arise as the facility is being brought up to operating conditions, both the E/I&C and the CSI teams will spring into action to help resolve the problem.

Hot commissioning is complete when the owner's performance test is complete, and product is being made within specification.

E. Adjustment of Document Package to Reflect Construction Modifications

During the deployment process, adjustments made to the design package during construction must be documented. Any deviations from the design should have been carefully documented during construction (see Management of Change (MOC) in Chapter 3). An MOC process for the constructor would include notification of the engineering entity that either a mistake was made in the engineering package, or there is a better way to execute. In either case, the engineering and design team should be involved in any deviations.

In a non-regulated industry (i.e., non-regulated in that DOE or the FDA is not involved), construction adjustments are generally documented by marking up the construction drawings to reflect the as-built condition. These are commonly referred to as "red-line drawings" since red pencil is usually used to make additions. (Green is used to delete). Construction redline drawings are usually copied since the redlines are the as-built version and only exist there and cannot be removed from the site. These copies are then sent to the engineering service provider to be formally incorporated into the documentation. This process is generally a simple CADD function, but there are occasionally more far-reaching modifications.

In a regulated industry (food, drug, and nuclear, predominately), design changes during construction must be fully vetted by the entire project team – constructor, owner, engineer – prior to implementation. The MOC process includes a validation step that is missing in most non-regulated industries.

F. Issue for Record

As soon after startup as practical, the updated drawings should be released back into the customer's possession, but before the documents are issued for record, the CADD supervisor usually makes one more pass to make sure there are no hidden problems that will be caught by the customer's organization. The drawing list should then be updated and a transmittal prepared. Finally, the entire project may be shipped to the customer.

This "issue for record" is an official turnover of responsibility for the project. Depending on the contract, this may conclude the engineering service provider's activity for this project. Usually, however, there is a period of free phone support followed by an option for additional ongoing support.

G. Phase 3 Summary

The construction period is when the design is fully validated, as the end result is a fully functional facility. If, at its conclusion, the facility is running as designed and the customer is happy, then the project was a success. Following an integrated design process, such as has been presented in this book, increases the likelihood of success.

Part III – Chapter 19: Phase 4—Support

After the facility is generating product, the engineering service provider enters the Service and Support phase, whereby warranty service and technical support is either implied or implicit in the contract with the owner.

A. Warranty Support

All materials supplied will be procured under a manufacturer's factory-direct warranty. Most of the time, on a large project, those warranty periods will be expired by the time the plant goes into operation. Frequently, the owner will stipulate extended warranties in the contract. These extended warranties generally are about 3% of the sale price for a one-year extension, and must be built into the service provider's bid. However, the cost of the warranty is not the only cost associated with warranty support. In order to effect a warranty repair, the service provider must sometimes travel to the site, spend time with the customer in troubleshooting, spend time working with the manufacturer to get the part shipped, and work with the customer to get the bad part shipped to the manufacturer. Sometimes, the manufacturer will not credit a warranty until the bad part is in hand and confirmed to be a warranty repair, and so the new part must be procured by the service provider until that evaluation is made.

As can be seen, warranty support can be a costly aspect of the project, and should be properly considered at the time of the bid preparation.

Another stipulation that can be made in the contract is a specific timeframe of consecutive days or weeks after commissioning in which there are no control system anomalies. The definition of an anomaly needs to be well defined in the contract, but this stipulation could cause the warranty period to extend indefinitely if the control system has problems. CSI support during this timeframe can usually be offsite unless a problem arises, at which time the clock will restart from zero. It behooves the CSI to address any issues that arise at high speed!

B. Continuing Service Support

Continuing services support is a paid service, usually performed by the CSI, and usually performed under contract.   This set of services is usually pre-negotiated, either as a fixed fee (subscription), a discounted hourly rate, or a blend of both. The following are some of the more common aspects of Continuing Services:

- Help Desk (Phone Support) – This service features case-tracking, in which each call is tracked against a case number until the case is resolved. Pricing for this service could be by the case, or by the call, or it could be provided at a flat subscription rate. The service can be provided 24/7, or just during normal working hours, with different costs associated with each. There is also a guaranteed return call time span which can also be graduated, depending on the price of the subscription.

- Offsite Storage – For a fee, some CSI providers will provide offsite storage for software and key documents. Usually this is done on a website that will give the customer direct access to the documents. The offsite servers are generally redundant and UPS-backed so that access to the information is assured at all times.

- Rapid Response Technical Support – This type of service is usually a blend of subscription plus a fee for each event. The CSI provider will agree to a quick response (response-time varies by contract) reaction to an emergency call by the customer. CSI personnel must usually be on call (hence the subscription). If called upon to react, then an hourly fee would kick in. In return, the customer is given a response guarantee.

- Periodic Plant Inspections – The CSI and/or E/I&C provider(s) would be engaged to perform a periodic (yearly, quarterly, monthly) inspection of the facility to look for problems, and to resolve any minor issues that may have arisen. This inspection would probably be covered by the subscription fee. If the inspection revealed issues that required more work, then those would be dealt with under work orders or projects.

- Procurement – Equipment breaks, wears out, needs to be replaced on a regular basis. The CSI can provide replacement parts on an as-needed basis. The contract usually stipulates the markup that would be applied for such a service. The advantage to the customer is that the CSI frequently has pricing agreements with the control systems manufacturer that partially or wholly counteracts the markup applied, so the customer can get this procurement activity done with little effort on his own part, and minimal expense.

Continuing services support is, for the CSI in particular, a great way to stay in close contact with the customer. Since the CSI has written the software, they will be the most knowledgeable about how the facility operates and will be best positioned to provide ongoing support. Whether the ongoing support is merely an agreed-upon hourly rate for services, or a hybrid set of services under contract, Phase 4 is important to the long-term viability of the facility.

REFERENCES – PART III

1. *Instrumentation Symbols and Identification*, ANSI/ISA-5.1-2009. Research Triangle Park: ISA, 2009.

2. *Graphic Symbols for Distributed Control/Shared Display*, ISA-5.3-1983. Research Triangle Park: ISA, 1983.

3. *Instrumentation Symbols and Identification*, ANSI/ISA-5.1-2009. Research Triangle Park: ISA, 2009.

4. Extracted from SAMA Standard PMC 22.1-1981, "Functional Diagramming of Instrument and Control Systems," with permission of Scientific Apparatus Makers Association, Process Measurement & Control Section (Washington, D.C., Scientific Apparatus Makers Association, 1981) – NOTE: SAMA is defunct. This standard has been replaced by ISA-5.1.

5. Lipták, B. G. *Instrument Engineers' Handbook, 4th Edition, Vol. 1: Process Measurement and Analysis*. ISA/CRC Press, 2003.

6. Lipták, B. G. *Instrument Engineers' Handbook, 4th Edition, Vol. 2: Process Control and Optimization*. ISA/CRC Press, 1995.

7. Baumann, H. D. *Control Valve Primer: A User's Guide, 4th Edition*. Research Triangle Park: ISA, 2009.

8. *Flow Equations for Sizing Control Valves*, ISA-75.01-1985. Research Triangle Park: ISA, 1985.

9. *TVA Nuclear Power Electrical Design*, STD DS-E18.1-24. Tennessee Valley Authority: Chattanooga, 1993.

10. Earley, M., Sheehan, J., Sargent, J., Caloggero, J., and Croushore, T. "NEC article 110-26(A)(1): Depth of Working Space" *NEC 2002 Handbook*. Quincy: National Fire Protection Association, Inc., 2002.

11. *Ibid.*, "NEC article 110-26(C)(2)(a): Unobstructed Exit."

12. *Ibid.*, "NEC article 110-26(C)(2)(b): Extra Working Space."

13. *Ibid.*, "NEC article 110-26: Spaces About Electrical Equipment."

14. *Instrument Loop Diagrams*, ISA-5.4-1991. Research Triangle Park: ISA, 1991.

Additional Resources

Software Tools for Successful Instrumentation and Control Systems Design

For additional tools that can help you with successful instrumentation and control systems design, see the CD-ROM entitled, "Software Tools for Successful Instrumentation and Control Systems Design," by Michael D. Whitt included in this book. This CD-ROM contains files for use in designing a project as well as the figures, charts, and forms presented in this book for use as training or presentation aids.

Section 1 – Estimate & Schedule

This section describes a method for generating a comprehensive, cross-referenced project estimate and execution schedule, using Microsoft® Excel. Microsoft® Excel is used due to its flexibility and usefulness and to its almost universal availability. An accompanying narrative provides instruction for building the material and labor estimates and a subsequent project schedule. Two Microsoft® Excel files are included—the first being fully implemented per the imaginary scenario described in the narrative. The second file is a blank file that may be used as a template for a new project. There are a total of three files:

- File_1A_Estimate&Schedule (the narrative text in Adobe format)
- File_1B_AddisonPlasticsEstimate&Schedule (the sample Microsoft® Excel workbook)
- File_1C_BlankEstimate&Schedule (the blank Microsoft® Excel workbook)

In this approach, teaching occurs through interactive involvement. You are forced to open Microsoft® Excel, for example, in order to follow the author's reasoning in building the estimate. New skills in the use of Microsoft® Excel are likely to be gained, as is a new understanding of the proper relationship that should exist between the project estimate and the schedule. All too often the link between the two is lost, thus making budget difficult to manage. Here, the estimating process is tied to the Work Breakdown Structure (WBS), which is then used to generate the schedule. And once the schedule is generated, a project planning and tracking tool facilitates labor management.

Section 2 – Forms, Figures, and Database

This section provides the figures, charts and forms discussed in each part of the book. The first four items (Files 2A through 2D) are the Figures that are in the book presented in their original Microsoft® Excel format. Having them available in their native format makes them suitable for use as training aids since they can be tailored to individual needs. File 2E is the Instrument Database (Microsoft® Access) developed in Part III. And finally, File 2F is a Microsoft® Visio file with the final deliverable drawings that are shown in the book.

- File_2A_Part1Figures (Book Part I charts and forms Excel file)
- File_2B_Part2Figures (Book Part II charts and forms Excel file)
- File_2C_Part3Figures_1-99 (Book Part III charts and forms Excel file)
- File_2D_Part3Figures_100-213 (Book Part III charts and forms Excel file)
- File_2E_Instrument Database (Access file)
- File_2F_Deliverables (Visio file)

If you wish to use the provided spreadsheet form, you must first copy the sheet from the Excel workbook and clear it of existing sample data. Then the form may be adapted to your purposes.

INDEX

A

A/M 203–204, 246, 334
A&E firm 74, 438
Access 132, 288, 291, 299, 372, 378, 394, 453, 524
accuracy 118, 120, 307
  resolution effects on 122
action list 93–96
actual hours 88–90
agenda 57, 92–93, 96–97
alarm 41, 48, 81, 120, 166–168, 176, 196–197, 201, 207–208, 211, 222, 225, 230, 232, 237, 239, 243, 245, 272, 276–277, 286, 324, 326, 331, 334–335, 338, 346, 349, 356, 358–359, 365, 367, 369, 386, 465
alarm manager 202, 211, 235, 243, 356, 366–367, 377–378
algorithm 119–120, 222, 238, 245–246
analog 47, 119, 122, 126, 128, 138, 178, 184–186, 188–189, 191, 195–198, 200–202, 206, 211, 214, 217, 220, 222, 225–226, 228, 230, 243, 245
analog alarm 221, 333, 337, 354, 357, 383–384
annunciator 28, 35, 167–168, 196, 204, 211, 225, 233, 384, 410, 444
auto/manual (A/M) 203–204, 246, 334
  mode switching 49, 245

B

benchboard 204
bid package 5–7, 20, 24, 71, 73–74, 76, 78, 242
bill of material (BOM) 13, 34, 37–39, 52, 74, 131, 268, 284, 429–430, 432, 434, 438, 479, 485, 487, 491, 493–497, 501–502, 506, 510
bit 119, 122–123, 126, 128, 157–158, 225, 233, 277, 333–334, 342, 345
bluetooth 234
BOM 13, 34, 37–39, 52, 74, 131, 268, 284, 429–430, 432, 434, 438, 479, 485, 487, 491, 493–497, 501–502, 506, 510
Boolean 196, 199, 225
break-before-make 162
bridges 236–237
budget 2–3, 11–12, 21, 58, 62, 68, 74, 78, 90, 92, 98, 228, 267, 317, 327, 382, 385, 401, 509, 523
budgetary 12, 14, 71, 73, 76, 514
budgetary estimate 26, 32, 71, 73, 79, 98, 309, 322
bulk material 13, 493, 496–497, 499
buyer 1, 10–16, 18, 32, 57–58, 71–73, 79–82, 96, 98

C

cable and conduit schedule 137, 155, 268, 282

calibrated
  range 118–119, 123, 397, 401
  span point 118
  zero point 118
capital projects 58, 78, 241
CAT5 236–237
CFC 220, 328, 333
change order 12, 16, 33–34, 100
chart 38, 81, 124, 131, 149, 151, 189, 197, 212, 219–220, 226, 268–270, 317, 321, 331, 368, 372–373, 426–427, 429, 450–451, 468–469, 478, 497, 499
checkout 3, 27, 43, 61, 69–70, 138–139, 166, 228, 312, 323, 346, 383–384, 395, 515
circuit
  four-wire 190
  two-wire 190
client 66, 129–130, 235, 237–238, 323
coil 120, 157, 159–160, 163, 167, 445–446, 448
cold-commissioning (also called "dry") 516
color plan 208, 210, 243, 370
colors 145, 208–209, 243, 366, 465, 477–478, 510
commercial off-the-shelf (COTS) 115
commissioning 17, 20, 27, 34, 43–44, 46, 50, 67, 141, 323, 383–385, 515–517, 519
communications 7, 12, 25, 45–46, 56, 198–199, 214–216, 229–231, 235–237, 239, 288, 364, 367, 379
comparators 221, 326
component schedule 154, 367–369, 372, 392, 394, 414–416, 426, 428, 435, 496
conduit 34, 38–39, 68, 115, 137–138, 149–152, 154–155, 171, 187, 268, 280, 282, 285, 307, 397, 400, 405, 414–417, 423, 425, 429–435, 473, 475, 493–494, 496–497, 500
  fill 150, 425
  routing 154–155, 391–392, 413, 425–426, 429, 471
connection diagram 156, 284, 442, 462–463, 466, 474, 476, 479
construction 1–2, 16, 20–22, 25–26, 34, 39, 43, 53, 63–64, 67–68, 70, 74, 83, 130, 138–141, 144, 148, 150, 154, 156, 164, 170, 196, 228, 267–268, 285, 312, 314, 385, 390, 395, 397–398, 401, 407, 415, 422, 429, 438, 441, 443, 452, 458, 462, 485, 493–494, 497, 509, 513–514, 517–518
contacts 120, 157–160, 162, 167, 182, 219, 276, 335, 346, 361, 403, 445, 448–449, 514
contingency 15, 17–18, 26, 33, 100
continuous control 245–246, 326, 328
continuous function chart (CFC) 220, 328, 333

control
    board 196–197, 201–202, 205
    narrative 24, 42, 47, 329, 332, 334–335
    panel 37, 41, 147, 149, 494, 496
    panel fabricator (CPF) 1, 4, 25, 41, 50, 53, 130
    room 40, 46, 129, 193, 196, 201–204, 208, 235–236, 286–287, 374, 409–411, 413, 426
    solutions providers (CSP) 19, 53
    variable (CV) 119–120
control system 1–2, 20, 22–23, 29–30, 35, 37, 40, 43, 47, 49, 115–116, 136, 139, 141, 145, 147, 154, 177, 186, 193–195, 197–202, 206–207, 213–214, 223–225, 227, 238, 241, 244–247, 272–273, 276, 278, 283, 285, 287, 325, 327, 334, 345, 362, 364–365, 370, 379, 381, 383, 385–386, 411, 414, 441, 444, 447–448, 461, 472, 474, 515–516, 519
    specification 42, 242, 322–324
control systems integrator (CSI) xxxii, 4, 40, 42–44, 47, 51, 53, 64, 70, 130, 210, 241, 321
cost 2, 6–14, 16, 22–24, 52–53, 66, 76–77, 79–80, 96, 156, 159, 183, 193, 200, 230, 302, 304, 309, 314, 316, 322, 415, 493, 504, 509, 511
cost-plus 2, 8, 10–15, 98
COTS 115
CPF xxxii, 1, 4, 25, 41, 50, 53, 130
CPU 214
CSI xxxii, 4, 40, 42–44, 47, 51, 53, 64, 70, 130, 210, 241, 321
CSP 19, 53
current source 125–126, 128–129, 189, 515
CV 119–120
cyber security 46, 213, 244

D

DAS 188, 197–198
data
    communications 45–46, 187, 193, 215, 232–234, 287, 321, 334
    highway plus 233
    integrity 46
    storm 237–238
data acquisition system (DAS) 188, 197
data transfer detail sheets (DTDS) 48
database 36, 48, 53, 64, 82, 96, 117, 127, 130, 132, 134–138, 140, 142, 144, 156, 198, 201–202, 206, 210–211, 216–218, 225–226, 283, 288–291, 293–297, 302, 304, 365, 372, 374, 377, 380, 383–384, 389–390, 392, 395, 397–398, 420, 428, 436, 438, 453, 460–461, 479, 483, 501, 512, 514–515, 524
DC power supply 184, 190, 451, 471, 474
DCDS 47, 136, 329, 345, 348, 350–351
DCS 23, 40, 66, 116, 197
definitive estimate 21, 26, 72, 76–77, 81, 283–284, 322
delay timer 354, 361
derived function blocks 219

design
    basis 17, 47, 72–74, 79, 280, 315
    check 36, 142, 394–395, 462, 509, 511–512
    range 118–119, 123
    supervisor 63, 76
detail sheet 47–48, 221–222, 329, 332, 339, 352, 354, 382
deterministic 4, 8, 15, 115, 230, 237–239
device control 243–244, 325
device control detail sheet (DCDS) 47, 136, 329, 345, 348, 350–351
device logic 49, 222, 329, 337, 340–341, 343–347, 350, 353–354, 356, 359, 383
devicenet 200
digital 46, 122, 159, 178, 188, 220, 225–226, 422, 443, 446
    input 168, 179, 181, 307, 448–449, 455–456, 464
    output 182–183, 228, 307, 346, 385, 448, 457–458, 465
DIO 115
discipline engineer 63
discipline lead engineer 63–64, 67
discrete 119, 157, 178, 181, 184, 195–196, 198–199, 201–202, 206, 215, 222, 225, 228, 230, 245–246, 274, 277, 334, 384, 405, 446, 455, 458–459, 462, 472, 479, 516
distributed control system (DCS) 23, 66, 116, 197
distributed I/O (DIO) 115
document control 48, 86, 142, 268, 289–293, 389, 395, 398, 479, 483
document list 35, 142
documentation 20, 27, 63, 95, 116, 136, 139, 217–218, 221, 246, 295, 323, 328–329, 364–365, 411, 444, 472, 518
DPDT 158
drawing list 32
dry contacts 158
DTDS 48

E

E/I&C xxxii, 1, 4, 20, 25, 30–35, 39–42, 47, 51, 53, 55, 61, 65, 70, 130, 151, 268, 271, 280, 398, 413, 515, 517, 520
earned hours 88, 90
ECDS 47
ECN 98
edge-triggered 158
efficiency ratio 88, 91
electrical/instrumentation and controls (E/I&C) 1, 4, 130, 268
elementary 156, 165, 194, 309–310, 347, 443, 446–447, 450, 452–453, 462, 465, 467, 472, 479
elevation 29, 277, 279, 281, 416, 419–420, 423, 429
engineering change notices (ECN) 98

INDEX

engineering units 117, 120, 122, 127, 138, 187, 207, 220, 333, 355–356
environmental 40, 121, 244, 409, 411, 413
EPC 16
equipment
    arrangement drawings 29
    arrangements 24, 39, 279, 417–418, 422–423
    lists 129
    specifications 29, 36, 278–279, 282
equipment control detail sheets (ECDS) 47
ergonomics 244, 416
estimate 3, 5, 7, 21, 24, 26, 71–73, 75–78, 84, 88, 98, 131, 143, 217, 302, 309–310, 315–318, 321–322, 325–326, 383, 450, 523
ethernet 42, 45, 65, 67, 115, 139, 198, 200, 234–241, 287, 379
ethernet master 237
execution strategy 8, 320
explosion 171, 174–175
explosionproof 174–176, 192, 414, 435

F

fabricator 2, 20–21, 51–52, 285, 321, 399, 442
factory acceptance test (FAT) 20, 34, 51, 61, 86, 321
failsafe 116, 147, 166–169, 346
FAT 20, 25
fault-tolerant 45, 166, 237
FEL 4, 20
field 36–37, 39, 49, 67–68, 81, 89, 115, 130, 132–133, 136, 138–142, 148, 156, 164, 169, 175, 177–179, 182–183, 186, 188, 190, 201–204, 215–216, 222, 226, 234, 238, 244, 279, 281, 290–291, 296, 317, 325–326, 370, 383, 392, 416, 422, 426, 444, 461
fixed price 10, 14–16
flag 157–158, 334–336, 339, 341–342, 344–346, 349, 353–354, 356, 359, 370
flip-flop 196
floor plans 29, 39, 279, 282, 419, 511
form 133
form-A 158
form-B 158
form-C 159
Foundation Fieldbus 198, 234
four-wire 187
    circuit 190
front-end loading (FEL) 4, 20, 135, 283
functional description 5, 24, 137, 323–324
fuse 179, 181, 184–185, 201, 400, 445, 449–450, 452, 456–458, 465, 471, 473–474, 477, 499

G

graphic user interface (GUI) 66, 196, 200, 205
green-field 17, 19, 22
GUI 66, 196, 200, 205

H

hand-off-auto (HOA) 204, 277
hazard and operability (HazOp) 24, 29
hazardous area classification 170–171
HazOp 24, 29–30
heat & material balance (HMB) 23, 28, 124, 271
historian 23, 197, 202, 205, 211–212, 235, 243, 326, 366, 378, 384, 516
historical
    data recorder 243
    trend 45, 212
HMB 23, 28, 124, 271
HMI 42, 48, 66, 193, 196, 205–206, 286, 365
    animation database 365
    animation plan 366, 368
    control graphics 365, 369
    device driver 202, 365, 374
    flash and beep 367
    messaging 366–367
    screen diagrams 367, 371, 376, 386
HOA 155, 204, 277, 324, 336, 342, 346, 349, 414, 447–449, 467–468
homerun 137–138, 148, 179, 228, 426, 457, 463, 465–466
homerun cables 148
hot-commissioning (also called "wet") 517
hot-standby 214, 238–239
human-machine interface (HMI) 42, 66, 193, 196, 205–206, 286, 365

I

I/O 23
    configuration 35–36, 309–310, 379
    count 136, 193, 226, 228, 230, 304, 307, 321, 325
    interface 116, 148, 177, 193, 217, 222, 226, 383
    list 24, 32–33, 35–36, 41–42, 53, 86, 129, 131, 134–135, 192, 283, 288–289, 295–297, 300, 303, 390–391, 453, 456–457
    map 193, 216, 224–226, 228, 309, 312, 374
    modules 36, 45, 115, 136, 177, 179, 181, 185, 188, 215–216, 222, 227–229, 239, 287, 310, 382, 450, 452, 472
    partitioning 41, 45, 51, 136–137, 185
    points 135, 177, 183, 189–190, 222, 227–228, 307, 452, 457
    rack 36, 136, 139, 177, 228, 230, 327
    termination room 40
ice-cube relay 159
ICR 98
ICS xxix, xxxii, 4
IEC 200
inhibit timers 245
input-output (I/O) 23

installation details 36, 38–39, 69, 295, 400, 402, 407, 409, 428–429, 433–434, 438, 501–502
instrument
    arrangement 155, 282, 427
    calibration 123
    database 36, 38, 192, 295, 303, 397, 416–417, 420, 433, 510, 524
    elementary 36, 444, 453, 464
    installation details 38, 433, 496
    list 32, 129, 142, 192, 394, 483
    range 123
    specifications 36, 138, 282, 284, 288, 397, 510
integrated control systems (ICS) xxix, xxxii, 4
integration v
    systems xxix, 19, 29, 33, 53, 55, 65, 241, 279, 286, 302, 322, 363–364, 379, 387
interconnection 36–37, 47, 148, 157, 199, 271, 442, 460, 492
interlocks 29, 35, 47, 49, 138, 161, 202, 204–205, 221, 245, 277, 307, 324, 335–336, 345–346, 349–350, 359, 383–384
internal change request (ICR) 98
interposing relay 159, 181, 183, 449
interval timer 162, 356
intrinsic safety 175
inverters 196
ISA v, xxxii, 120, 147, 208, 272–274, 302, 318, 362, 397

J
junction boxes 37, 39, 137, 149, 154, 175, 179, 279–281, 296, 400, 413, 416, 422–423, 425, 442, 457, 463–464, 466, 469, 475–476, 479, 487, 492–493, 497, 499

K
kickoff meeting 32, 56, 513

L
labor 7, 9, 11, 13, 47, 65, 75–76, 202, 241, 302, 310, 312, 314–316, 387, 504, 523
ladder logic 200, 218–219, 286, 363
LAN 45, 230–236, 238–239, 287, 326, 380
leakage current 161, 180, 183
legacy 219, 246–247
linearization 124
list
    action 94
    document 35
    drawing 32
    I/O 32, 35
    instrument 32, 129, 142, 192, 394, 483
    needs 94
    suggestion 95
local-off-remote (LOR) 204
    mode switching 244

logic 28, 35, 47, 49, 136–137, 140, 159–160, 180, 195–196, 199, 201, 205, 215–217, 246, 272, 291, 296, 329–330, 334, 345, 353, 355, 359, 366, 382, 445, 447, 452, 461, 465
    diagrams 47–48, 136, 286–287, 329, 350, 352, 356, 362, 367, 462
loop checks 27, 43
loop sheets 36, 39, 78, 156, 284, 295, 309–310, 443, 446, 459, 461, 471, 473, 479
LOR 204, 244
lump sum 10, 14–15, 17–18

M
maintenance xxix, 1–2, 12, 23, 27, 35, 44, 46, 55, 57–59, 63, 82, 101, 116, 132, 136, 156, 172, 195, 198, 228, 241, 246–247, 267–268, 286–287, 321, 385–387, 389, 394, 401, 407, 410, 413, 417, 423, 438, 441, 443, 452, 458, 461–462, 467, 479, 515, 517
make-before-break 162
managed ethernet switches 237
management of change (MOC) 14, 16, 57, 73, 96, 99
man hours 9
manuals 46
marshalling 37, 40, 179, 184–186, 201, 411, 426, 453, 457, 472, 479
material classification 496
MCC 287, 411, 447, 467–468, 483
memorandum of understanding (MOU) 32, 83
memory map 217–218
milestone 5, 11, 13, 15, 25, 60, 75, 79, 81–84, 88, 140, 143–144, 198, 285, 290, 385, 389, 514
MOC 14, 16, 57, 73–74, 96, 98–99
modbus 198, 233, 378
modbus-plus 45, 200, 233, 237, 378–379
motor 24, 27, 34, 36, 43, 45, 47, 49, 140, 161, 167–168, 185–186, 200, 204, 219, 225, 243–245, 271–272, 278, 280, 307, 325–327, 335–336, 346, 349, 357, 360–361, 385, 444, 446–449, 452, 467, 479, 515–517
motor bucket 204, 447, 449
motor control center 200, 204, 287, 421
MOU 32, 83

N
National Electrical Code (NEC) 147, 413
National Fire Protection Association (NFPA) 147, 150
navigation plan 44, 243
NEC 147, 150–151, 153, 170–172, 174–175, 228, 413, 423, 432
needs list 93–95

INDEX

network 42, 45, 47–48, 64, 66–67, 116, 126–127, 129, 134, 139, 193, 210–211, 222, 229–231, 233–239, 267, 286–287, 326–327, 356, 367, 375, 378, 380, 411, 457, 515
network block diagram 32
NFPA 147–148, 150
noise 179, 185–188, 208, 232, 245, 405, 407, 410, 423
non-incendive 175
normally closed 157–159, 167–168, 445
normally open 157–159, 167, 169, 346
notice to proceed (NTP) 32, 82, 285
NTE 14
NTP 32, 82, 285

O

OE 4, 24–26, 28, 34, 60–61
OEM 46, 55, 61, 86, 214
operations 1, 20, 27, 33, 41–44, 46, 49, 58–59, 61, 64, 128–129, 131–132, 172, 201–202, 239, 286, 321, 337, 376, 385–386, 517
operator interface 23, 197–198, 200, 202, 205, 213, 215, 222, 232, 334, 365
opto-isolators 215
orthographics 29, 279, 281, 418
owner 1, 4, 6, 10, 19–21, 24–25, 27–28, 30, 40, 42, 44, 55, 57, 73, 78, 143, 170, 175, 271, 515–516, 518–519
owner's engineer (OE) 4, 19–20, 22, 25–29, 34, 41–42, 55, 57, 60, 74

P

P&ID 24, 28–29, 33, 35–36, 42, 78, 130, 136, 208, 228, 242, 271–273, 275–276, 278–280, 282, 284–285, 291, 295–296, 300, 303–304, 309, 312, 321–322, 326, 329, 357–358, 369–370, 401–402, 419, 428, 443, 445, 460, 510
panel arrangement 37, 484–485, 492
partitioning 35, 135, 223, 226–228, 379
PCDS 48
PEP 7, 84
permissives 47, 138, 201, 205, 245, 334–336, 342, 359
PFD 23, 28, 205, 242, 271
phasing 211
physical address 223–224
PID 234
piping 28–29, 39, 49, 64, 69, 271–272, 279, 281–282, 370, 375, 398–400, 402, 406, 412, 417–418, 438, 460, 497, 499–500, 510
piping and instrumentation diagram (P&ID) 24, 28, 35, 136, 271
PLC 23, 40, 66, 115, 199, 214, 285
PLC cabinet 286, 479
PMT 44
poles 159–160, 162

post-modification test (PMT) 44
power distribution 37, 70, 177–178, 181, 226, 247, 270, 400, 407, 446, 449–452, 456, 465, 470, 474, 479
pre-commissioning 49, 515–516
preliminary engineering 283, 285, 312, 319, 327–329, 392
prime 16, 52, 60–61, 515
probabilistic 4, 8, 15
procedure 24–25, 29, 32–33, 42, 46, 49, 59, 63, 69, 84, 86, 98, 140, 142, 212, 286, 346, 381, 383–384, 386, 410, 418, 486, 515–516
process
    alarm 168–169, 341, 344
    controller 66, 193, 196–197, 365
    engineering 3, 28–29, 35, 271–272
    variable (PV) 117, 119–120, 122, 124, 126, 167, 169, 221, 333, 356, 358, 363
process control detail sheets (PCDS) 47
process flow diagram (PFD) 23, 28, 205, 242, 271
procurement 16–17, 33, 42, 51, 138, 268, 284, 398, 493, 495, 497, 520
profibus 233
programmable logic controller (PLC) 23, 66, 115, 199, 214, 285
programming languages 218
project 44
    engineer 3, 60, 63, 65, 84–85
    management 1–2, 14, 52, 58, 60–61, 65, 73, 77, 79, 90, 115, 144, 289, 317
    manager 25, 32, 53, 56–58, 60–63, 65, 71, 74, 76, 79, 89–90, 170, 193, 293, 318
    specification 5–6, 24, 73–74, 76, 85–86, 147, 268
project execution plan (PEP) 7, 21, 24, 32–33, 40, 42, 51, 73–74, 78, 84, 87
propagation delay 160
proposal 4, 34, 79
proprietary 66, 230, 232, 238–239, 378
purchasing 38–39, 59–60, 145, 215, 286–287, 401, 407, 415, 429, 493–495, 497
purging 174–177
PV 117, 119–120, 122, 124, 126, 167, 169, 221, 333, 356, 358, 363

Q

quality 2, 7, 9, 11, 13, 27, 59–60, 63, 68, 96, 130, 134, 141, 288, 302, 365, 383, 387, 395, 511, 517
query 133–134, 136, 138, 247, 291–300, 392–394, 420, 453, 501

R

race condition 160
range 117–119, 122–123, 165, 187–188, 397, 401, 487
rangeability 119, 123

rathole 352
raw units 120
real-time trend 212
record 36, 43, 96, 130, 132, 134, 145, 149, 212, 288, 291, 378, 391, 394, 415, 509, 518
redlines 144, 518
relay 37, 120, 156–158, 160, 166–168, 175, 179, 182–183, 195–196, 199–200, 214, 218, 225, 346, 349, 364, 445, 449, 457
relay, timing 160, 163
remote I/O 45, 138–139, 201, 230, 238–239, 287, 327, 378, 380
remote terminal units (RTU) 233
repeatability 59, 119–121, 197, 202
repeaters 237
report 2, 43, 45, 52, 84, 87, 131, 133, 138, 140–141, 196, 202, 213, 295, 300–301, 318, 366, 378, 384, 393, 394, 416
request for proposal (RFP) 4, 15, 20, 24, 71, 323
resolution 53, 88, 94, 119, 122–123, 126, 128, 139, 205, 217, 243, 333
resolution effects on accuracy 122
retrofit 15, 17, 19, 143, 389, 415
RFP 4, 15, 18, 20, 24, 71, 323
ride-through timer 161, 163–164, 330, 348–349, 359
risk 3–4, 8, 10–11, 13–15, 17, 46, 80, 134, 170, 289, 320, 328, 514
routers 236–237, 326
RS-232 232–233
RTD 187
RTU 233

S

safe-state 341, 344, 361
safety 8, 29, 46, 58–59, 68, 75, 81, 83, 85, 147, 168, 170, 175, 181, 272, 274, 279, 284, 328, 338, 448
safety interrupt 338, 341, 344
SAMA 246, 362–363
SAT 20, 27, 43, 49, 140–141, 383–384, 516
SCADA 42
scale 10, 39, 117–121, 123, 138, 187, 356, 415, 486, 488
scaling 49, 115–117, 124, 129, 189, 201, 207, 211, 217–218, 245, 383, 459, 492
scan 181, 210–212, 222, 243
SCDS 329, 359
schedule 2, 5, 9, 12, 16, 21, 24, 26, 32, 39, 62, 68, 73–74, 76, 78, 92, 115, 148–149, 154, 267–268, 285, 302, 316–317, 322, 367, 401, 466, 496, 514, 523
schematic diagrams 36
scope 3, 5–6, 16, 18, 57, 62, 72, 75, 78, 92, 95, 100, 267, 269, 284–285, 430, 510

scope of work (SOW) 7, 11, 15–18, 21, 24, 32, 40, 51, 60, 63, 72–74, 76–78, 81, 84, 98, 283–284, 317, 319, 327, 514
screen design guidelines 208
SDCS 48
security 46, 134, 244
seller 1, 6–7, 10–11, 13–15, 18, 20, 29, 31, 71–73, 80–81, 98
sequence 220
sequence control detail sheet (SCDS) 48, 329, 359
sequence of events 142, 330
sequence of events (SOE) recorder 243
sequence step detail sheet (SSDS) 337–338, 342, 345, 351, 386
sequential control 246, 325
sequential function chart (SFC) 219–220, 328–330, 332
serial communications 232
server 45, 48, 66, 235–238, 240–241, 520
setpoint 48, 119, 138, 245
SFC 219–220, 328–330, 332
shelf state 159–161, 167–169
simulation 42, 49, 246, 321, 383, 516
single-line 47, 378–379
sinking and sourcing 179, 182, 185, 456
site acceptance tests (SAT) 20, 27, 43, 49, 140–141, 383–384, 516
software & logic assignments 136
solid-state relay 161
SOW 5, 7, 74–75, 84–85
spools 399
spreadsheet 130–132, 134, 217, 225, 230, 295, 302, 309, 372, 392, 416, 420, 422, 428, 463, 465–466, 469–470, 486, 504, 506
squad check 510
SSDS 337–338, 342, 345, 351, 386
startup 3, 17, 27, 48–49, 61, 67, 70, 130, 156, 242, 246, 312, 325, 328, 383, 386, 415, 511, 517–518
status report 74, 87, 91
structured text 219
subnet 235, 237–241
suggestion list 93, 95
supervisory control and data acquisition (SCADA) 42, 139, 235
switch 49, 156–159, 162, 168–169, 178–179, 182, 184, 197, 204, 231, 237–238, 240, 244, 274, 276–277, 324, 334, 336, 346, 357–359, 366, 370, 376, 397, 399, 403, 405, 414, 427–428, 445, 447–449, 465, 467–468, 516
symbols library 243
systems integration xxix, 19, 29, 33, 53, 55, 65, 241, 279, 286, 302, 322, 363–364, 379, 387
systems integrator 5, 20, 25, 33, 50, 193, 242, 322–323, 367, 379, 384–386, 517

T

T&M 10, 13–14, 17–18, 302
T&M/NTE 10, 14, 16, 18
table 81, 132, 138, 142, 149, 152, 154, 224, 289–290, 295–297, 299–300, 304, 309, 312, 372, 391, 394, 406, 426, 429, 432, 453, 457, 483, 510
tagname 137, 206, 218, 277, 296–299, 345–346, 352, 374, 377
takeoff 136, 295, 300, 303–304, 392-393, 420, 429, 432, 438, 494, 497, 500, 506
terminal server 45
termination cabinet 179, 449–450, 457, 463–465, 469, 499
terminations 178, 462, 497
thermocouple 124–125, 175, 187–188, 226
thin client 45, 48, 236, 241
time 8
time and material (T&M) 10, 302
time and material, not-to-exceed (T&M, NTE) 10
timer, interval 162, 356
timer, off-delay 160, 163
timer, on-delay 163
timer, ride-through 161, 163–164, 330, 348–349, 359
training 27, 33, 42, 44, 83, 196, 198, 241–242, 247, 269, 312, 321, 328, 376, 386–387, 415, 516–517, 523–524
training plan 46
transmittal 127, 289, 291–293, 518
turndown 119–120
turnkey 2, 16, 19, 241–242
twisted pair 179, 184, 186, 237, 426, 430, 472
two-wire 124, 185, 187, 189–190, 226, 398–399
  circuit 190

U

unit conversions 115, 117, 124
units, engineering 120, 122, 127, 138, 187, 207, 220, 333, 355–356
units, raw 120
unmanaged ethernet switches 236
update by exception 211
update by view 211

V

V&V 42–43, 49, 51, 139–140
validation & verification (V&V) 42, 49
variable frequency drives (VFD) 42
variable, control (CV) 119–120
variable, process (PV) 117, 119–120, 122, 124, 126, 167, 169, 221, 333, 356, 358, 363
vessel trim 434, 436, 493
VFD 42, 385
voltage source 125, 189–190

W

walkdown 25, 32, 34, 71, 86, 269, 312, 438, 514
WBS 26, 34, 43, 45, 73–74, 78, 88–89, 142, 290, 294, 494, 513, 523
wetted voltage 177
Wheatstone bridge 187
wi-fi 235
wire color scheme 478
wire numbering 116, 164–165, 441, 457
wiring 26, 34, 37, 43, 47, 49, 51, 68, 129, 138, 147–149, 155–156, 159, 165–166, 168, 178–179, 184, 188, 193, 199, 201, 226, 228, 267, 272, 394, 400–401, 411–412, 428, 441–442, 444, 446, 452, 455, 457–458, 461–462, 466, 472–474, 479, 485–486, 488, 490, 514
wiring diagram 36–37, 39, 148, 156, 399, 407, 443, 465–466, 469, 471, 479, 488, 510
work breakdown structure (WBS) 26, 74, 78, 142, 290, 513, 523
work-order projects 58